INTERDISCIPLINARY ANCHOR TITLE

The City Reader 3rd ed
Richard LeGates and Frederic Stout eds

URBAN DISCIPLINARY READERS

The Urban Geography Reader
Nick Fyfe and Judith Kenny eds

The Urban Sociology Reader
Jan Lin and Christopher Mele eds

The Urban Politics Reader
Elizabeth Strom and John Mollenkopf eds

The Urban and Regional Planning Reader
Eugenie Birch

TOPICAL URBAN READERS

The City Cultures Reader, 2nd ed
Malcolm Miles, Tim Hall and Iain Borden eds

The Cybercities Reader
Stephen Graham ed

The Sustainable Urban Development Reader
Stephen M. Wheeler and Timothy Beatley eds

The Global Cities Reader
Neil Brenner and Roger Keil eds

For further Information on The Routledge Urban Series
Please Contact:

Richard LeGates
Urban Studies Program
San Francisco State University
1600 Holloway Avenue
San Francisco, California 94132
(415) 338-2875
dlegates@sfsu.edu

or

Frederic Stout
Urban Studies Program
Stanford University
Palo Alto, California
(650) 725-6321
fstout@stanford.edu

The Urban Design Reader

Edited by

Michael Larice

and

Elizabeth Macdonald

Routledge
Taylor & Francis Group

LONDON AND NEW YORK

First published 2007 by Routledge
2 Park Square, Milton Park, Abingdon, Oxon OX14 4RN

Simultaneously published in the USA and Canada
by Routledge
270 Madison Avenue, New York, NY 10016

Routledge is an imprint of the Taylor & Francis Group, an informa business

© 2007 Michael Larice and Elizabeth Macdonald for selection and editorial matter:
the contributors and publishers for individual chapters

Typeset in Amasis and Univers by
Graphicraft Limited, Hong Kong
Printed and bound in Great Britain by
Bell & Bain Ltd, Glasgow

British Library Cataloguing in Publication Data
A catalogue record for this book is available from the British Library

Library of Congress Cataloging in Publication Data

ISBN 10: 0-415-33386-5 (hbk)
ISBN 10: 0-415-33387-3 (pbk)

ISBN 13: 978-0-415-33386-3
ISBN 13: 978-0-415-33387-0

To my students – who continue to inspire me (ML)

and

To Jake (EM)

Contents

Plates

Acknowledgments

We would like to thank the many people who made this book possible. We are extremely grateful for the assistance we received from many of the contributors whose works are included herein and which have inspired us in our professional and academic careers.

Series editor Richard T. LeGates has been a great source of guidance during the process of putting this volume together. Routledge's *The City Reader*, edited by LeGates and Frederick Stout, and *The Sustainable Urban Development Reader*, edited by Stephen M. Wheeler and Timothy Beatley, served as instructive and inspiring models for this reader. We particularly wish to thank Andrew Mould, our editor at Routledge, who has encouraged this project from the start. Thanks go as well to Zoe Kruze and Melanie Attridge, our editorial assistants at Routledge, and Steve Thompson, our Production Editor. In addition, we owe a large debt to the several anonymous reviewers who gave us helpful and wise comments on the book's content. We are grateful to the Routledge team for their constant advice and patience in the production of this volume.

We would like to thank Tony Dorcey, Director of the School of Community and Regional Planning at the University of British Columbia, for generous financial assistance in helping to produce this book. Several people helped us with technical support throughout the permissions, editing, and writing process. People we would like to thank include our permissions manager and editorial assistant Molly O'Neill, and our assistants Marcela de la Peña, Bryan Sherrell, Garlen Capita, and Corinne Stewart. Michael Larice is grateful to the many people who reviewed written work and provided personal support in the preparation of this volume, including Liana Evans, Holly Alyea, Scott Manson, his parents Tony and Bonnie, and his sister Lisa Larice Barber. Elizabeth Macdonald would like to thank her husband, Allan Jacobs, for helpful editorial comments during the writing process and most particularly for his constant support and sense of humor.

GENERAL INTRODUCTION

In the late 1950s, criticism of mid-century city planning and urban design demanded a return to origins within the built environment professions and prompted the rediscovery of the principles of classic urbanism: walkable streets, human-scaled buildings, an active public realm, and meaningful and context-relevant places. This reawakening resulted in a flurry of research and writing that led to the development of new university-level urban design programs and ignited popular interest in designing better neighborhoods and cities. Around the globe, the number of urban design students and professionals continues to grow with each passing year. Design and planning firms search for employees who possess a well-developed understanding of urban spatial form and who can communicate design ideas both visually and verbally. Within an increasingly project-driven planning environment, municipal planning agencies look for staff who can effectively interface with both design professionals and community stakeholders – and can take leadership roles in design decision-making and policy formation. Urban design is increasingly seen as an economic driver in place marketing, tourism development, and business attraction. It is becoming an integral part of urban amenity provision – important for a growing horizontal labor market that allows workers to move according to desired lifestyle choices. Recent concerns in preparing for a less secure and resource constrained future are also driving new interests in public safety, sustainability, ecological design, and livability. These bursts of energy are impacting the way we make cities and offering great opportunities for innovative urban design practice.

In the tradition of other volumes of the Routledge Urban Reader Series, this anthology of literature brings together some of the most influential and seminal material in the field of urban design. Included in this reader are both classic and newer selections that help to describe both historical and contemporary activity in urban design thought and practice. Without an anthology such as this, collecting this material together would be an arduous and time-consuming undertaking. We hope this collection of essential literature will open windows for students and others interested in urban design.

The forty-one readings in *The Urban Design Reader* focus on important historical, theoretical and practical material within several organizational themes. In choosing the selections we sought highly readable and accessible material that focuses directly on urban design. Because this is a multidisciplinary field, the readings are written from a variety of perspectives: architecture, activism and advocacy, landscape architecture, planning, theory and the academies, among others. The material is drawn from historical texts, popular and academic books, advocacy writing, and journal articles. The vast majority of the readings were written since the 1950s, with many drawn from the 1980s and 1990s. *The Urban Design Reader* is international in scope and focuses on material from North America, Europe, and Australia. Not surprisingly however, the great bulk of material is from the United States, Canada, and Great Britain, where urban design is widely taught and researched.

Due to space limitations we have sought a balance in the material, avoiding duplication within each part of the reader. Thus, we have not included every last word or work on every topic. In editing the readings we were compelled to omit some of the visual material because of space considerations. References to material in other sections of the original works were also omitted. Our primary intention in editing the selections was to give readers the essential and most important material from each original piece. Each reading is introduced by way of a short essay that summarizes significant content, the importance and context of the piece, information on the author, and supplemental reading related to the topic material. We have selected readings and introduced each selection objectively, often presenting conflicting debates and arguments over the issues presented therein.

This anthology begins with a selection of writings from signature moments and movements in the history of urban design and city-making. The readings are presented chronologically from the period of the Renaissance up to early-twentieth-century Modernism. Although we could have easily filled a single reader with important historical material, we found these selections to be particularly resonant with the design concerns of our own day. This material offers lessons about collaboration, leadership, unexpected outcomes, public health, satisfaction, and community design – and includes some of the true classics of the urban design canon from Olmsted, Perry, and Le Corbusier. In several instances we present critical material from the perspective of those who came later than the topics of which they write. In these works by Bacon, Berman, and Mumford, the original design ideas have been filtered with the benefit of hindsight so that critical lessons can be presented for current day consumption. These readings serve as precedents for current urban design practice and remain examples of living history. Part One ends with Le Corbusier's theories of modern city design, which provided the impetus for mid-century city-making – and subsequently spawned the criticism that would lead to the reconceptualization of the urban design field.

The selections in Part Two of *The Urban Design Reader* begin with these criticisms of modern planning and design from the late 1950s, and illustrate the normative design theories that were voiced in the ensuing decades. The signature voice of this period is Jane Jacobs' clarion call for planning and design reform in *The Death and Life of Great American Cities*. It is likely the most widely published and republished piece of literature in planning history and continues to inspire new urban devotees to this day. In addition to presenting the book's powerful "Introduction," we also include one of the lesser known chapters, "The Uses of Sidewalks: Contact," which we feel illustrates an important design perspective about the social life of urban space. The other three readings in Part Two provide very different normative filters on how cities should be designed. Christopher Alexander's contribution suggests a single "timeless way" of place-making that stems from traditional and vernacular processes. Allan B. Jacobs and Donald Appleyard provide a recipe for city-making, aimed at creating livable and authentic urbanism. And in the first of two readings by Kevin Lynch, we present his very objectively derived "Dimensions of Performance" that can be applied nearly universally in affecting "*Good City Form*." These pieces are some of the earliest and most important writings by the great minds of the nascent urban design field.

Part Three provides an overview of the theoretical concepts of *place* and rationales for why it has become important in contemporary urban design discourse. In these selections place is understood as physical space that is largely identified with social and personal meaning. The three readings all share distaste for the lack of accessible meaning found in many examples of modern architecture and urban development. Edward Relph's selection from *Place and Placelessness* is a direct attack on the homogenous and meaningless places of the modern urban landscape. In a similar manner, Ray Oldenburg is critical of contemporary development and its dearth of public places, subsequently recommending creation and support for "third places" – places where people can voluntarily come together in informal public settings. In much of the literature on place, phenomenology has proven to be a valuable theoretical starting point – primarily for its elevation of sensory material in making necessary connections between people, physical reality and meaning. These ideas are reinforced in the writing of Christian Norberg-Schulz, who introduces the idea of "spirit of place," which stems from the Roman concept *genus loci*.

Part Four introduces readers to various approaches and practices of place-making. Unlike the previous part, which was theoretical in tone, the selections from Part Four are more prescriptive and provide useful strategies for reinforcing place qualities in urban design. The readings are roughly presented in chronological order, with the first two writings by Kevin Lynch and Gordon Cullen dealing with issues of legibility, imageability, orientation, and experiential movement in understanding the built environment. The next three readings involve the important use of context in the design process. These authors discuss the use of local resources, plant materials, site conditions, histories and craft traditions in evoking places that are filled with meaning and local or regional relevance. The last two readings discuss the characteristics of contemporary Postmodern Urbanism and the Post-Urban city – frequently in a highly critical manner. While not necessarily championing these perspectives, these authors clearly delineate current trends in urban design and city-making and the sometimes unexpected products that result.

Part Five describes a set of very different design strategies and research methods that focus on typology and morphology in urban design. Stemming from architectural theories that suggest the use of well-understood and categorized building types, typology in urban design relies instead on the categorizing of public realm and urban form types that help in shaping space. The three readings in Part Five all share a respect for traditional urbanism and the lessons that can be learned from existing built form – otherwise known within research discourse as "empirical induction." Morphology, on the other hand, is simply the study of urban form. While it is a relatively easy concept to grasp, its practice in different contexts has resulted in a particularly rich diversity of uses, which are highlighted in the reading by Anne Vernez Moudon. Morphological practice is increasingly becoming an important method of "knowing" in the urban design field and is used widely and differently by varied researchers.

Part Six focuses on multidisciplinary challenges and responses in physical planning and urban form-making. While not necessarily design movements in themselves, these advocacy concerns relate to problems with sprawl, higher density development, growth management, sustainability, and compact development. All of these were borne of the negative environmental externalities associated with pollution, threats of resource depletion, and auto-dependent urbanism. Many people in North America are familiar with the contemporary imagery this brings to mind: sprawling low-density suburbia, multi-lane freeways congested with traffic, big-box retail outlets surrounded by seas of asphalt, and pedestrian-unfriendly arterials lined with strip malls – to name a few of these ills. This part opens with a discussion of sprawl and its many definitions, and follows with the Charter of the New Urbanism – a key text related to both the Smart Growth movement and trends in neo-traditional urban design. The next reading highlights the importance of density, not only for the sake of more compact urban form, but also for its role in community building. The final reading of Part Six presents various debates associated with compact cities and urban sustainability. Culling this burgeoning material to just four selections was difficult and remains eminently debatable.

Part Seven introduces the reader to physical elements of the urban public realm and approaches to their design. The readings include observations and suggestions for plazas, neighborhoods, streets, and transit. The authors are concerned with more than the formal and aesthetic aspects of urban design. Each in their own way is seeking more sociable, active and enjoyable places in the city. The first three readings involve various aspects of plaza and urban open space design. One of these is material excerpted from William Whyte's classic *The Social Life of Small Urban Spaces*, which not only provides suggestions for making better urban plazas, but also guides researchers to new methods of urban spatial analysis. Randolph Hester's seminal piece on neighborhood design not only summarizes historical notions of what makes a neighborhood, but also provides surprising insights into the use and design of valued neighborhood space. The readings on street design include the conclusions of Allan B. Jacobs' masterwork *Great Streets*, and also best practice lessons for traffic calming and green infrastructure. Part Seven closes with Robert Cervero's wisdom on successful urban transit and land use linkages found in *The Transit Metropolis*. The readings in this part are the "meat and potatoes" of urban design – opportunities for both problem solving and public realm creativity.

The final set of readings concerns urban design practice and process. Ensuring that urban design intentions see the light of day requires knowledge, skills, and savvy – but also a healthy pragmatism that remains true to inspirational vision. In some ways the readings in Part Eight are the most generally applicable in *The Urban Design Reader*. The opening selection reviews the vast amount of knowledge required in urban design research and practice. In this second reading by Anne Vernez Moudon, she reminds us of the integrative nature of the urban design field and asks "what urban designers should know?" The selection provides a categorization of several knowledge areas, but also an extensive bibliography that readers will find a useful reference. Jon Lang's chapter "Urban Design as a Discipline and as a Profession" poses similar questions about integration and explores the nature of professional activity, its responsibilities, its working contexts and the roles of the urban designer. As a multi-stakeholder process, communication between urban designers and other interested parties poses potential difficulties both in clear presentation and sensitive listening. The chapter on communication from *Public Places – Urban Spaces* discusses not only various communication gaps, but also the power of the designer in manipulating, obfuscating and seducing audiences with visual imagery, written material and oral presentations. The last two readings illustrate issues in urban design regulation and design review – areas of design influence that are growing in use, but not always in effectiveness. The potential material for inclusion in this part of the reader was significant. For space reasons, we were unable to include important material on urban design education, ethics, visual representation, design charrettes, the use of computer techniques, urban design plan writing, and other aspects of professional practice. What is presented here should suffice only as a brief introduction, and we encourage interested readers to immerse themselves in the larger urban design literature.

In producing this reader, we remained aware of its potential to help shape the next generation of urban designers. Although aimed primarily at new professionally oriented students seeking to enter the urban design field, it should also be of interest to professors, established design professionals, urban connoisseurs, and fans of urbanism as a reference text on prevailing urban design thought. *The Urban Design Reader* will provide a comprehensive introduction to the various aspects of the field, its history, its subject matter, and its practice. For those pursuing advanced knowledge, beyond an introductory sampling, we strongly encourage you to dig deeper into the supplemental readings we reference throughout the anthology. In our introductory urban design courses, we often begin with a discussion and definition of the field – those characteristics that make it a different endeavor from other disciplines. These include a deep love of urbanism, the excitement of the urban experience, and the thrill of learning from the great built examples we discover in our travels and research. In our own personal development as designers and educators, we continue to be inspired by the literature and many diverse voices in this rich field. We hope this material helps inspire and introduce readers to the great joys of city-making, as much as it has for us.

PART ONE

Historical Precedents for the Urban Design Field

Plate 1 The Champs Elysées from the Place de l'Etoile in Paris: Haussmann's transformation of Paris provided aesthetic harmony, greater transport access, and visual clarity to the city's medieval street pattern. (Photo: M. Larice)

INTRODUCTION TO PART ONE

Sooner or later in the design of cities, historical precedents for design approaches will come into play: design ideas have to come from somewhere and in the activity of arranging urban physical environments this often involves precedents. Precedents can help us understand why the physical environments that people live in are the way they are, can tell us what has been tried and worked or failed and why, and can give us a range of possibilities to emulate or avoid. They can inspire at the same time as they mislead: a medieval city with narrow, crooked streets and many small buildings crowded up against each other may appear pleasing to the visitor, but what is it like to live in such a place? An urban designer may well evoke an image from a valued historic place, such as Philadelphia's tight grid of narrow residential streets or Paris's grand boulevards, to justify a design proposal. Such associations are often helpful for conveying an idea to an urban designer's many clients, be they communities, developers, public planning agencies, city councils, or design review commissions.

The world's many historic and more recent cities are rich with physical forms that have allowed people to live in settlements. Different eras, times, geographies, and economies have produced a variety of urban physical forms that come to us as precedents. So, too, have people from different times and places produced various theories of what makes "the good city."

Through books and personal experience, present-day urban designers have access to a wealth of design theory and urban form precedents to draw upon. That being said, we learn that the history of urban form and ideas about good city form do not follow a single, steady path. New ideas – grand diagonal avenues, curvilinear residential streets, garden cities, traffic-protected neighborhood enclaves, high-rise towers – come and have their impacts, then retreat or move in another direction, only to be reborn later or to disappear. It is important to know and understand the origins and impacts of the many different urban forms at a range of scales that make up today's cities, including their theoretical foundations, in order to have a sense of what to design for the future.

In this opening part, we focus on theoretical ideas and urban design approaches that serve as the main influences on modern cities and on today's field of urban design, which is commonly understood to have had its birth in the 1960s. This consists of precedents coming primarily from the mid-nineteenth to mid-twentieth centuries, although we also look briefly at the substantial influences of the Renaissance period. This is not to undervalue earlier design ideas and physical forms, most particularly the ubiquitous grid and its many manifestations, but rather to present the most useful combination of readings for present-day students and practitioners. For sure, more reading on the rich history of city form is recommended. Here we are concerned primarily with the ideas and physical forms that have their roots in the modernization process that started with what is commonly referred to as the industrial revolution.

The social and economic changes that accompanied the industrialization of European and North American cities in the late eighteenth and early nineteenth centuries brought forth some of the worst living conditions seen before or since in the Western world. Central areas of cities became densely built, overcrowded, heavily polluted, and highly unsanitary. New physical form answers as well as socio-economic reforms had to be found for those conditions, and a host of people, as individuals and in groups, came forth with answers.

We start with a reading from Edmund N. Bacon's seminal *Design of Cities*, a path-breaking book in the urban design field, a reading that connects us with design ideas and professional practices of the Renaissance. We learn about how, with the invention of perspective drawing and the rise of a humanist desire for visual order, architects took on a major new role as arrangers of urban space. The urban modernization process, in physical terms, can be said to have started with Baron Haussmann's reconstruction of Paris in the 1850s and 1860s. The two poetically written selections from Marshall Berman's excellent book, *All That is Solid Melts into Air*, evoke the feeling of the time and the enormous social changes wrought when the city was opened up with wide new boulevards and public spaces. The glittering café life that developed along the tree-lined boulevards stood in stark contrast to the poverty of surrounding working-class areas, and people for the first time had to contend with fast-moving city traffic. The next reading is a classic writing by Frederick Law Olmsted, the father of the American Parks Movement. Written at the same time that the remodeling of Paris was going on, and taking partial inspiration from Haussmann's parks and boulevards, it extols the virtues of large picturesquely designed urban parks for bringing together diverse urban populations and providing relief from the stresses of urban life. Woven within the narrative is a vision for suburban expansion that includes separated land uses and picturesquely designed residential districts. The reading from Camillo Sitte, two chapters taken from his book *City Planning According to Artistic Principles*, which was written just before the beginning of the twentieth century, directs attention to the aesthetic deficiencies of the modern rectilinear urban spatial forms being implemented at the time, and urges a reappreciation of picturesque medieval forms. Next we turn to a piece by Lewis Mumford that analyzes Ebenezer Howard's turn-of-the-century Garden City idea and how it transformed the field of town planning. Mumford's discussion helps set the Garden City idea within a larger historical context because he focuses on addressing and correcting what he felt were the misconceptions that had come to be associated with it. From Clarence Perry, we learn about the neighborhood unit concept. First presented within the 1929 *Regional Plan of New York and Its Environs*, Perry's concept was to have an enormous influence on the form of future residential areas in the United States. Finally, two pieces from *The City of To-Morrow and Its Planning* present in Le Corbusier's own words his disparagement of pre-modern urban forms and his vision for the rationalized modern city that would become such a paramount force in urban planning and architecture throughout the twentieth century and beyond. His vision compelled the massive urban clearance and redevelopment schemes and urban highway building programs of the 1950s and 1960s, which destroyed the fabrics of so many American city centers.

"Upsurge of the Renaissance"
from *Design of Cities* (1967)

Edmund N. Bacon

Editors' Introduction

The roles and filters of urban designers in the process of city making evolved dramatically during the Renaissance with the humanist desire for visual order, the discovery of mathematical perspective, and new conceptions of space related to time and experiential movement. Early in *Design of Cities*, Edmund Bacon notes that a fundamental shift occurred from the medieval city, which was composed intuitively, perceived simultaneously from different viewpoints, and well integrated to its environment – to a perception of the Renaissance city that was dependent on the personal filters of designers at specific geographic locations and moments in time. The rise of one-point perspective in design practice elevated the individual eyes of designers and their particular focus to new importance (typically targeted at works of art or ecclesiastical and civic buildings of the powerful). In addition to reinforcing the power of capital and elite interests, Renaissance designs based in visual order and one-point perspective emphasized harmony in building design, the linearity of streets, and faster movement through the city. As a result, designers occupied a heightened role in urban decision-making processes, either in predicating new designs or in responding to the design of others (see the section on "Principle of the Second Man" herein). Despite Bacon's suggestion that the form of a city is "determined by the multiplicity of decisions made by the people who live in it," the influence of the Renaissance designer to evoke a singular vision suggests anything but a participatory multi-stakeholder process of design. Renaissance reliance on the designer's perspective elevated the role of the design eye and created a new elite class that was able to direct the focus of others and create perceptual harmonies where none previously existed.

Bacon's intention in writing *Design of Cities* was to investigate the many decisions that influence urban form and to expose the various individual *acts of will* used in making the noble cities of the past. This particular reading from the book is notable not only for the specific innovations of the Renaissance, but also because it echoes debates that were occurring at the time of its publication in the mid-1960s over the role of designers and planners in making cities. Emanating from Jane Jacobs' critique of elite planners and designers in *The Death and Life of Great American Cities*, Bacon politely refutes the notion of incremental growth and the ad-hoc city that was gaining popularity in the mid-1960s. He rejects the notion "that cities are a kind of grand accident, beyond the control of human will," and instead contends that designers should assume responsibility for expressing "the highest aspirations of our civilization." In addition to highlighting key theories of city design, it is of little surprise that examples throughout the book accentuate individual *acts of will* by well-known designers through history, including Hippodamus of Miletus, Michelangelo, Sixtus V, the Woods at Bath, Haussmann in Paris, and Le Corbusier. As a well-positioned practitioner and academic who recognized the ground-shift beginning to occur in the planning field toward more participatory design, Bacon was sensitive to the need for "democratic feedback" and participatory project review to ensure that people's needs were met. Yet at the same time, he could not deny his own professional standing as a leader in the field, his faith in government to improve society on behalf of the public interest, and the power of individual design ideas that are "necessary to create noble cities in our own day." Although public participation in design practice is increasingly valorized, the role of personal agency in creativity and design leadership is still debated.

Edmund N. Bacon (1910–2005) was a planner and educator, who studied at Cornell University and the Cranbrook Academy of Art under the renowned Finnish architect Eliel Saarinen. He worked as an architect and planner in Flint (Michigan), China, and Philadelphia, before becoming managing director of the Philadelphia Housing Authority and later executive director of the City Planning Commission from 1949 to his retirement in 1970. He taught for a time at the University of Illinois and then the University of Pennsylvania between 1950 and 1987. He focused much of his design and planning career on the City of Philadelphia, where he achieved fame in the popular press in the mid-1960s, including a cover article in *Time* and a feature in *Life* magazine. Alongside significant efforts to preserve and restore the city's historic core, his visions and projects transformed the city. While some, such as Penn's Landing, Market East, the redevelopment of Society Hill, and improvements to Independence Mall, were hailed as successes, other modernist proposals for downtown expressways and various mega-projects were met with opposition. This experience as a practitioner undoubtedly helped shape his perspective on urban history and the role of the urban designer as a pivotal participant in the larger urban development process.

Other general urban form histories include: Leonardo Benevolo, *The History of the City* (London: Scolar Press, 1980); Sir Peter Hall, *Cities in Civilization* (New York: Pantheon, 1998); Mark Girouard, *Cities and People: A Social and Architectural History* (New Haven, CT: Yale University Press, 1985); Spiro Kostoff, *The City Shaped: Urban Patterns and Meanings Through History* (Boston, MA: Little, Brown, 1991) and *The City Assembled: The Elements of Urban Form Through History* (Boston, MA: Little, Brown, 1992); A.E.J. Morris, *History of Urban Form Before the Industrial Revolution*, 3rd edn (New York: John Wiley, 1994); Lewis Mumford, *The City in History: Its Origins, Its Transformation and Its Prospects* (New York: Harcourt, Brace, Jovanovich, 1961); Steen Eiler Rasmussen, *Towns and Buildings* (Cambridge, MA: MIT Press, 1994, original 1949); Aidan Southall, *The City in Time and Space* (London: Cambridge University Press, 1998); and Paul Zucker, *Town and Square: From the Agora to the Village Green* (New York: Columbia University Press, 1959).

Important books focusing on Renaissance urbanism, architecture and urban design include: Leonardo Benevolo, *Architecture of the Renaissance* (Boulder, CO: Westview Press, 1978); Peter Murray, *Architecture of the Italian Renaissance* (New York: Schocken, 1997); Rudolf Wittkower, *Architectural Principles in the Age of Humanism* (Chichester, UK: Academy Editions, 1998, original 1949); James S. Ackerman, *The Architecture of Michelangelo* (New York: Viking Press, 1961); Leon Battista Alberti, *The Ten Books on Architecture: The 1755 Leoni Edition* (New York: Dover, 1987); and the important text that was instrumental in shaping Renaissance architecture and urbanism, Vitruvius, *The Ten Books on Architecture* (New York: Dover, 1960).

The coming of the Renaissance brought new energy, new ideas, and a new rational basis for city extension in accord with the new scale of city growth. It was in Florence that the Renaissance first found full expression.

In 1420 the building of the dome over the octagonal walls at the crossing of the cathedral of Florence, designed by the architect Brunelleschi, was far more than a brilliant achievement of building technology. It provided Florence with a psychological and visual center which became the orientation point for much of the later work.

When the Servite monks decided to lay out a new street through property they owned, from the cathedral to their church of Santissima Annunziata, probably during the second half of the thirteenth century, they set into motion a process of orderly city extension which culminated in the great expression of the emerging ideas of the Renaissance, Piazza della Santissima Annunziata. The design of Brunelleschi for the arcade of the Foundling Hospital set a level of architectural excellence that was continued around the square by later designers, so creating a spectacular architectural termination for the much earlier plan of movement from the cathedral.

Figure 1 showing the cathedral dome illustrates the direct physical relationship to it of both the Piazza della Santissima Annunziata and the Uffizi extension from Piazza della Signoria to the Arno River. This shows the partial network of interconnecting streets and squares and the principal church buildings that suggests the beginning of a

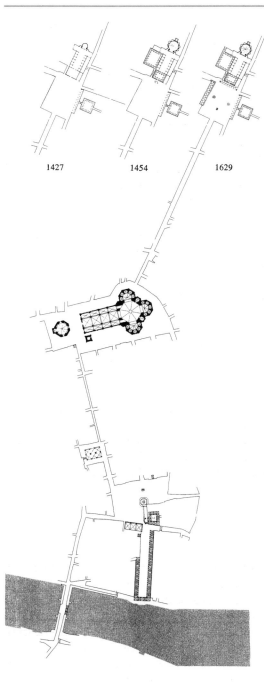

1427   1454   1629

Figure 1 Florence – showing three stages in the development of the Piazza della Santissima Annunziata at top; the Duomo and Baptistry at center; and the Piazza della Signoria, the narrow courtyard of the Uffizi Palace to the River Arno, and street connections to the Ponte Vecchio over the River Arno at the bottom. Drawing by Alois K. Strobl.

city-wide design structure on a new scale, an idea that reached full magnificence in the later development of Rome.

PRINCIPLE OF THE SECOND MAN

Any really great work has within it seminal forces capable of influencing subsequent development around it, and often in ways unconceived of by its creator. The great beauty and elegance of Brunelleschi's arcade of the Foundling Hospital found expression elsewhere in the Piazza della Santissima Annunziata, whether or not Brunelleschi intended this to be so.

The first significant change in the square, following the completion of the arcade in 1427, was the construction of a central bay of the Santissima Annunziata church. This was designed by Michelozzo in 1454 and is harmonious with Brunelleschi's work. However, the form of the square remained in doubt until 1516, when architects Antonio da Sangallo the Elder and Baccio d'Agnolo were commissioned to design the building opposite to Brunelleschi's arcade. It was the great decision of Sangallo to overcome his urge toward self-expression and follow, almost to the letter, the design of the then eighty-nine-year-old building of Brunelleschi. This design set the form of Piazza della Santissima Annunziata and established, in the Renaissance train of thought, the concept of a space created by several buildings designed in relation to one another. From this the "principle of the second man" can be formulated: it is the second man who determines whether the creation of the first man will be carried forward or destroyed.

Sangallo was well prepared for the decision he faced, having worked as a pupil of Bramante, possibly on the plan for the Vatican Cortile, which was the first great effort of the Renaissance in space-planning. Sangallo's arcade is at the left (in the map of 1629), and in the center are the fountains and the equestrian statue of Grand Duke Ferdinand I sculptured by Giambologna (placed there as a directional accent in imitation of Michelangelo's siting of Marcus Aurelius in the Campidoglio). Behind these are the architect Caccini's extensions of Michelozzo's central bay, forming the arcade of Santissima Annunziata, which was completed about 1600. Figure 1 shows three stages of the development of this piazza in relation to the design structure of Florence.

The quality of Piazza della Santissima Annunziata is largely derived from the consummate architectural expression that Brunelleschi gave the first work, the Innocenti arcade, but it is really to Sangallo that we owe the piazza in its present form. He set the course of continuity that has been followed by the designers there ever since.

IMPOSITION OF ORDER

It is impossible to enter Piazza della Signoria (at the base of the previous map) at any point without being confronted with a complete and organized design composition. The powerful impression received is largely due to the interplay of points in space defined by the sculpture with the formal façades of the medieval and Renaissance buildings behind them, a Renaissance ordering of the space of a medieval square.

If one enters by Via Calimaruzza in the northwest corner of the square, looking east, one sees a view of Bartolommeo Ammanati's massive white statue of Neptune, which is silhouetted against the shadowed north wall of the Palazzo Vecchio, and the dark equestrian figure of Cosimo I by Giambologna stands sharply outlined in the center of the sun-bathed Palazzo della Tribunale di Mercanzia. The view from the northeast shows the buildings on each side of the narrow street framing the steeply vertical composition of the Palazzo Vecchio and its tower. The equestrian figure and the figure of Neptune almost overlap, forming a plane in space, which reinforces the direction of movement of this approach to the square.

The view suddenly opens up at the entry point of the Via Vaccherecia in the southwest corner. Neptune now appears at the center of the Palazzo della Tribunale di Mercanzia's façade, and Cosimo I has moved to the center of the richly rusticated façade of the palazzo on the north side of the square. The Loggia dei Lanzi, on the south side of the square, acts as a fulcrum at the point of juncture with the Uffizi.

As one walks about the square, the variously placed sculptural groups appear to move in different directions in relation to their backgrounds and to one another, involving the onlooker in continual orientation, disorientation, and reorientation to a new set of relationships.

DESIGN IN DEPTH

One function of architecture is to create spaces to intensify the drama of living. [. . .] Figure 1 shows the way in which the pavilion of the Uffizi projects out into the street by the Arno, giving the effect of seizing the flow of space along the river's course, and pulling it into the Piazza della Signoria.

The principle of the recession plane and of design in depth is illustrated by . . . the shaft of space contained by the Uffizi walls and framed by the arch at the end (which) links the planes together and focuses on the cathedral dome, with the result that its importance is drawn into the space of the square. [. . .]

One of the most remarkable aspects of the Piazza della Signoria, the plane in space established by the line of sculpture from Hercules and Cacus to the right of the Palazzo Vecchio's entrance, on to the copy of Michelangelo's David, to Ammanati's Neptune fountain, and ending in the figure of Cosimo I on horseback, starts and ends in the physical sense, but in spirit extends in each direction, exercising extraordinary influence in all parts of the square.

MICHELANGELO'S ACT OF WILL

Only by reconstructing the Capitoline Hill as it existed before Michelangelo went to work on it can we comprehend the magnitude of this artist's genius in creating the Campidoglio. This masterwork forms a link between the early Renaissance expressions of urban design in Florence and the great Baroque developments in Rome.

The drawing by J.H. Aronson, see Figure 2, based on various sketches by artists of the period, is an attempt to reconstruct the area as it existed in 1538, when Michelangelo began work. It shows the Palazzo del Senatore at the top and the Palazzo dei Conservatori at the right. [. . .] The formless, unplanned relationship between the medieval Palazzo del Senatore and the Palazzo dei Conservatori was complicated by mounds of earth, columns, and an obelisk. There were also the

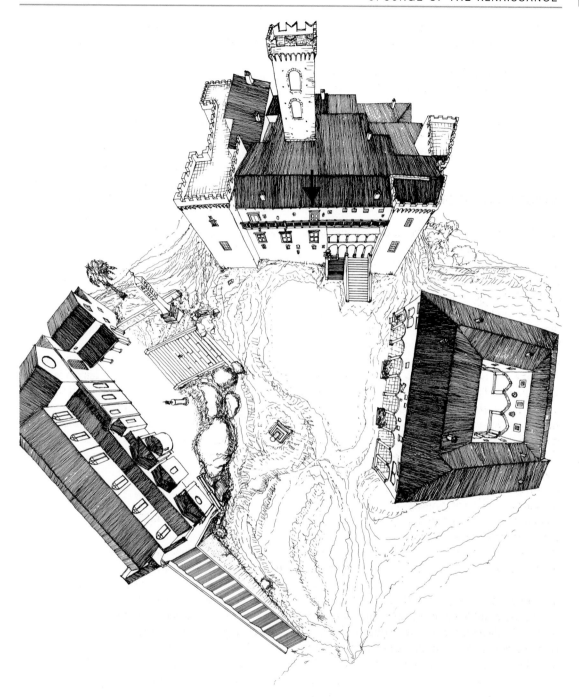

Figure 2 The Capitoline Hill before reconstruction by Michelangelo. Drawing by Joseph H. Aronson.

statues of the two Roman river gods, on each side of the entrance to the Palazzo dei Conservatori. This, then, was the physical situation with which Michelangelo was faced when he reluctantly

acceded to Pope Paul III's orders to recreate the Campidoglio as the heart of Rome.

The approach Michelangelo took produced one of the great masterpieces of all time. The actual basis for the artist's design was an enormous intellectual achievement. By a single act of will he established a line of force on the axis of the Palazzo del Senatore, a line which in effect became the organizing element that pulled chaos into order. [...]

DEVELOPMENT OF ORDER

In their discussion of the angle between the two flanking buildings of the Campidoglio, and its significance in relation to perspective as diminishing or increasing apparent distance, antiquarians sometimes seem to lose sight of the fact that this angle was determined long before Michelangelo started work. What Michelangelo did was to repeat the angle already set by the Palazzo dei Conservatori, symmetrically on the other side of the axis of the Palazzo del Senatore. Accepting this angle as a point of departure, he set about to treat the space it created. The decision he made is remarkable because it contained elements of two violently contradictory points of view.

On one hand, Michelangelo saved the basic structure of the two old palaces which he found on the site by confining his efforts to the building of new façades. On the other hand, what he did was to create a totally new effect. One might have thought a man of such drive toward order and beauty would have swept away the old buildings in order to give free rein to his own creative efforts, or, conversely, that such modesty would have led to a hodgepodge compromise. Michelangelo has proved that humility and power can coexist in the same man, that it is possible to create a great work without destroying what is already there.

On orders of the Pope, and against the advice of Michelangelo, the statue of Marcus Aurelius was moved from San Giovanni in Laterano to the Campidoglio, and Michelangelo positioned the figure and designed a base for it. The 2nd drawing by Aronson (Figure 3) shows that the first act was the setting of this figure, and by that single act the integrity of the total idea was established. Aronson's first drawing (Figure 2) shows, in the extremely disorganized front that the Palazzo del Senatore presents, the degree of imagination necessary to conceive the order that eventually would arise. [...] The completion of the stairway of the Palazzo del Senatore and the positioning of the Marcus Aurelius statue establish a relationship between two architectural elements in space. Each of these is modest in extent, yet of such power that the feeling of order is already present and the drive toward the larger order is irreversibly set in motion.

Michelangelo had designed a new tower to replace the unsymmetrical medieval one, but his design was only vaguely followed when the tower was finally, in 1578, replaced by another.

A comparison (between Aronson's two drawings) reveals the admirable skill with which Michelangelo introduced a totally new scale into this space. He modulated the façade of the Palazzo del Senatore by establishing a firm line defining the basement, and above this he placed a monumental order of Corinthian pilasters. These interact effectively with the colossal two-story order of the flanking palaces which sweep from the base to the cornice in one mighty surge.

ORDER ARRIVES ON THE CAPITOL

One of the greatest attributes of the Campidoglio composition is the modulation of the land. Without the shape of the oval, and its two-dimensional star-shaped paving pattern, as well as its three-dimensional projection in the subtly designed steps that surround it, the unity and coherence of the design would not have been achieved. The paved area stands as an element in its own right, in effect creating a vertical oval shaft of space which greatly reinforces the value of the larger space defined by the three buildings. [...] The product is a space which, apart from its beauty, still serves as the symbolic heart of Rome.

The Campidoglio was designed some thirty-five years after Bramante made his great plan for the Vatican Cortile, and it followed by just over twenty years Sangallo's plan for the second arcade in Piazza della Santissima Annunziata in Florence. While the Campidoglio incorporated ideas contained in each of these earlier works, it went far beyond them in the degree of integration between the architecture of the buildings, the placement of sculpture, and the modulation of the land. Furthermore, it established more powerfully than any

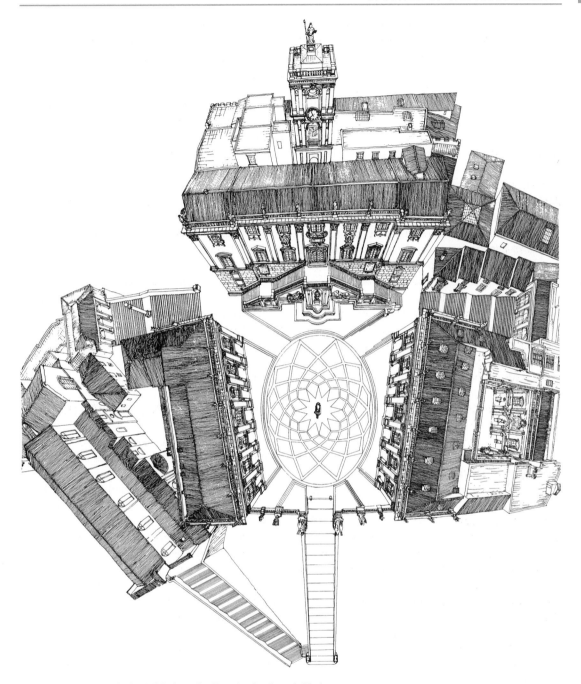

Figure 3 Campidoglio by Michelangelo. Drawing by Joseph H. Aronson

previous example the fact that space itself could be the subject of design. In the richness of its forms, the Campidoglio heralded the arrival of the Baroque.

[…]

STIRRINGS OF THE NEW ORDER

The first hint of the system that would lead to the new order came from the artists using the new and glittering tools of scientific perspective. We have already seen how these stemmed the intuitive flow of experience which led to the organic design of cities in the medieval period, and, how they led to positive advocacy of organized confusion.

In a drawing of Antonio Pisanello, who lived in the first half of the fifteenth century, we see the beginning of a new idea of design. [. . .] In Pisanello's drawing, he is fascinated, not with the shape of mass, but with the shape of space. He has created a tunnel of space articulated by the series of recession planes, through which his figures move in depth toward the pull of the vanishing point.

This sets into motion the idea of architectural design, not as the manipulation of mass but as artic-ulation of experience along an axis of movement through space. It was provided by exactly the same basic scientific technology, but this was employed in a different way, which led to a liber-ation of the designer's thinking, and set into motion a new ordering principle in city design. Over the next two hundred years, one can observe a continuous growth and development of the seminal idea contained here, in its acceptance by designers and its application on a vast scale in actual construction on the ground.

[. . .]

INSIDE–OUTSIDE RELATIONSHIPS

Until the beginning of the seventeenth century, the energy of designers was absorbed in the problem of applying new-found Renaissance principles to the solution of the interior form of the building, and then to the façade. In the seventeenth century, after almost two hundred years of experimentation, the flow of energy was reversed. The design vit-ality began to spill out of the building into the streets of the city around it. The designer, having mastered the internal building problem, now cast his eye on the building's environment, and expended his extra energy in a euphoric flow of design activity to create a setting for his structure. [. . .]

The lines of energy (in Baroque design) radiate outward from a central source in a manner similar to that of Baroque design. The energies of the design expired in the depths of the city, the points of expiration themselves creating a form – as, for example, a piazza connected with a Baroque church. Out of this grew the deliberate planning of a network of lines of design energy on a city-wide basis, providing channels for the transmission of the design energy of buildings already built and at the same time creating locations calling for a new design energy in buildings yet to come.

It was the extra energy of the Baroque period, resulting from the confidence inspired by the mastery of design technique, which produced the great interaction between structure and setting. [. . .] Similar exuberance was expressed in the Roman plan of Sixtus V for a city design structure binding the points of design energy into a total system.

"The Family of Eyes" and "The Mire of the Macadam"

from *All That is Solid Melts into Air: The Experience of Modernity* (1982)

Marshall Berman

Editors' Introduction

By the mid-nineteenth century, following the industrial revolution, many cities had become extremely crowded, especially those contained by protective city walls, such as Paris and Barcelona. From 1853 through 1870, Georges Eugène Haussmann, Prefect of the Seine during the rule of Napoleon III, implemented large-scale public works in Paris that included carving wide boulevards through the dense medieval city center, laying out boulevard systems in peripheral areas to promote urban development, turning former royal hunting grounds at the edge of the city into large new public parks, and redesigning existing city parks. Drawing on Baroque axial planning ideas, the new boulevards linked important public places – train stations, public markets, civic buildings, and parks. A municipal sewer system was built under them and many thousands of trees were planted along them. The boulevards were lined with mandated six-story buildings sporting uniform empire-style façades and mansard roofs. Ground floors of these buildings held cafés and restaurants. At the time, Haussmann's work was looked upon as a model of city modernization, and the boulevards were much admired worldwide. However, particularly following the excesses and failures of the modern urban renewal projects of the 1950s and the self-reflection this engendered in the planning profession, Haussmann's Paris has been much criticized on the rightful grounds that it displaced large numbers of mostly poor people, and that it was an expression of political power meant to clear out working-class neighborhoods that were hotbeds of resistance and made it easier for Napoleon's troops to move through the city and break up political unrest.

In *All That is Solid Melts into Air: The Experience of Modernity* (New York: Simon & Schuster, 1982) Marshall Berman looks at Haussmann's Paris and the new boulevards from a different perspective. Writing at a time when the positivism associated with modernity was being highly criticized and dismissed, Berman dared to call attention to the expanding nature of the new urban forms Haussmann introduced into Paris. Within the chapters on Paris (other chapters discuss St. Petersburg and New York City), which use as their starting points poems written by the nineteenth-century Parisian poet and flâneur Charles Baudelaire, Berman discusses the reactions and social changes that came from people's encounters with the new boulevards. He characterizes the boulevards as the most important invention of the nineteenth century because they physically opened up the city, giving breathing room and creating spaces – the cafés, restaurants, and wide sidewalks – where new forms of urban public social life could develop, while at the same time they forced the middle and upper classes to confront the reality of urban working-class poverty because people from the dense surrounding neighborhoods spilled out onto the boulevards. The boulevards also introduced the beginnings of modern traffic into the city, because their wide and unencumbered roadways allowed private carriages to move at much faster speeds than previously possible.

As well as giving a vivid impression of the changes wrought by the new boulevards, Berman's account of Haussmann's Paris gives us insight into the complexity of the modern public realm and its physical forms, and the often neither-black-nor-white outcomes of urban design projects – there are generally both positive and negative effects associated with any urban design undertaking. One of the lessons to be taken from study of Haussmann's transformation of Paris is the power of an appointed civil servant to transform a cityscape. Latter-day civil servants who have had enormous influence on their city's built form include Robert Moses, chairman of New York City's Triborough Bridge Authority and park commissioner during the mid-twentieth century, Ed Logue, the director of Boston's redevelopment agency in the 1960s, Allan B. Jacobs, San Francisco's planning director during the late 1960s and early 1970s, and Larry Beasley, Vancouver's co-director of planning since the 1990s.

Marshall Berman is a Professor of Political Science at the City University of New York (CUNY), where he teaches political philosophy and urbanism. His other writings include *The Politics of Authenticity* (New York: Athenaeum, 1970) and *On the Town: One Hundred Years of Spectacle in Times Square* (New York: Random House, 2006).

Other resources on the subject of Haussmann's Paris include Nicholas Papayanis, *Planning Paris Before Haussmann* (Baltimore, MD: Johns Hopkins Press, 2004), Howard Saalman, *Haussmann: Paris Transformed* (New York: G. Braziller, 1971), and Michel Carmona, *Haussmann: His Life and Times*, translated by Patrick Camiller (Chicago, IL: I.R. Dee, 2002). A book by Haussmann's landscape architect Adolphe Alphand that contains plans and sections of Paris's new boulevards and parks has been reprinted: *Les Promenades de Paris* (Princeton, NJ: Princeton Architectural Press, 1984).

It is interesting to note that while Haussmann's classic boulevards were considered the height of modern street building in the latter half of the nineteenth century, by the 1930s they were considered outdated in the United States. These boulevards have a wide roadway in the center for fast-moving through traffic and narrow roadways on the side for slow-moving local traffic. The roadways are separated by tree-lined medians, or malls. The three-roadway configuration came to be considered dangerous by traffic engineers and they ceased to be built. However, three-roadway boulevards (or multiway boulevards, to use a recently coined term) are currently being looked to as a possible solution for handling large amounts of traffic in cities, where it is necessary to do so, without deadening the local environment. A book that documents extensive research into multiway boulevards is Allan B. Jacobs, Elizabeth Macdonald, and Yodan Rofé, *The Boulevard Book: History, Evolution, Design of Multiway Boulevards* (Boston, MA: MIT Press, 2002).

THE FAMILY OF EYES

OUR FIRST primal scene emerges in "The Eyes of the Poor." (Paris Spleen #26) This poem takes the form of a lover's complaint: the narrator is explaining to the woman he loves why he feels distant and bitter toward her. He reminds her of an experience they recently shared. It was the evening of a long and lovely day that they had spent alone together. They sat down on the terrace "in front of a new cafe that formed the corner of a new boulevard." The boulevard was "still littered with rubble," but the cafe "already displayed proudly its unfinished splendors." Its most splendid quality was a flood of new light: "The cafe was dazzling. Even the gas burned with the ardor of a debut; with

all its power it lit the blinding whiteness of the walls, the expanse of mirrors, the gold cornices and moldings." Less dazzling was the decorated interior that the gaslight lit up: a ridiculous profusion of Hebes and Ganymedes, hounds and falcons; "nymphs and goddesses bearing piles of fruits, pâtés and game on their heads," a mélange of "all history and all mythology pandering to gluttony." In other circumstances the narrator might recoil from this commercialized grossness; in love, however, he can laugh affectionately, and enjoy its vulgar appeal – our age would call it Camp.

As the lovers sit gazing happily into each other's eyes, suddenly they are confronted with other people's eyes. A poor family dressed in rags – a graybearded father, a young son, and a baby – come to a stop directly in front of them and gaze raptly at the bright new world that is just inside. "The three faces were extraordinarily serious, and those six eyes contemplated the new cafe fixedly with an equal admiration, differing only according to age." No words are spoken, but the narrator tries to read their eyes. The father's eyes seem to say, "How beautiful it is! All the gold of the poor world must have found its way onto these walls." The son's eyes seem to say, "How beautiful it is! But it is a house where only people who are not like us can go." The baby's eyes "were too fascinated to express anything but joy, stupid and profound." Their fascination carries no hostile undertones; their vision of the gulf between the two worlds is sorrowful, not militant, not resentful but resigned. In spite of this, or maybe because of it, the narrator begins to feel uneasy, "a little ashamed of our glasses and decanters, too big for our thirst." He is "touched by this family of eyes," and feels some sort of kinship with them. But when, a moment later, "I turned my eyes to look into yours, dear love, to read *my* thoughts there" (Baudelaire's italics), she says, "Those people with their great saucer eyes are unbearable! Can't you go tell the manager to get them away from here?"

This is why he hates her today, he says. He adds that the incident has made him sad as well as angry: he sees now "how hard it is for people to understand each other, how incommunicable thought is" – so the poem ends – "even between people in love."

What makes this encounter distinctively modern? What marks it off from a multitude of earlier Parisian scenes of love and class struggle? The difference lies in the urban space where our scene takes place. "Toward evening you wanted to sit down in front of a new cafe that formed the corner of a new boulevard, still piled with rubble but already displaying its unfinished splendors." The difference, in one word, is the *boulevard*: the new Parisian boulevard was the most spectacular urban innovation of the nineteenth century, and the decisive breakthrough in the modernization of the traditional city.

In the late 1850s and through the 1860s, while Baudelaire was working on *Paris Spleen*, Georges Eugène Haussmann, the Prefect of Paris and its environs, armed with the imperial mandate of Napoleon III, was blasting a vast network of boulevards through the heart of the old medieval city.[1] Napoleon and Haussmann envisioned the new roads as arteries in an urban circulatory system. These images, commonplace today, were revolutionary in the context of nineteenth-century urban life. The new boulevards would enable traffic to flow through the center of the city, and to move straight ahead from end to end – a quixotic and virtually unimaginable enterprise till then. In addition, they would clear slums and open up "breathing space" in the midst of layers of darkness and choked congestion. They would stimulate a tremendous expansion of local business at every level, and thus help to defray the immense municipal demolition, compensation and construction costs. They would pacify the masses by employing tens of thousands of them – at times as much as a quarter of the city's labor force – on long-term public works, which in turn would generate thousands more jobs in the private sector. Finally, they would create long and broad corridors in which troops and artillery could move effectively against future barricades and popular insurrections.

The boulevards were only one part of a comprehensive system of urban planning that included central markets, bridges, sewers, water supply, the Opera and other cultural palaces, a great network of parks. "Let it be said to Baron Haussmann's eternal credit" – so wrote Robert Moses, his most illustrious and notorious successor, in 1942 – "that he grasped the problem of step-by-step large-scale city modernization." The new construction wrecked hundreds of buildings, displaced uncounted thousands of people, destroyed whole neighborhoods that had lived for centuries. But it opened up the whole of the city, for the first time in its history, to all its inhabitants. Now, at last, it was possible to move not only within neighborhoods, but through them. Now, after centuries of life as a cluster of isolated cells, Paris was becoming a unified physical and human space.[2]

The Napoleon-Haussmann boulevards created new bases – economic, social, aesthetic – for bringing enormous numbers of people together. At the street level they were lined with small businesses and shops of all kinds, with every corner zoned for restaurants and terraced sidewalk cafes. These cafes, like the one where Baudelaire's lovers and his family in rags come to look, soon came to be seen all over the world as symbols of *la vie parisienne*. Haussmann's sidewalks, like the boulevards themselves, were extravagantly wide, lined with benches, lush with trees.[3] Pedestrian islands were installed to make crossing easier, to separate local from through traffic and to open up alternate routes for promenades. Great sweeping vistas were designed, with monuments at the boulevards' ends, so that each walk led toward a dramatic climax. All these qualities helped to make the new Paris a uniquely enticing spectacle, a visual and sensual feast. Five generations of modern painters, writers and photographers (and, a little later, filmmakers), starting with the impressionists in the 1860s, would nourish themselves on the life and energy that flowed along the boulevards. By the 1880s, the Haussmann pattern was generally acclaimed as the very model of modern urbanism. As such, it was soon stamped on emerging and expanding cities in every corner of the world, from Santiago to Saigon.

What did the boulevards do to the people who came to fill them? Baudelaire shows us some of the most striking things. For lovers, like the ones in "The Eyes of the Poor," the boulevards created a new primal scene: a space where they could be private in public, intimately together without being physically alone. Moving along the boulevard, caught up in its immense and endless flux, they could feel their love more vividly than ever as the still point of a turning world. They could display their love before the boulevard's endless parade of strangers – indeed, within a generation Paris would be world-famous for this sort of amorous display – and draw different forms of joy from them all. They could weave veils of fantasy around the multitude of passers-by: who were these people, where did they come from and where were they going, what did they want, whom did they love? The more they saw of others and showed themselves to others – the more they participated in the extended "family of eyes" – the richer became their vision of themselves.

In this environment, urban realities could easily become dreamy and magical. The bright lights of street and cafe only heightened the joy; in the next generations, the coming of electricity and neon would heighten it still more. Even the most blatant vulgarities, like those café nymphs with fruits and pâtés on their heads, turned lovely in this romantic glow. Anyone who has ever been in love in a great city knows the feeling, and it is celebrated in a hundred sentimental songs. In fact, these private joys spring directly from the modernization

of public urban space. Baudelaire shows us a new private and public world at the very moment when it is coming into being. From this moment on, the boulevard will be a vital boudoir in the making of modern love.

But primal scenes, for Baudelaire as later on for Freud, cannot be idyllic. They may contain idyllic material, but at the climax of the scene a repressed reality creaks through, a revelation or discovery takes place: "a new boulevard, still littered with rubble . . . displayed its unfinished splendors." Alongside the glitter, the rubble: the ruins of a dozen inner-city neighborhoods – the city's oldest, darkest, densest, most wretched and most frightening neighborhoods, home to tens of thousands of Parisians – razed to the ground. Where would all these people go? Those in charge of demolition and reconstruction did not particularly concern themselves. They were opening up vast new tracts for development on the northern and eastern fringes of the city; in the meantime, the poor would make do, somehow, as they always did. Baudelaire's family in rags step out from behind the rubble and place themselves in the center of the scene. The trouble is not that they are angry or demanding. The trouble is simply that they will not go away. They, too, want a place in the light.

This primal scene reveals some of the deepest ironies and contradictions in modern city life. The setting that makes all urban humanity a great extended "family of eyes" also brings forth the discarded stepchildren of that family. The physical and social transformations that drove the poor out of sight now bring them back directly into everyone's line of vision. Haussmann, in tearing down the old medieval slums, inadvertently broke down the self-enclosed and hermetically sealed world of traditional urban poverty. The boulevards, blasting great holes through the poorest neighborhoods, enable the poor to walk through the holes and out of their ravaged neighborhoods, to discover for the first time what the rest of their city and the rest of life is like. And as they see, they are seen: the vision, the epiphany, flows both ways. In the midst of the great spaces, under the bright lights, there is no way to look away. The glitter lights up the rubble, and illuminates the dark lives of the people at whose expense the bright lights shine.[4] Balzac had compared those old neighborhoods to the darkest jungles of Africa; for Eugène Sue they epitomized

"The Mysteries of Paris." Haussmann's boulevards transform the exotic into the immediate; the misery that was once a mystery is now a fact.

The manifestation of class divisions in the modern city opens up new divisions within the modern self. How should the lovers regard the ragged people who are suddenly in their midst? At this point, modern love loses its innocence. The presence of the poor casts an inexorable shadow over the city's luminosity. The setting that magically inspired romance now works a contrary magic, and pulls the lovers out of their romantic enclosure, into wider and less idyllic networks. In this new light, their personal happiness appears as class privilege. The boulevard forces them to react politically. The man's response vibrates in the direction of the liberal left: he feels guilty about his happiness, akin to those who can see but cannot share it; he wishes, sentimentally, to make them part of his family. The woman's affinities – in this instant, at least – are with the right, the Party of Order: we have something, they want it, so we'd better "*prier le maître*," call somebody with the power to get rid of them. Thus the distance between the lovers is not merely a gap in communication, but a radical opposition in ideology and politics. Should the barricades go up on the boulevard – as in fact they will in 1871, seven years after the poem's appearance, four years after Baudelaire's death – the lovers could well find themselves on opposite sides.

That a loving couple should find themselves split by politics is reason enough to be sad. But there may be other reasons: maybe, when he looked deeply into her eyes, he really did, as he hoped to do, "read *my* thoughts there." Maybe, even as he nobly affirms his kinship in the universal family of eyes, he shares her nasty desire to deny the poor relations, to put them out of sight and out of mind. Maybe he hates the woman he loves because her eyes have shown him a part of himself that he hates to face. Maybe the deepest split is not between the narrator and his love but within the man himself. If this is so, it shows us how the contradictions that animate the modern city street resonate in the inner life of the man on the street.

Baudelaire knows that the man's and the woman's responses, liberal sentimentality and reactionary ruthlessness, are equally futile. On one hand, there is no way to assimilate the poor into any family of the comfortable; on the other hand, there is no form of repression that can get rid of them for long – they'll always be back. Only the most radical reconstruction of modern society could even begin to heal the wounds – personal as much as social wounds – that the boulevards bring to light. And yet, too often, the radical solution seems to be dissolution: tear the boulevards down, turn off the bright lights, expel and resettle the people, kill the sources of beauty and joy that the modern city has brought into being. We can hope, as Baudelaire sometimes hoped, for a future in which the joy and beauty, like the city lights, will be shared by all. But our hope is bound to be suffused by the self-ironic sadness that permeates Baudelaire's city air.

THE MIRE OF THE MACADAM

OUR NEXT archetypal modern scene is found in the prose poem "Loss of a Halo" (Paris Spleen #46), written in 1865 but rejected by the press and not published until after Baudelaire's death. Like "The Eyes of the Poor," this poem is set on the boulevard; it presents a confrontation that the setting forces on the subject; and it ends (as its title suggests) in a loss of innocence. Here, however, the encounter is not between one person and another, or between people of different social classes, but rather between an isolated individual and social forces that are abstract yet concretely dangerous. Here, the ambience, imagery and emotional tone are puzzling and elusive; the poet seems intent on keeping his readers off balance, and he may be off balance himself. "Loss of a Halo" develops as a dialogue between a poet and an "ordinary man" who bump into each other in *un mauvais lieu*, a disreputable or sinister place, probably a brothel, to the embarrassment of both. The ordinary man, who has always cherished an exalted idea of the artist, is aghast to find one here:

> "What! you here, my friend? you in a place like this? you, the eater of ambrosia, the drinker of quintessences! I'm amazed!"

The poet then proceeds to explain himself:

> "My friend, you know how terrified I am of horses and vehicles? Well, just now as I was crossing the

boulevard in a great hurry, splashing through the mud, in the midst of a moving chaos, with death galloping at me from every side, I made a sudden move (un mouvement brusque), *and my halo slipped off my head and fell into the mire of the macadam. I was much too scared to pick it up. I thought it was less unpleasant to lose my insignia than to get my bones broken. Besides, I said to myself, every cloud has a silver lining. Now I can walk around incognito, do low things, throw myself into every kind of filth* (me livrer à la crapule), *just like ordinary mortals* (simples mortel). *So here I am, just as you see me, just like yourself!"*

The straight man plays along, a little uneasily:

"But aren't you going to advertise for your halo? or notify the police?"

No: the poet is triumphant in what we recognize as a new self-definition:

"God forbid! I like it here. You're the only one who's recognized me. Besides, dignity bores me. What's more, it's fun to think of some bad poet picking it up and brazenly putting it on. What a pleasure to make somebody happy! especially somebody you can laugh at. Think of X! Think of Z! Don't you see how funny it will be?"

It is a strange poem, and we are apt to feel like the straight man, knowing something's happening here but not knowing what it is.

One of the first mysteries here is that halo itself. What's it doing on a modern poet's head in the first place? It is there to satirize and to criticize one of Baudelaire's own most fervent beliefs: belief in the holiness of art. We can find a quasi-religious devotion to art throughout his poetry and prose. Thus, in 1855: "The artist stems only from himself. . . . He stands security only for himself. . . . He dies childless. He has been his own king, his own priest, his own God."[5] "Loss of a Halo" is about how Baudelaire's own God fails. But we must understand that this God is worshipped not only by artists but equally by many "ordinary people" who believe that art and artists exist on a plane far above them. "Loss of a Halo" takes place at the point at which the world of art and the ordinary world converge. This is not only a spiritual point but a physical one, a point in the landscape of the modern city. It is the point where the history of modernization and the history of modernism fuse into one.

Walter Benjamin seems to have been the first to suggest the deep affinities between Baudelaire and Marx. Although Benjamin does not make this particular connection, readers familiar with Marx will notice the striking similarity of Baudelaire's central image here to one of the primary images of the *Communist Manifesto*: "The bourgeoisie has stripped off its halo every activity hitherto honored and looked up to with reverent awe. It has transformed the doctor, the lawyer, the priest, the poet, the man of science, into its paid wage-laborers."[6] For both men, one of the crucial experiences endemic to modern life, and one of the central themes for modern art and thought, is *desanctification*. Marx's theory locates this experience in a world-historical context; Baudelaire's poetry shows how it feels from inside. But the two men respond to this experience with rather different emotions. In the *Manifesto*, the drama of desanctification is terrible and tragic: Marx looks back to, and his vision embraces, heroic figures like Oedipus at Colonnus, Lear on the heath, contending against the elements, stripped and scorned but not subdued, creating a new dignity out of desolation. "Eyes of the Poor" contains its own drama of desanctification, but there the scale is intimate rather than monumental, the emotions are melancholy and romantic rather than tragic and heroic. Still, "Eyes of the Poor" and the *Manifesto* belong to the same spiritual world. "Loss of a Halo" confronts us with a very different spirit: here the drama is essentially comic, the mode of expression is ironic, and the comic irony is so successful that it masks the seriousness of the unmasking that is going on. Baudelaire's denouement, in which the hero's halo slips off his head and rolls through the mud – rather than being torn off with a violent *grand geste*, as it was for Marx (and Burke and Blake and Shakespeare) – evokes vaudeville, slapstick, the metaphysical pratfalls of Chaplin and Keaton. It points forward to a century whose heroes will come dressed as anti-heroes, and whose most solemn moments of truth will be not only described but actually experienced as clown shows, music-hall or nightclub routines-shticks. The setting plays the same sort of decisive role in Baudelaire's black comedy that it will play in Chaplin's and Keaton's later on.

"Loss of a Halo" is set on the same new boulevard as "Eyes of the Poor." But although the two poems are separated physically by only a few feet, spiritually they spring from different worlds. The gulf that separates them is the step from the sidewalk into the gutter. On the sidewalk, people of all kinds and all classes know themselves by comparing themselves to each other as they sit or walk. In the gutter, people are forced to forget what they are as they run for their lives. The new force that the boulevards have brought into being, the force that sweeps the hero's halo away and drives him into a new state of mind, is modern *traffic*.

When Haussmann's work on the boulevards began, no one understood why he wanted them so wide: from a hundred feet to a hundred yards across. It was only when the job was done that people began to see that these roads, immensely wide, straight as arrows, running on for miles, would be ideal speedways for heavy traffic. Macadam, the surface with which the boulevards were paved, was remarkably smooth, and provided perfect traction for horses' hooves. For the first time, riders and drivers in the heart of the city could whip their horses up to full speed. Improved road conditions not only speeded up previously existing traffic but – as twentieth-century highways would do on a larger scale – helped to generate a volume of new traffic far greater than anyone, apart from Haussmann and his engineers, had anticipated. Between 1850 and 1870, while the central city population (excluding newly incorporated suburbs) grew by about 25 percent, from about 1.3 million to 1.65 million, inner-city traffic seems to have tripled or quadrupled. This growth exposed a contradiction at the heart of Napoleon's and Haussmann's urbanism. As David Pinkney says in his authoritative study, *Napoleon III and the Rebuilding of Paris*, the arterial boulevards "were from the start burdened with a dual function: to carry the main stream of traffic across the city and to serve as major shopping and business streets; and as the volume of traffic increased, the two proved to be ill-compatible." The situation was especially trying and terrifying to the vast majority of Parisians who walked. The macadam pavements, a source of special pride to the Emperor – who never walked – were dusty in the dry months of summer, and

muddy in the rain and snow. Haussmann, who clashed with Napoleon over macadam (one of the few things they ever fought about), and who administratively sabotaged imperial plans to cover the whole city with it, said that this surface required Parisians "either to keep a carriage or to walk on stilts."[7] Thus the life of the boulevards, more radiant and exciting than urban life had ever been, was also more risky and frightening for the multitudes of men and women who moved on foot.

This, then, is the setting for Baudelaire's primal modern scene: "I was crossing the boulevard, in a great hurry, in the midst of a moving chaos, with death galloping at me from every side." The archetypal modern man, as we see him here, is a pedestrian thrown into the maelstrom of modern city traffic, a man alone contending against an agglomeration of mass and energy that is heavy, fast and lethal. The burgeoning street and boulevard traffic knows no spatial or temporal bounds, spills over into every urban space, imposes its tempo on everybody's time, transforms the whole modern environment into a "moving chaos." The chaos here lies not in the movers themselves – the individual walkers or drivers, each of whom may be pursuing the most efficient route for himself – but in their interaction, in the totality of their movements in a common space. This makes the boulevard a perfect symbol of capitalism's inner contradictions: rationality in each individual capitalist unit, leading to anarchic irrationality in the social system that brings all these units together.[8]

The man in the modern street, thrown into this maelstrom, is driven back on his own resources – often on resources he never knew he had – and forced to stretch them desperately in order to survive. In order to cross the moving chaos, he must attune and adapt himself to its moves, must learn to not merely keep up with it but to stay at least a step ahead. He must become adept at *soubresauts* and *mouvements brusques*, at sudden, abrupt, jagged twists and shifts – and not only with his legs and his body, but with his mind and his sensibility as well.

Baudelaire shows how modern city life forces these new moves on everyone; but he shows, too, how in doing this it also paradoxically enforces new modes of freedom. A man who knows how to move in and around and through the traffic can go anywhere, down any of the endless urban corridors where traffic itself is free to go. This mobility opens up a great wealth of new experiences and activities for the urban masses.

Moralists and people of culture will condemn these popular urban pursuits as low, vulgar, sordid, empty of social or spiritual value. But when Baudelaire's poet lets his halo go and keeps moving, he makes a great discovery. He finds to his amazement that the aura of artistic purity and sanctity is only incidental, not essential, to art, and that poetry can thrive just as well, and maybe even better, on the other side of the boulevard, in those low, "unpoetic" places like un mauvais lieu where this poem itself is born. One of the paradoxes of modernity, as Baudelaire sees it here, is that its poets will become more deeply and authentically poetic by becoming more like ordinary men. If he throws himself into the moving chaos of everyday life in the modern world – a life of which the new traffic is a primary symbol – he can appropriate this life for art. The "bad poet" in this world is the poet who hopes to keep his purity intact by keeping off the streets, free from the risks of traffic. Baudelaire wants works of art that will be born in the midst of the traffic, that will spring from its anarchic energy, from the incessant danger and terror of being there, from the precarious pride and exhilaration of the man who has survived so far. Thus "Loss of a Halo" turns out to be a declaration of something gained, a rededication of the poet's powers to a new kind of art. His mouvements brusques, those sudden leaps and swerves so crucial for everyday survival in the city streets, turn out to be sources of creative power as well. In the century to come, these moves will become paradigmatic gestures of modernist art and thought.[9]

Ironies proliferate from this primal modern scene. They unfold in Baudelaire's nuances of language. Consider a phrase like La fange du macadam, "the mire of the macadam;" La fange in French is not only a literal word for mud; it is also a figurative word for mire, filth, vileness, corruption, degradation, all that is foul and loathsome. In classical oratorical and poetic diction, it is a "high" way of describing something "low." As such, it entails a whole cosmic hierarchy, a structure of norms and values not only aesthetic but metaphysical, ethical, political. La fange might be the nadir of the moral universe whose summit is signified by l'auréole. The irony here is that, so long as the poet's halo

falls into "La fange," it can never be wholly lost, because, so long as such an image still has meaning and power – as it clearly has for Baudelaire – the old hierarchical cosmos is still present on some plane of the modern world. But it is present precariously. The meaning of macadam is as radically destructive to La fange as to l'auréole: it paves over high and low alike.

We can go deeper into the macadam: we will notice that the word isn't French. In fact, the word is derived from John McAdam of Glasgow, the eighteenth-century inventor of modern paving surface. It may be the first word in that language that twentieth-century Frenchmen have satirically named Franglais: it paves the way for le parking, le shopping, le weekend, le drugstore, le mobile-home, and far more. This language is so vital and compelling because it is the international language of modernization. Its new words are powerful vehicles of new modes of life and motion. The words may sound dissonant and jarring, but it is as futile to resist them as to resist the momentum of modernization itself. It is true that many nations and ruling classes feel – and have reason to feel – threatened by the flow of new words and things from other shores.[10] There is a wonderful paranoid Soviet word that expresses this fear: infiltrazya. We should notice, however, that what nations have normally done, from Baudelaire's time to our own, is, after a wave (or at least a show) of resistance, not only to accept the new thing but to create their own word for it, in the hope of blotting out embarrassing memories of underdevelopment. (Thus the Académie Française, after refusing all through the 1960s to admit le parking meter to the French language, coined and quickly canonized le parcmetre in the 1970s.)

Baudelaire knew how to write in the purest and most elegant classical French. Here, however, with the "Loss of a Halo," he projects himself into the new, emerging language, to make art out of the dissonances and incongruities that pervade – and, paradoxically, unite – the whole modern world. "In place of the old national seclusion and self-sufficiency," the Manifesto says, modern bourgeois society brings us "intercourse in every direction, universal interdependence of nations. And, as in material, so in intellectual production. The spiritual creations of nations become" – note this image, paradoxical in a bourgeois world – "common property." Marx goes on: "National one-sidedness and narrow-mindedness become more and more impossible, and from the numerous local and national literatures, there arises a world literature." The mire of the macadam will turn out to be one of the foundations from which this new world literature of the twentieth century will arise.[11]

There are further ironies that arise from this primal scene. The halo that falls into the mire of the macadam is endangered but not destroyed; instead, it is carried along and incorporated into the general flow of traffic. One salient feature of the commodity economy, as Marx explains, is the endless metamorphosis of its market values. In this economy, anything goes if it pays, and no human possibility is ever wiped off the books; culture becomes an enormous warehouse in which everything is kept in stock on the chance that someday, somewhere, it might sell. Thus the halo that the modern poet lets go (or throws off) as obsolete may, by virtue of its very obsolescence, metamorphose into an icon, an object of nostalgic veneration for those who, like the "bad poets" X and Z, are trying to escape from modernity. But alas, the anti-modern artist – or thinker or politician – finds himself on the same streets, in the same mire, as the modernist one. This modern environment serves as both a physical and a spiritual lifeline – a primary source of material and energy – for both.

The difference between the modernist and the anti-modernist, so far as they are concerned, is that the modernist makes himself at home here, while the anti-modern searches the streets for a way out. So far as the traffic is concerned, however, there is no difference between them at all: both alike are hindrances and hazards to the horses and vehicles whose paths they cross, whose free movement they impede. Then, too, no matter how closely the anti-modernist may cling to his aura of spiritual purity, he is bound to lose it, more likely sooner than later, for the same reason that the modernist lost it: he will be forced to discard balance and measure and decorum and to learn the grace of brusque moves in order to survive. Once again, however opposed the modernist and the anti-modernist may think they are, in the mire of the macadam, from the viewpoint of the endlessly moving traffic, the two are one.

Ironies beget more ironies. Baudelaire's poet hurls himself into a confrontation with the "moving chaos" of the traffic, and strives not only to survive

but to assert his dignity in its midst. But his mode of action seems self-defeating, because it adds yet another unpredictable variable to an already unstable totality. The horses and their riders, the vehicles and their drivers, are trying at once to outpace each other and to avoid crashing into each other. If, in the midst of all this, they are also forced to dodge pedestrians who may at any instant dart out into the road, their movements will become even more uncertain, and hence more dangerous than ever. Thus, by contending against the moving chaos, the individual only aggravates the chaos.

But this very formulation suggests a way that might lead beyond Baudelaire's irony and out of the moving chaos itself. What if the multitudes of men and women who are terrorized by modern traffic could learn to confront it *together*? This will happen just six years after "Loss of a Halo" (and three years after Baudelaire's death), in the days of the Commune in Paris in 1871, and again in Petersburg in 1905 and 1917, in Berlin in 1918, in Barcelona in 1936, in Budapest in 1956, in Paris again in 1968, and in *dozens* of cities all over the world, from Baudelaire's time to our own – the boulevard will be abruptly transformed into the stage for a new primal modern scene. This will not be the sort of scene that Napoleon or Haussmann would like to see, but nonetheless one that their mode of urbanism will have helped to make.

As we reread the old histories, memoirs and novels, or regard the old photos or newsreels, or stir our own fugitive memories of 1968, we will see whole classes and masses move into the street together. We will be able to discern two phases in their activity. At first the people stop and overturn the vehicles in their path, and set the horses free: here they are avenging themselves on the traffic by decomposing it into its inert original elements. Next they incorporate the wreckage they have created into their rising barricades: they are recombining the isolated, inanimate elements into vital new artistic and political forms. For one luminous moment, the multitude of solitudes that make up the modern city come together in a new kind of encounter, to make a *people*. "The streets belong to the people": they seize control of the city's elemental matter and make it their own. For a little while the chaotic modernism of solitary brusque moves gives way to an ordered modernism of mass movement. The "heroism of modern life" that

Baudelaire longed to see will be born from his primal scene in the street. Baudelaire does not expect this (or any other) new life to last. But it will be born again and again out of the street's inner contradictions. It may burst into life at any moment, often when it is least expected. This possibility is a vital flash of hope in the mind of the man in the mire of the macadam, in the moving chaos, on the run.

NOTES

1 My picture of the Napoleon III–Haussmann transformation of Paris has been put together from several sources: Siegfried Giedion, *Space, Time and Architecture* (1941; 5th edition, Harvard, 1966), 744–775; Robert Moses, "Haussmann," in *Architectural Forum*, July 1942, 57–66; David Pinkney, *Napoleon III and the Rebuilding of Paris* (1958; Princeton, 1972); Leonardo Benevolo, *A History of Modern Architecture* (1960, 1966; translated from the Italian by H.J. Landry, 2 volumes, MIT, 1971), I, 61–65; Françoise Choay, *The Modern City: Planning in the Nineteenth Century* (George Braziller, 1969), especially 15–26; Howard Saalman, *Haussmann: Paris Transformed* (Braziller, 1971); and Louis Chevalier, *Laboring Classes and Dangerous Classes: Paris in the First Half of the Nineteenth Century*, 1970, translated by Frank Jellinek (Howard Fertig, 1973). Haussmann's projects are skillfully placed in the context of long-term European political and social change by Anthony Vidler, "The Scenes of the Street: Transformations in Ideal and Reality, 1750–1871," in *On Streets*, edited by Stanford Anderson (MIT, 1978), 28–111. Haussmann commissioned a photographer, Charles Marville, to photograph dozens of sites slated for demolition and so preserve their memory for posterity. These photographs are preserved in the Musée Carnavalet, Paris. A marvelous selection was exhibited in New York and other American locations in 1981. The catalogue, French Institute/Alliance Française, *Charles Marville: Photographs of Paris, 1852–1878*, contains a fine essay by Maria Morris Hamburg.

2 In *Laboring Classes and Dangerous Classes*, cited in note 1, Louis Chevalier, the venerable historian of Paris, gives a horrific, excruciatingly detailed account of the ravages to which the old central neighborhoods in the pre-Haussmann decades were subjected: demographic bombardment, which doubled the population while the erection of luxury housing and government buildings sharply reduced the overall housing stock; recurrent mass unemployment, which in a pre-welfare era led directly to starvation; dreadful epidemics of typhus and cholera, which took their greatest toll in the old *quartiers*. All this suggests why the Parisian poor, who fought so bravely on so many fronts in the nineteenth century, put up no resistance to the destruction of their neighborhoods: they may well have been willing to go, as Baudelaire said in another context, anywhere out of their world.

The little-known essay by Robert Moses, also cited in note 1, is a special treat for all those who savor the ironies of urban history. In the course of giving a lucid and balanced overview of Haussmann's accomplishments, Moses crowns himself as his successor, and implicitly bids for still more Haussmann-type authority to carry out even more gigantic projects after the war. The piece ends with an admirably incisive and trenchant critique that anticipates, with amazing precision and deadly accuracy, the criticism that would be directed a generation later against Moses himself, and that would finally help to drive Haussmann's greatest disciple from public life.

3 Haussmann's engineers invented a tree-lifting machine that enabled them to transplant thirty-year-old trees in full leaf, and thus to create shady avenues overnight, seemingly ex nihilo. Giedion, *Space, Time and Architecture*, 757–59.

4 See Engels, in his pamphlet *The Housing Question* (1872), on

the method called "Haussmann" . . . I mean the practice, which has now become general, of making breaches in working-class quarters of our big cities, especially in those that are centrally situated . . . The result is everywhere the same: the most scandalous alleys and lanes disappear, to the accompaniment of lavish self-glorification by the bourgeoisie on account of this tremendous success – but ashamed of it, I have looked studiously but vainly among them for a single face completely unsympathetic with the prevailing expression of good nature and light-heartedness.

Is it doubtful that it does men good to come together in this way in pure air and under the light of heaven, or that it must have an influence directly counteractive to that of the ordinary hard, hustling working hours of town life?

You will agree with me, I am sure, that it is not, and that opportunity, convenient, attractive opportunity, for such congregation, is a very good thing to provide for, in planning the extension of a town.

[. . .]

Think that the ordinary state of things to many is at this beginning of the town. The public is reading just now a little book in which some of your streets of which you are not proud are described. Go into one of those red cross streets any fine evening next summer, and ask how it is with their residents. Oftentimes you will see half a dozen sitting together on the door-steps or, all in a row, on the curb-stones, with their feet in the gutter; driven out of doors by the closeness within; mothers among them anxiously regarding their children who are dodging about at their play, among the noisy wheels on the pavement.

Again, consider how often you see young men in knots of perhaps half a dozen in lounging attitudes rudely obstructing the sidewalks, chiefly led in their little conversation by the suggestions given to their minds by what or whom they may see passing in the street, men, women, or children, whom they do not know and for whom they have no respect or sympathy. There is nothing among them or about them which is adapted to bring into play a spark of admiration, of delicacy, manliness, or tenderness. You see them presently descend in search of physical comfort to a brilliantly lighted basement, where they find others of their sort, see, hear, smell, drink, and eat all manner of vile things.

Whether on the curb-stones or in the dram-shops, these young men are all under the influence of the same impulse which some satisfy about the tea-table with neighbors and wives and mothers and children, and all things clean and wholesome, softening, and refining.

If the great city to arise here is to be laid out little by little, and chiefly to suit the views of land-

owners, acting only individually, and thinking only of how what they do is to affect the value in the next week or the next year of the few lots that each may hold at the time, the opportunities of so obeying this inclination as at the same time to give the lungs a bath of pure sunny air, to give the mind a suggestion of rest from the devouring eagerness and intellectual strife of town life, will always be few to any, to many will amount to nothing.

But is it possible to make public provision for recreation of this class, essentially domestic and secluded as it is?

It is a question which can, of course, be conclusively answered only from experience. And from experience in some slight degree I shall answer it. There is one large American town, in which it may happen that a man of any class shall say to his wife, when he is going out in the morning:

"My dear, when the children come home from school, put some bread and butter and salad in a basket, and go to the spring under the chestnut-tree where we found the Johnson's last week. I will join you there as soon as I can get away from the office. We will walk to the dairy-man's cottage and get some tea, and some fresh milk for the children, and take our supper by the brook-side"

and this shall be no joke, but the most refreshing earnest.

There will be room enough in the Brooklyn Park, when it is finished, for several thousand little family and neighborly parties to bivouac at frequent intervals through the summer, without discommoding one another, or interfering with any other purpose, to say nothing of those who can be drawn out to make a day of it, as many thousand were last year. And although the arrangements for the purpose were yet very incomplete, and but little ground was at all prepared for such use, besides these small parties, consisting of one or two families, there came also, in companies of from thirty to a hundred and fifty, somewhere near twenty thousand children with their parents, Sunday-school teachers, or other guides and friends, who spent the best part of a day under the trees and on the turf, in recreations of which the predominating element was of this neighborly receptive class. Often they would bring a fiddle, flute, and harp, or other music.

Tables, seats, shade, turf, swings, cool spring-water, and a pleasing rural prospect, stretching off half a mile or more each way, unbroken by a carriage road or the slightest evidence of the vicinity of the town, were supplied them without charge and bread and milk and ice-cream at moderate fixed charges. In all my life I have never seen such joyous collections of people. I have, in fact, more than once observed tears of gratitude in the eyes of poor women, as they watched their children thus enjoying themselves.

The whole cost of such neighborly festivals, even when they include excursions by rail from the distant parts of the town, does not exceed for each person, on an average, a quarter of a dollar; and when the arrangements are complete, I see no reason why thousands should not come every day where hundreds come now to use them; and if so, who can measure the value, generation after generation, of such provisions for recreation to the over-wrought, much-confined people of the great town that is to be?

For this purpose neither of the forms of ground we have heretofore considered are at all suitable. We want a ground to which people may easily go after their day's work is done, and where they may stroll for an hour, seeing, hearing, and feeling nothing of the bustle and jar of the streets, where they shall, in effect, find the city put far away from them. We want the greatest possible contrast with the streets and the shops and the rooms of the town which will be consistent with convenience and the preservation of good order and neatness. We want, especially, the greatest possible contrast with the restraining and confining conditions of the town, those conditions which compel us to walk circumspectly, watchfully, jealously, which compel us to look closely upon others without sympathy. Practically, what we most want is a simple, broad, open space of clean greensward, with sufficient play of surface and a sufficient number of trees about it to supply a variety of light and shade. This we want as a central feature. We want depth of wood enough about it not only for comfort in hot weather, but to completely shut out the city from our landscapes.

The word *park*, in town nomenclature, should, I think, be reserved for grounds of the character and purpose thus described.

[...]

A park fairly well managed near a large town, will surely become a new center of that town. With the determination of location, size, and boundaries should therefore be associated the duty of arranging new trunk routes of communication between it and the distant parts of the town existing and forecasted.

These may be either narrow informal elongations of the park, varying say from two to five hundred feet in width, and radiating irregularly from it, or if, unfortunately, the town is already laid out in the unhappy way that New York and Brooklyn, San Francisco and Chicago, are, and, I am glad to say, Boston is not, on a plan made long years ago by a man who never saw a spring-carriage, and who had a conscientious dread of the Graces, then we must probably adopt formal Park-ways. They should be so planned and constructed as never to be noisy and seldom crowded, and so also that the straightforward movement of pleasure-car carriages need never be obstructed, unless at absolutely necessary crossings, by slow-going heavy vehicles used for commercial purposes. If possible, also, they should be branched or reticulated with other ways of a similar class, so that no part of the town should finally be many minutes' walk from some one of them; and they should be made interesting by a process of planting and decoration, so that in necessarily passing through them, whether in going to or from the park, or to and from business, some substantial recreative advantage may be incidentally gained. It is a common error to regard a park as something to be produced complete in itself, as a picture to be painted on canvas. It should rather be planned as one to be done in fresco, with constant consideration of exterior objects, some of them quite at a distance and even existing as yet only in the imagination of the painter.

I have thus barely indicated a few of the points from which we may perceive our duty to apply the means in our hands to ends far distant, with reference to this problem of public recreations. Large operations of construction may not soon be desirable, but I hope you will agree with me that there is little room for question, that reserves of ground for the purposes I have referred to should be fixed upon as soon as possible, before the difficulty of arranging them, which arises from private building, shall be greatly more formidable than now.

"The Meager and Unimaginative Character of Modern City Plans" and "Artistic Limitations of Modern City Planning"

from *City Planning According to Artistic Principles* (1898)

Camillo Sitte

Editors' Introduction

By the late nineteenth century, processes of modernization had speeded up and industrial mechanization of the building trades and bureaucratization of planning were taking hold. Rapid urban growth continued and people were witnessing immense changes to their cities as old areas were "modernized" and new ones were built to the modern aesthetic of straight lines and rectilinear, unornamented buildings. Within the emerging profession of city planning, an engineering approach was dominant. Public spaces were increasingly designed around the main goal of efficiently moving ever increasing amounts of carriage traffic. A faster, larger, less detailed, and less finely nuanced way of life was coming into being and was widely embraced.

Within this context, when to be modern meant looking completely to the future and ignoring the past, Viennese architect and city planner Camillo Sitte (1843–1903) argued that the past had much to offer. Coming from an arts and crafts tradition, he emerged as a strong voice for human scale in architecture, lamenting its lack in modern architecture along with the pervasive functional approach to urban design. Over a hundred years later, his laments sound familiar to our ears: the loss of public life from public spaces, bland environments, lack of detail, standardization, and excessively wide streets.

In his 1898 treatise *City Planning According to Artistic Principles*, from which these readings are taken, Sitte called for an "artistic renaissance" of city-building, arguing that a balance could be found between art and function. While giving full due to modern city planning for improving the sanitary conditions of cities, he severely criticized the profession's lack of ability to create new urban public spaces that were as good or as inviting to people as the old. He admired the picturesque qualities and human scale of pre-industrial European cities and argued for an approach to modernism that built on these traditions rather than discarded them. He advocated looking at the good public spaces of the past, determining their essential physical qualities, and applying those qualities to modern conditions. His ideas were grounded in extensive empirical observations, and he presented them using figure ground plans that analyzed and compared the spatial qualities of old and new plazas.

Sitte's ideas briefly found an audience and influenced numerous town planning ordinances throughout Europe. In the 1920s, however, modernist architects soundly renounced Sitte's theory – the architect Le Corbusier referring to Sitte's ideal of crooked streets as the pack-donkey's way. In 1965, Sitte's treatise was republished (Random House) and rediscovered by a new generation of humanistic architects and city planners. The ideas influenced the current New Urbanism movement, whose adherents look to Sitte as a seminal historic reference.

Two books in the same spirit as Sitte's *City Planning According to Artistic Principles* are Werner Hegemann and Elbert Peets, *American Vitruvius* (New York: B. Blom, 1972) and Andres Duany, Elizabeth Plater-Zyberk, and Robert Alminana, *The New Civic Art: Elements of Town Planning* (New York: Rizzoli, 2003).

THE MEAGER AND UNIMAGINATIVE CHARACTER OF MODERN CITY PLANS

How in recent times the history of the art of city building has failed to synchronize with the history of architecture and with that of the other creative arts is indeed astonishing. City planning stubbornly goes its own way, unconcerned with what transpires around it. This difference was already striking in the Renaissance and Baroque periods, but it has become even more pronounced in modern times as old styles have been revived once again. This time, of course, exactitude in imitation was taken much more seriously, the example of the Ancients being faithfully adhered to in every way possible. Actual copies of old structures were erected in monumental and costly fashion without answering any real need or practical purpose – merely out of enthusiasm for the splendor of ancient art. The Walhalla at Regensburg was created in the exact image of a Greek temple, the Loggia dei Lanzi found its imitation (the Feldherrn-Halle) at Munich, Early Christian basilicas were erected again, Greek propylaea and Gothic cathedrals were built, but what became of the plazas that belonged with them? Agora, forum, market place, acropolis – nobody remembered them.

The modern city-builder has been deprived in alarming fashion of the resources of his art. The precisely straight house-line and the cubic building-block are all that he can offer to compete with the wealth of the past. The architect is allowed millions to construct balconies, towers, gables, caryatids or anything else that his sketchbook might contain, and his sketchbook contains everything the past has ever created in any corner of the world. The town planner, on the other hand, is not given a penny for the installation of colonnades, porticoes, triumphal arches, or any other motifs that are essential to his art; not even the voids between the building blocks are put at his disposal for artistic use, because even the open air already belongs to someone else: the highway or sanitation engineer.

So it has come to pass that all the good features of artistic city-building were dropped by the wayside one after the other, until nothing is left of them, not even a memory. Although we can see clearly the tremendous difference that exists between the old plazas, still charming as they are, and the monotonous modern ones, yet, despite this, we unfortunately consider it self-evident that churches and monuments must stand in the center of their plazas, that all streets must intersect at right angles and open wide all around a plaza, that the buildings need not close up about a square, and that monumental structures need not form part of such a closure. We are well aware of the effect of an old plaza, but how to produce it under modern conditions is not understood because we are no longer cognizant of the relation between cause and effect in these matters.

The theorist of modern city planning, R. Baumeister, says in his book about city expansion, '. . . the various elements which produce a pleasing architectural impression (as regards plazas) *are hardly reducible to universal rules.*' Does not this statement confirm what we have just said? Do not the results of what has been presented so far add up to precisely such general rules? – Enough rules, in fact, to compose a whole textbook on city planning, as well as a history of this art, if they are worked out in detail? A thorough study of the variations that the Baroque masters alone carried out would suffice to fill volumes. If, however, our first and thus far our only theorist in this field can express the above opinion, does this not demonstrate that we are now no longer aware of the relationship between cause and effect?

Today nobody is concerned with city planning as an art – only as a technical problem. When, as a result, the artistic effect in no way lives up to our expectations, we are left bewildered and helpless; nevertheless, in dealing with the next project it is again treated wholly from the technical point of view, as if it were the layout of a railroad in which artistic questions are not involved.

Even in modern histories of art, which discuss every insignificant thing, city planning has not been granted the humblest little spot, whereas bookbinding, pewter work, and costume design are readily allowed space next to Phidias and Michelangelo. From this it might be understood that we have lost the thread of artistic tradition in city planning, although it is not clear why. But now back to our analysis of the matter at hand.

There exist an infinite variety of derogatory opinions about modern planning. In the daily press and in professional publications they are repeated again and again. However, the most they do is to attribute the cause of bad effects to an overly pedantic straightness of line in our house fronts. Even Baumeister says, 'one rightly laments the boredom of modern streets,' and he then criticizes the 'unwieldy massive effect' of modern blocks of buildings. With regard to the siting of monuments it is only reported that several major 'monumental' catastrophes can be listed; yet no reason for the bad effects is ever given, since it is as irrevocable as natural law that every monument can only be placed in the middle of its plaza, in order that one also may have a good look at the celebrity from the rear. One of the most discerning opinions, which Baumeister mentioned, can be quoted here. It is from the Paris *Figaro* of August 23, 1874, and it says in a report about the trip of Marshal MacMahon:

> Rennes does not actually feel an antipathy toward the Marshal, but the town is totally incapable of any enthusiasm. I have noticed that this is the case with all towns that are laid out along straight lines and in which the streets intersect at right angles. The straight line prevents any excitement from arising. Thus one could also observe in the year 1870 how the completely regularly designed towns could be captured by three lancers, while really old and twisted towns were ready to defend themselves to the utmost.

Straight lines and right angles are certainly characteristic of insensitive planning, but are apparently not decisive in this matter, because Baroque planning also used straight lines and right angles, achieving powerful and truly artistic effects in spite of them. In the layout of streets it is true that rectilinearity is a weakness. An undeviating boulevard, miles long, seems boring even in the most beautiful surroundings. It is unnatural, it does not adapt itself to irregular terrain, and it remains uninteresting in effect, so that, mentally fatigued, one can hardly await its termination. An ordinary street, if excessively long, has the same effect. But as the more frequent shorter streets of modern planning also produce an unfortunate effect, there must be some other cause for it. It is the same as in the plazas, namely *faulty closure of the sides of the street*. The continual breaching by wide cross streets, so that on both sides nothing is left but a row of separated blocks of buildings, is the main reason why no unified impression can be attained. This may be demonstrated most clearly by comparing old arcades with their modern imitations. Ancient arcades, nothing short of magnificent in their architectural detail, run uninterruptedly along the whole curve of a street as far as the eye can see; or they encircle a plaza enclosing it completely; or at least they run unbroken along one side of it. Their whole effect is based on continuity, for only by it can the succession of arches become a large enough unity to create an impact. The situation is completely different in modern planning. Although occasional outstanding architects have, in their enthusiasm for this magnificent old motif, succeeded in providing us with such covered walks – as, for instance, in Vienna around the Votive Church and at the new Rathaus – these hardly remind us of the ancient models, because their effect is totally different. The separate sections are larger and much more sumptuously carried out than almost any ancient predecessors. Yet the intended effect is absent. Why? Because each separate loggia is attached to its own building-block, and the cuts made by the numerous broad cross streets prevent the slightest effect of continuity. Only if the openings of these intersecting streets were spanned by a continuation of the arcade could any coherence result that might then create a grandiose impression. Lacking this, the dismembered motif is like a hoe without a handle.

For the same reason a coherent effect does not come about in our streets. A modern street is made up primarily of corner buildings. A row of isolated blocks of buildings is going to look bad under any circumstances, even if placed in a curved line.

These considerations bring us close to the crux of the matter. In modern city planning the ratio between the built-up and the open spaces is exactly reversed. Formerly the empty spaces (streets and plazas) were a unified entity of shapes calculated for their impact; today building lots are laid out as regularly-shaped closed forms, and what is left over between them become streets or plazas. Formerly all that was crooked and ugly lay hidden in the built-up areas; today in the process of laying out the various building lots all irregular wedges that are left over become plazas, since the prime rule is that '*architecturally* speaking, a street pattern should first of all provide convenient house plans. Therefore street crossings at right angles are an advantage. And it is certainly wrong to adopt irregular angles as a principle of parcelling' (Baumeister). Well, but what architect is afraid of an irregularly-shaped building lot? Indeed, that would be a man who has not advanced beyond the most elementary principles of planning. Irregular building lots are just the ones that allow, without exception, the most interesting solutions and usually the better ones; not only because they demand a more careful study of the plan and prevent mechanical, run-of-the-mill design, but because, in the interior of such a building, wedge-shaped pieces are repeatedly left over and are splendidly suited for all sorts of little extra rooms (elevators, spiral-staircases, storage rooms, toilets, etc.), a feature which we miss in regular plans. To recommend rectangular building lots for their presumed architectural advantage is completely wrong. This could only be done by those who do not understand how to lay out ground plans. Is it possible that all the attractiveness of streets and plazas could fall victim to such a trivial misconception? It would almost seem so.

Studying the ground plan of a complicated building on an irregular building lot, one finds, if it is well designed, that all halls, chambers, and other principal rooms are of excellent proportion. Here again, irregularities are concealed in the thickness of the walls or in the shape of the service rooms described above. Nobody likes a triangular room, because the sight of it is unbearable and because furniture can never be well placed in it. Yet the circle or the ellipse of a spiral staircase can be accommodated nicely in it by varying the thickness of the wall. It is quite similar to what we find in ancient city plans. The hall-like forums were of regular shape, their voids calculated for their visible

effect, while all their irregularities were absorbed in the mass of surrounding structures. This was carried out down to the smallest detail, and in the end every irregularity of the site seems to dissolve away and be hidden in the thickness of the walls; it is simple and very clever. Today the exact opposite of this takes place. As illustrations, we take three plazas from the same city, Trieste: the Piazza della Caserma, the Piazza della Legna, and the Piazza della Borsa. Artistically speaking, these are not really plazas at all, but only triangularly-shaped remnants of empty space, left over in the cutting out of right-angled city blocks. When one then notices the frequency of broad, unfavorable street openings, it becomes immediately clear that it is just as impossible to position a monument on such plazas as to show a building off to advantage. Such a plaza is as unbearable as a triangular room.

Regarding this, one thing needs further discussion. A special chapter has already been devoted to proving the appropriateness of the irregularity of old plazas. One might expect it also to be applicable here. But such is not the case, for between the two kinds of irregularity there is a crucial difference: the irregularity present in (the above mentioned plazas) is obvious and immediately observable to the eye, and it becomes the more awkward, the more regular the adjoining building façades and nearby town sections are shaped; on the contrary, the other irregularities were of a kind that deceived the eye, being noticeable on the drawing board, of course, but not in actuality.

Something similar occurs in ancient structures. In the ground plans of Romanesque and Gothic churches one rarely finds the various axes at right angles to each other, since the old masters were unable to gauge this accurately enough. It does not matter in this case, because it goes unnoticed. Similarly, there are great irregularities in the ground plans of ancient temple structures as regards the intercolumniation, etc. All this one only detects with precise measurement, not by the naked eye; it mattered little, since they were building for visual effect and not for the sake of the plan on paper. On the other hand, one has discovered almost incredible refinements in the curvatures of entablature, etc., refinements which, although they almost elude measurement, were carried out because their absence would have been noticed by the eye, and it was the eye that counted. The more com-

Le Corbusier was a founding member of the Congrès Internationaux d'Architecture Moderne (CIAM), whose modernist philosophy was proclaimed in its 1933 Manifesto, *The Charter of Athens*. Allan B. Jacobs and Donald Appleyard's "Toward an Urban Design Manifesto," *Journal of the American Planning Association* (1987), reprinted in Part Two of this volume, takes on *The Charter of Athens*, rejecting its utopian program and calling instead for an urbanism based upon social objectives for urban development and upon how people actually live and experience cities and space. Earlier, Jane Jacobs soundly refuted Le Corbusier's ideas and the modernist movement in architecture and city planning in the introduction to her book *The Life and Death of Great American Cities*, also reprinted in Part Two.

Many of Le Corbusier's most provocative writings are included in *The City of To-morrow and Its Planning* (New York: Dover, 1987) (translated by Frederich Etchells from *Urbanisme* [1929]). His later, less-well-known books include *Concerning Town Planning* (New Haven, CT: Yale University Press, 1948) (translated by Cliver Entwistle from *Propos d'Urbanisme* [1946]) and *L'Urbanisme des trois etablissements humaines* (Paris: Editions de Minuit, 1959).

Discussions and criticisms of Le Corbusier's ideas and their influences can be found in Robert Fishman, *Urban Utopias in the Twentieth Century* (New York: Basic Books, 1977), Peter Hall, *Cities of Tomorrow* (Oxford: Blackwell, 1988), Henry Russell Hitchcock, *Architecture: Nineteenth and Twentieth Centuries* (Baltimore, MD: Penguin, 1967); and Siegfried Giedion, *Space, Time and Architecture: The Growth of a New Tradition* (Cambridge, MA: Harvard University Press, 1963).

■ ■ ■ ■ ■ ■

THE PACK-DONKEY'S WAY AND MAN'S WAY

MAN walks in a straight line because he has a goal and knows where he is going; he has made up his mind to reach some particular place and he goes straight to it.

The pack-donkey meanders along, meditates a little in his scatter-brained and distracted fashion, he zigzags in order to avoid the larger stones, or to ease the climb, or to gain a little shade; he takes the line of least resistance.

But man governs his feelings by his reason; he keeps his feelings and his instincts in check, subordinating them to the aim he has in view. He rules the brute creation by his intelligence. His intelligence formulates laws which are the product of experience. His experience is born of work; man works in order that he may not perish. In order that production may be possible, a line of conduct is essential, the laws of experience must be obeyed. Man must consider the result in advance.

But the pack-donkey thinks of nothing at all, except what will save himself trouble.

The Pack-Donkey's Way is responsible for the plan of every continental city; including Paris, unfortunately.

In the areas into which little by little invading populations filtered, the covered wagon lumbered along at the mercy of bumps and hollows, of rocks or mire; a stream was an intimidating obstacle. In this way were born roads and tracks. At cross roads or along river banks the first huts were erected, the first houses and the first villages; the houses were planted along the tracks, along the Pack-Donkey's Way. The inhabitants built a fortified wall round and a town hall inside it. They legislated, they toiled, they lived, and always they respected the Pack-Donkey's Way. Five centuries later another and larger enclosure was built, and five centuries later still a third yet greater. The places where the Pack-Donkey's Way entered the town became the City Gates and the Customs officers were installed there. The village has become a great capital; Paris, Rome, and Stamboul are based upon the Pack-Donkey's Way.

The great capitals have no arteries; they have only capillaries: further growth, therefore, implies sickness or death. In order to survive, their existence has for a long time been in the hands of surgeons who operate constantly.

The Romans were great legislators, great colonizers, great administrators. When they arrived at a place, at a cross roads or at a river bank, they took a square and set out the plan of a rectilinear town, so that it should be clear and well-arranged, easy to police and to clean, a place in which you could find your way about and stroll with comfort – the working town or the pleasure town (Pompeii). The square plan was in conformity with the dignity of the Roman citizen.

But at home, in Rome itself, with their eyes turned towards the Empire, they allowed themselves to be stifled by the Pack-Donkey's Way. What an ironical situation! The wealthy, however, went far from the chaos of the town and built their great and well-planned villas, such as Hadrian's villa.

They were, with Louis XIV, the only great town-planners of the West.

In the Middle Ages, overcome by the year 1000, men accepted the leading of the pack-donkey, and long generations endured it after. Louis XIV, after trying to tidy up the Louvre (i.e. the Colonnade), became disgusted and took bold measures: he built Versailles, where both town and chateau were created in every detail in a rectilinear and well-planned fashion; the Observatoire, the Invalides and the Esplanade, the Tuileries and the Champs Elysées, rose far from the chaos, outside the town – all these were ordered and rectilinear.

The overcrowding had been exorcised. Everything else followed, in a masterly way: the Champs de Mars, l'Etoile, the avenues de Neuilly, de Vincennes, de Fontainebleau, etc., for succeeding generations to exploit.

But imperceptibly, as a result of carelessness, weakness and anarchy, and by the system of "democratic" responsibilities, the old business of overcrowding began again.

And as if that were not enough, people began to desire it; they have even created it in invoking the laws of beauty! The Pack-Donkey's Way has been made into a religion.

The movement arose in Germany as a result of a book by Camillo Sitte on town-planning, a most wilful piece of work; a glorification of the curved line and a specious demonstration of its unrivalled beauties. Proof of this was advanced by the example of all the beautiful towns of the Middle Ages; the author confounded the picturesque with the conditions vital to the existence of a city. Quite recently whole quarters have been constructed in Germany based on this *aesthetic*. (For it was purely a question of aesthetics.)

This was an appalling and paradoxical misconception in an age of motor-cars. "So much the better," said a great authority to me, one of those who direct and elaborate the plans for the extension of Paris; "motors will be completely held up!"

But a modern city lives by the straight line, inevitably; for the construction of buildings, sewers and tunnels, highways, pavements. The circulation of traffic demands the straight line; it is the proper thing for the heart of a city. The curve is ruinous, difficult and dangerous; it is a paralyzing thing.

The straight line enters into all human history, into all human aim, into every human act.

We must have the courage to view the rectilinear cities of America with admiration. If the aesthete has not so far done so, the moralist, on the contrary, may well find more food for reflection than at first appears.

The winding road is the Pack-Donkey's Way, the straight road is man's way.

The winding road is the result of happy-go-lucky heedlessness, of looseness, lack of concentration and animality.

The straight road is a reaction, an action, a positive deed, the result of self-mastery. It is sane and noble.

A city is a centre of intense life and effort.

A heedless people, or society, or town, in which effort is relaxed and is not concentrated, quickly becomes dissipated, overcome and absorbed by a nation or a society that goes to work in a positive way and controls itself.

It is in this way that cities sink to nothing and that ruling classes are overthrown.

A CONTEMPORARY CITY

The use of technical analysis and architectural synthesis enabled me to draw up my scheme for a contemporary city of three million inhabitants. The result of my work was shown in November 1922 at the Salon d' Automne in Paris. It was greeted with a sort of stupor; the shock of surprise caused rage in some quarters and enthusiasm in others. The solution I put forward was a rough one and completely uncompromising. There were no notes to accompany the plans, and, alas! not everybody can read a plan. I should have had to be constantly on the spot in order to reply to the fundamental questions which spring from the very depths of human feelings. Such questions are of profound interest and cannot remain unanswered. When at a later date it became necessary that this book should be written, a book in which I could formulate the new principles of Town Planning, I resolutely decided *first of all* to find answers to these fundamental questions. I have used two kinds of argument: first, those essentially human ones which start from the mind or the heart or the physiology of our sensations as a basis; secondly, historical and statistical arguments. Thus I could keep in touch with what is fundamental and at the same time be master of the environment in which all this takes place.

In this way I hope I shall have been able to help my reader to take a number of steps by means of which he can reach a sure and certain position. So that when I unroll my plans I can have, the happy assurance that his astonishment will no longer be stupefaction nor his fears mere panic.

A CONTEMPORARY CITY OF THREE MILLION INHABITANTS

Proceeding in the manner of the investigator in his laboratory, I have avoided all special cases, and all that may be accidental, and I have assumed an ideal site to begin with. My object was not to overcome the existing state of things, but *by constructing a theoretically water-tight formula to arrive at the fundamental principles of modern town planning*. Such fundamental principles, if they are genuine, can serve as the skeleton of any system of modern town planning; being as it were the *rules* according to which development will take place. We shall then be in a position to take a special case, no matter what: whether it be Paris, London, Berlin, New York or some small town. Then, as a result of what we have learnt, we can take control and decide in what direction the forthcoming battle is to be waged. For the desire to rebuild any great city in a modern way is to engage in a formidable battle. Can you imagine people engaging in a battle without knowing their objectives? Yet that is exactly what is happening. The authorities are compelled to do something, so they give the police white sleeves or set them on horseback, they invent sound signals and light signals, they propose to put bridges over streets or

moving pavements under the streets; more garden cities are suggested, or it is decided to suppress the tramways, and so on. And these decisions are reached in a sort of frantic haste in order, as it were, to hold a wild beast at bay. That BEAST is the great city. It is infinitely more powerful than all these devices. And it is just beginning to wake. What will to-morrow bring forth to cope with it?

We must have some rule of conduct.[1]

We must have fundamental principles for modern town planning.

Site

A level site is the ideal site. In all those places where traffic becomes over-intensified the level site gives a chance of a normal solution to the problem. Where there is less traffic, differences in level matter less.

The river flows far away from the city. The river is a kind of liquid railway, a goods station and a sorting house. In a decent house the servants' stairs do not go through the drawing room – even if the maid is charming (or if the little boats delight the loiterer leaning on a bridge).

Population

This consists of the citizens proper; of suburban dwellers; and of those of a mixed kind.

(a) Citizens are of the city: those who work and live in it.
(b) Suburban dwellers are those who work in the outer industrial zone and who do not come into the city: they live in garden cities.
(c) The mixed sort are those who work in the business parts of the city but bring up their families in garden cities.

To classify these divisions (and so make possible the transmutation of these recognized types) is to attack the most important problem in town planning, for such a classification would define the areas to be allotted to these three sections and the delimitation of their boundaries. This would enable us to formulate and resolve the following problems:

1 The *City* as a business and residential centre.
2 The *Industrial City* in relation to the *Garden Cities* (i.e. the question of transport).
3 The *Garden Cities* and the *daily transport* of the workers. Our first requirement will be an organ that is compact, rapid, lively and concentrated: this is the City with its well-organized centre. Our second requirement will be another organ, supple, extensive and elastic; this is the *Garden City* on the periphery.

Lying between these two organs, we must *require the legal establishment* of that absolute necessity, a protective zone which allows of extension, *a reserved zone* of woods and fields, a fresh-air reserve.

Density of population

The more dense the population of a city is the less are the distances that have to be covered. The moral, therefore, is that we must *increase the density of the centres of our cities, where business affairs are carried on.*

Lungs

Work in our modern world becomes more intensified day by day, and its demands affect our nervous system in a way that grows more and more dangerous. Modern toil demands quiet and fresh air, not stale air.

The towns of to-day can only increase in density at the expense of the open spaces which are the lungs of a city.

We must *increase the open spaces and diminish the distances to be covered.* Therefore the centre of the city must be constructed *vertically*.

The city's residential quarters must no longer be built along "corridor-streets," full of noise and dust and deprived of light.

It is a simple matter to build urban dwellings away from the streets, without small internal courtyards and with the windows looking on to large parks; and this whether our housing schemes are of the type with "set-backs" or built on the "cellular" principle.

The street

The street of to-day is still the old bare ground which has been paved over, and under which a few tube railways have been run.

The modern street in the true sense of the word is a new type of organism, a sort of stretched-out workshop, a home for many complicated and delicate organs, such as gas, water and electric mains. It is contrary to all economy, to all security, and to all sense to bury these important service mains. They ought to be accessible throughout their length. The various storeys of this stretched-out workshop will each have their own particular functions. If this type of street, which I have called a "workshop," is to be realized, it becomes as much a matter of *construction* as are the houses with which it is customary to flank it, and the bridges which carry it over valleys and across rivers.

The modern street should be a masterpiece of civil engineering and no longer a job for navvies.

The "corridor-street" should be tolerated no longer, for it poisons the houses that border it and leads to the construction of small internal courts or "wells."

Traffic

Traffic can be classified more easily than other things.

To-day traffic is not classified – it is like dynamite flung at hazard into the street, killing pedestrians. Even so, *traffic does not fulfill its function.* This sacrifice of the pedestrian leads nowhere.

If we classify traffic we get:

(a) Heavy goods traffic.
(b) Lighter goods traffic, i.e. vans, etc., which make short journeys in all directions.
(c) Fast traffic, which covers a large section of the town.

Three kinds of roads are needed, and in superimposed storeys:

(a) Below-ground there would be the street for heavy traffic.[2] This storey of the houses would consist merely of concrete piles, and between them large open spaces which would form a sort of clearing-house where heavy goods traffic could load and unload.

(b) At the ground floor level of the buildings there would be the complicated and delicate network of the ordinary streets taking traffic in every desired direction.

(c) Running north and south, and east and west, and forming the two great axes of the city, there would be great *arterial roads for fast one-way traffic* built on immense reinforced concrete bridges 120 to 180 yards in width and approached every half-mile or so by subsidiary roads from ground level. These arterial roads could therefore be joined at any given point, so that even at the highest speeds the town can be traversed and the suburbs reached without having to negotiate any cross-roads.

The number of existing streets *should be diminished by two-thirds*. The number of crossings depends directly on the number of streets; and *cross-roads are an enemy to traffic*. The number of existing streets was fixed at a remote epoch in history. The perpetuation of the boundaries of properties has, almost without exception, preserved even the faintest tracks and footpaths of the old village and made streets of them, and sometimes even an avenue (see "The Pack-Donkey's Way and Man's Way").

The result is that we have cross-roads every fifty yards, even every twenty yards or ten yards. And this leads to the ridiculous traffic congestion we all know so well.

The distance between two bus stops or two tube stations gives us the necessary unit for the distance between streets, though this unit is conditional on the speed of vehicles and the walking capacity of pedestrians. So an average measure of about 400 yards would give the normal separation between streets, and make a standard for urban distances. My city is conceived on the gridiron system with streets every 400 yards, though occasionally these distances are subdivided to give streets every 200 yards.

This triple system of superimposed levels answers every need of motor traffic (lorries, private cars, taxis, buses) because it provides for rapid and *mobile* transit.

Traffic running on fixed rails is only justified if it is in the form of a convoy carrying an immense load; it then becomes a sort of extension of the underground system or of trains dealing with suburban traffic. *The tramway has no right to exist in the heart of the modern city.*

If the city thus consists of plots about 400 yards square, this will give us sections of about 40 acres in area, and the density of population will vary from 50,000 down to 6,000, according as the "lots" are developed for business or for residential purposes. The natural thing, therefore, would be to continue to apply our unit of distance as it exists in the Paris tubes to-day (namely, 400 yards) and to put a station in the middle of each plot.

Following the two great axes of the city, two "storeys" below the arterial roads for fast traffic, would run the tubes leading to the four furthest points of the garden city suburbs, and linking up with the metropolitan network [. . .]. At a still lower level, and again following these two main axes, would run the one-way loop systems for suburban traffic, and below these again the four great main lines serving the provinces and running north, south, east and west. These main lines would end at the Central Station, or better still might be connected up by a loop system.

The station

There is only one station. The only place for the station is in the centre of the city. It is the natural place for it, and there is no reason for putting it anywhere else. The railway station is the hub of the wheel.

The station would be an essentially subterranean building. Its roof, which would be two storeys above the natural ground level of the city, would form the aerodrome for aero-taxis. This aerodrome (linked up with the main aerodrome in the protected zone) must be in close contact with the tubes, the suburban lines, the main lines, the main arteries and the administrative services connected with all these.

THE PLAN OF THE CITY

The basic principles we must follow are these:

1 We must de-congest the centres of our cities.
2 We must augment their density.
3 We must increase the means for getting about.
4 We must increase parks and open spaces.

At the very centre we have the STATION with its landing stage for aero-taxis.

Running north and south, and east and west, we have the MAIN ARTERIES for fast traffic, forming elevated roadways 120 feet wide.

At the base of the sky-scrapers and all round them we have a great open space 2,400 yards by 1,500 yards, giving an area of 3,600,000 square yards, and occupied by gardens, parks and avenues. In these parks, at the foot of and round the sky-scrapers, would be the restaurants and cafes, the luxury shops, housed in buildings with receding terraces: here too would be the theatres, halls and so on; and here the parking places or garage shelters.

The sky-scrapers are designed purely for business purposes.

On the left we have the great public buildings, the museums, the municipal and administrative offices. Still further on the left we have the "Park" (which is available for further logical development of the heart of the city).

On the right, and traversed by one of the arms of the main arterial roads, we have the warehouses, and the industrial quarters with their goods stations.

All round the city is the *protected zone* of woods and green fields.

Further beyond are the *garden cities*, forming a wide encircling band.

Then, right in the midst of all these, we have the *Central Station*, made up of the following elements:

(a) The landing-platform; forming an aerodrome of 200,000 square yards in area.
(b) The entresol or mezzanine; at this level are the raised tracks for fast motor traffic: the only crossing being gyratory.
(c) The ground floor where are the entrance halls and booking offices for the tubes, suburban, main line and air traffic.
(d) The "basement": here are the tubes which serve the city and the main arteries.
(e) The "sub-basement": here are the suburban lines running on a one-way loop.

(f) The "sub-sub-basement": here are the main lines (going north, south, east and west).

The city

Here we have twenty-four sky-scrapers capable each of housing 10,000 to 50,000 employees; this is the business and hotel section, etc., and accounts for 400,000 to 600,000 inhabitants.

The residential blocks, of the two main types already mentioned, account for a further 600,000 inhabitants.

The garden cities give us a further 2,000,000 inhabitants, or more.

In the great central open space are the cafes, restaurants, luxury shops, halls of various kinds, a magnificent forum descending by stages down to the immense parks surrounding it, the whole arrangement providing a spectacle of order and vitality.

Density of population

(a) The sky-scraper: 1,200 inhabitants to the acre.
(b) The residential blocks with set-backs: 120 inhabitants to the acre. These are the luxury dwellings.
(c) The residential blocks on the "cellular" system, with a similar number of inhabitants.

This great density gives us our necessary shortening of distances and ensures rapid intercommunication.

Note – The average density to the acre of Paris in the heart of the town is 146, and of London 63; and of the over-crowded quarters of Paris 213, and of London 169.

Open spaces

Of the area (a), 95 percent of the ground is open (squares, restaurants, theatres).

Of the area (b), 85 percent of the ground is open (gardens, sports grounds).

Of the area (c), 48 percent of the ground is open (gardens, sports grounds).

Educational and civic centres, universities, museums of art and industry, public services, county hall

The "Jardin anglais." (The city can extend here, if necessary.)

Sports grounds: Motor racing track, Racecourse, Stadium, Swimming baths, etc.

The Protected Zone (which will be the property of the city), with its *Aerodome*, a zone in which all building would be prohibited; reserved for the growth of the city as laid down by the municipality: it would consist of woods, fields, and sports grounds. The forming of a "protected zone" by continual purchase of small properties in the immediate vicinity of the city is one of the most essential and urgent tasks which a municipality can pursue. It would eventually represent a tenfold return on the capital invested.

Industrial quarters[3]
Types of buildings to be employed

For business: sky-scrapers sixty storeys high with no internal wells or courtyards.

Residential buildings with "set-backs," of six double storeys; again with no internal wells: the flats looking on either side on to immense parks.

Residential buildings on the "cellular" principle, with "hanging gardens," looking on to immense parks; again no internal wells. These are "service-flats" of the most modern kind.

GARDEN CITIES

Their aesthetic, economy, perfection and modern outlook

A simple phrase suffices to express the necessities of tomorrow: WE MUST BUILD IN THE OPEN. The lay-out must be of a purely geometrical kind, with all its many and delicate implications.

The city of to-day is a dying thing because it is not geometrical. To build in the open would be to replace our present haphazard arrangements, *which are all we have to-day*, by a *uniform* lay-out. Unless we do this *there is no salvation*.

The result of a true geometrical lay-out is *repetition*.

The result of repetition is a *standard*, the perfect form (i.e. the creation of standard types). A geometrical lay-out means that mathematics play their part. There is no first-rate human production but has geometry at its base. It is of the very essence of Architecture. To introduce uniformity into the building of the city we must *industrialize building*. Building is the one economic activity which has so far resisted industrialization.

It has thus escaped the march of progress, with the result that the cost of building is still abnormally high.

The architect, from a professional point of view, has become a twisted sort of creature. He has grown to love irregular sites, claiming that they inspire him with original ideas for getting round them. Of course he is wrong. For nowadays the only building that can be undertaken must be either for the rich or built at a loss (as, for instance, in the case of municipal housing schemes), or else by jerry-building and so robbing the inhabitant of all amenities. A motor-car which is achieved by mass production is a masterpiece of comfort, precision, balance and good taste. A house built to order (on an "interesting" site) is a masterpiece of incongruity – a monstrous thing.

If the builder's yard were reorganized on the lines of standardization and mass production we might have gangs of workmen as keen and intelligent as mechanics.

The mechanic dates back only twenty years, yet already he forms the highest caste of the working world.

The mason dates . . . from time immemorial! He bangs away with feet and hammer. He smashes up everything round him, and the plant entrusted to him falls to pieces in a few months. The spirit of the mason must be disciplined by making him part of the severe and exact machinery of the industrialized builder's yard.

The cost of building would fall in the proportion of 10 to 2.

The wages of the labourers would fall into definite categories; to each according to his merits and service rendered.

The "interesting" or erratic site absorbs every creative faculty of the architect and wears him out.

What results is equally erratic: lopsided abortions; a specialist's solution which can only please other specialists.

We must build *in the open*: both within the city and around it. Then having worked through every necessary technical stage and using absolute ECONOMY, we shall be in a position to experience the intense joys of a creative art which is based on geometry.

THE CITY AND ITS AESTHETIC

The plan of a city which is here presented is a direct consequence of purely geometric considerations.

A new unit *on a large scale* (400 yards) inspires everything. Though the gridiron arrangement of the streets every 400 yards (sometimes only 200) is uniform (with a consequent ease in finding one's way about), no two streets are in any way alike. This is where, in a magnificent contrapuntal symphony, the forces of geometry come into play.

Suppose we are entering the city by way of the Great Park. Our fast car takes the special elevated motor track between the majestic sky-scrapers: as we approach nearer there is seen the repetition against the sky of the twenty-four sky-scrapers; to our left and right on the outskirts of each particular area are the municipal and administrative buildings; and enclosing the space are the museums and university buildings.

Then suddenly we find ourselves at the feet of the first sky-scrapers. But here we have, not the meagre shaft of sunlight which so faintly illumines the dismal streets of New York, but an immensity of space. The whole city is a Park. The terraces stretch out over lawns and into groves. Low buildings of a horizontal kind lead the eye on to the foliage of the trees. Where are now the trivial *Procuracies?* Here is the CITY with its crowds living in peace and pure air, where noise is smothered under the foliage of green trees. The chaos of New York is overcome. Here, bathed in light, stands the modern city.

Our car has left the elevated track and has dropped its speed of sixty miles an hour to run gently through the residential quarters. The "set-backs" permit of vast architectural perspectives.[4] There are gardens, games and sports grounds. And sky everywhere, as far as the eye can see. The square silhouettes of the terraced roofs stand clear against the sky, bordered with the verdure of the hanging gardens. The uniformity of the units that compose the picture throw into relief the firm lines on which the far-flung masses are constructed. Their outlines softened by distance, the sky-scrapers raise immense geometrical façades all of glass, and in them is reflected the blue glory of the sky. An overwhelming sensation. Immense but radiant prisms.

And in every direction we have a varying spectacle: our "gridiron" is based on a unit of 400 yards, but it is strangely modified by architectural devices! (The "set-backs" are in counterpoint, on a unit of 600 × 400.)

The traveler in his airplane, arriving from Constantinople, or Peking it may be, suddenly sees appearing through the wavering lines of rivers and patches of forests that clear imprint which marks a city which has grown in accordance with the spirit of man: the mark of the human brain at work.

As twilight falls the glass sky-scrapers seem to flame. This is no dangerous futurism, a sort of literary dynamite flung violently at the spectator. It is a spectacle organized by an Architecture which uses plastic resources for the modulation of forms seen in light.

NOTES

1   New suggestions shower on us. Their inventors and those who believe in them have their little thrill. It is so easy for them to believe in them. But what if they are based on grave errors? How are we to distinguish between what is reasonable and an over-poetical dream? The leading newspapers accept everything with enthusiasm. One of them said, "The cities of to-morrow must be built on new virgin soil." But no, this is not true! We must go to the old cities, all our inquiries confirm it. One of our leading papers supports the suggestion made by one of our greatest and most reasonable architects, who for once gives us bad counsel in proposing to erect round about Paris a ring of sky-scrapers. The idea is romantic enough, but it cannot be defended. The sky-scrapers must be built *in the centre* and not on the periphery.

2 I say "below-ground," but *it* would be more exact to say at what we call *basement level*, for *if* my town, built on concrete piles, were realized, this "basement" would no longer be *buried* under the earth.

3 In this section I make new suggestions in regard to the industrial quarters: they have been content to exist too long in disorder, dirt and in a hand-to-mouth way. And this is absurd, for Industry, when it is on a properly ordered basis, should develop in an orderly fashion. A portion of the industrial district could be constructed of ready-made sections by using standard units for the various kinds of buildings needed. Fifty percent of the site would be reserved for this purpose. In the event of considerable growth, provision would thus be made for moving them into a different district where there was more space. Bring about "*standardization*" in the building of a works and you would have mobility instead of the crowding which results when factories become impossibly congested.

4 As before, this refers to set-backs *on plan*; buildings "à redents," i.e. with projecting salients.

real life, have been incurious about the reasons for unexpected success, and are guided instead by principles derived from the behavior and appearance of towns, suburbs, tuberculosis sanatoria, fairs, and imaginary dream cities – from anything but cities themselves.

If it appears that the rebuilt portions of cities and the endless new developments spreading beyond the cities are reducing city and countryside alike to a monotonous, unnourishing gruel, this is not strange. It all comes, first-, second-, third- or fourth-hand, out of the same intellectual dish of mush, a mush in which the qualities, necessities, advantages and behavior of great cities have been utterly confused with the qualities, necessities, advantages and behavior of other and more inert types of settlements.

There is nothing economically or socially inevitable about either the decay of old cities or the fresh-minted decadence of the new unurban urbanization. On the contrary, no other aspect of our economy and society has been more purposefully manipulated for a full quarter of a century to achieve precisely what we are getting. Extraordinary governmental financial incentives have been required to achieve this degree of monotony, sterility and vulgarity. Decades of preaching, writing and exhorting by experts have gone into convincing us and our legislators that mush like this must be good for us, as long as it comes bedded with grass.

Automobiles are often conveniently tagged as the villains responsible for the ills of cities and the disappointments and futilities of city planning. But the destructive effects of automobiles are much less a cause than a symptom of our incompetence at city building. Of course planners, including the highwaymen with fabulous sums of money and enormous powers at their disposal, are at a loss to make automobiles and cities compatible with one another. They do not know what to do with automobiles in cities because they do not know how to plan for workable and vital cities anyhow – with or without automobiles.

The simple needs of automobiles are more easily understood and satisfied than the complex needs of cities, and a growing number of planners and designers have come to believe that if they can only solve the problems of traffic, they will thereby have solved the major problem of cities. Cities have much more intricate economic and social concerns than automobile traffic. How can you know what to try with traffic until you know how the city itself works, and what else it needs to do with its streets? You can't.

[...]

THE USES OF SIDEWALKS: CONTACT

Reformers have long observed city people loitering on busy corners, hanging around in candy stores and bars and drinking soda pop on stoops, and have passed a judgment, the gist of which is: "This is deplorable! If these people had decent homes and a more private or bosky outdoor place, they wouldn't be on the street!"

This judgment represents a profound misunderstanding of cities. It makes no more sense than to drop in at a testimonial banquet in a hotel and conclude that if these people had wives who could cook, they would give their parties at home.

The point of both the testimonial banquet and the social life of city sidewalks is precisely that they are public. They bring together people who do not know each other in an intimate, private social fashion and in most cases do not care to know each other in that fashion.

Nobody can keep open house in a great city. Nobody wants to. And yet if interesting, useful and significant contacts among the people of cities are confined to acquaintanceships suitable for private life, the city becomes stultified. Cities are full of people with whom, from your viewpoint, or mine, or any other individual's, a certain degree of contact is useful or enjoyable; but you do not want them in your hair. And they do not want you in theirs either.

In speaking about city sidewalk safety, I mentioned how necessary it is that there should be, in the brains behind the eyes on the street, an almost unconscious assumption of general street support when the chips are down – when a citizen has to choose, for instance, whether he will take responsibility, or abdicate it, in combating barbarism or protecting strangers. There is a short word for this assumption of support: trust. The trust of a city street is formed over time from many, many little public sidewalk contacts. It grows out of people stopping by at the bar for a beer, getting advice from the grocer and giving advice to the newsstand man, comparing opinions with other customers at the bakery and nodding hello to the two boys drinking pop on the stoop, eying the girls while waiting to be called for dinner, admonishing the children, hearing about a job from the hardware man and borrowing a dollar from the druggist, admiring the new babies and sympathizing over the way a coat faded. Customs vary: in some neighborhoods people compare notes on their dogs; in others they compare notes on their landlords.

Most of it is ostensibly utterly trivial but the sum is not trivial at all. The sum of such casual, public contact at a local level – most of it fortuitous, most of it associated with errands, all of it metered by the person concerned and not thrust upon him by anyone – is a feeling for the public identity of people, a web of public respect and trust, and a resource in time of personal or neighborhood need. The absence of this trust is a disaster to a city street. Its cultivation cannot be institutionalized. And above all, *it implies no private commitments.*

I have seen a striking difference between presence and absence of casual public trust on two sides of the same wide street in East Harlem, composed of residents of roughly the same incomes and same races. On the old-city side, which was full of public places and the sidewalk loitering so deplored by Utopian minders of other people's leisure, the children were being kept well in hand. On the project side of the street across the way, the children, who had a fire hydrant open beside their play area, were behaving destructively, drenching the open windows of houses with water, squirting it on adults who ignorantly walked on the project side of the street, throwing it into the windows of cars as they went by. Nobody dared to stop them. These were anonymous children, and the identities behind them were an unknown. What if you scolded or stopped them? Who would back you up over there in the blind-eyed Turf? Would you get, instead, revenge? Better to keep out of it. Impersonal city streets make anonymous people, and this is not a matter of esthetic quality nor of a mystical emotional effect in architectural scale. It is a matter of what kinds of tangible enterprises sidewalks have, and therefore of how people use the sidewalks in practical, everyday life.

The casual public sidewalk life of cities ties directly into other types of public life, of which I shall mention one as illustrative, although there is no end to their variety.

Formal types of local city organizations are frequently assumed by planners and even by some social workers to grow in direct, common-sense fashion out of announcements of meetings, the presence of meeting rooms, and the existence of problems of obvious public concern. Perhaps they grow so in suburbs and towns. They do not grow so in cities.

Formal public organizations in cities require an informal public life underlying them, mediating between them and the privacy of the people of the city. We catch a hint of what happens by contrasting, again, a city area possessing a public sidewalk life with a city area lacking it, as told about in the report of a settlement-house social researcher who was studying problems relating to public schools in a section of New York City:

> Mr. W— (principal of an elementary school) was questioned on the effect of J— Houses on the school, and the uprooting of the community around the school. He felt that there had been many effects and of these most were negative. He mentioned that the project had torn out numerous institutions for socializing. The present atmosphere of the project was in no way similar to the gaiety of the streets before the project was built. He noted that in general there seemed fewer people on the streets because there were fewer places for people to gather. He also contended that before the projects were built the Parents Association had been very strong, and now there were only very few active members.

Mr. W— was wrong in one respect. There were not fewer places (or at any rate there was not less space) for people to gather in the project, if we count places deliberately planned for constructive socializing. Of course there were no bars, no candy stores, no hole-in-the-wall *bodegas*, no restaurants in the project. But the project under discussion was equipped with a model complement of meeting rooms, craft, art and game rooms, outdoor benches, malls, etc., enough to gladden the heart of even the Garden City advocates.

Why are such places dead and useless without the most determined efforts and expense to inveigle users – and then to maintain control over the users? What services do the public sidewalk and its enterprises fulfill that these planned gathering places do not? And why? How does an informal public sidewalk life bolster a more formal, organizational public life?

To understand such problems – to understand why drinking pop on the stoop differs from drinking pop in the game room, and why getting advice from the grocer or the bartender differs from getting advice from either your next-door neighbor or from an institutional lady who may be hand-in-glove with an institutional landlord – we must look into the matter of city privacy.

Privacy is precious in cities. It is indispensable. Perhaps it is precious and indispensable everywhere, but most places you cannot get it. In small settlements everyone knows your affairs. In the city everyone does not. Only those you choose to tell will know much about you. This is one of the attributes of cities that is precious to most city people, whether their incomes are high or their incomes are low, whether they are white or colored, whether they are old inhabitants or new, and it is a gift of great-city life deeply cherished and jealously guarded.

Architectural and planning literature deals with privacy in terms of windows, overlooks, sight lines. The idea is that if no one from outside can peek into where you live – behold, privacy. This is simple-minded. Window privacy is the easiest commodity in the world to get. You just pull down the shades or adjust the blinds. The privacy of keeping one's personal affairs to those selected to know them, and the privacy of having reasonable control over who shall make inroads on your time and when, are rare commodities in most of this world, however, and they have nothing to do with the orientation of windows.

Anthropologist Elena Padilla, author of *Up from Puerto Rico*, describing Puerto Rican life in a poor and squalid district of New York, tells how much people know about each other – who is to be trusted and who not, who is defiant of the law and who upholds it, who is competent and well informed and who is inept and ignorant – and how these things are known from the public life of the sidewalk and its associated enterprises. These are matters of public character. But she also tells how select are those permitted to drop into the kitchen for a cup of coffee, how strong are the ties, and how limited the number of a person's genuine confidants, those who share in a person's private life and private affairs. She tells how it is not considered dignified for everyone to know one's affairs. Nor is it considered dignified to snoop on others beyond the face presented in public. It does violence to a person's privacy and rights. In this, the people she describes are essentially the same as the people of the mixed, Americanized city street on which I live, and essentially the same as the people who live in high-income apartments or fine town houses, too.

A good city street neighborhood achieves a marvel of balance between its people's determination to have essential privacy and their simultaneous wishes for differing degrees of contact, enjoyment or help from the people around. This balance is largely made up of small, sensitively managed details, practiced and accepted so casually that they are normally taken for granted.

Perhaps I can best explain this subtle but all-important balance in terms of the stores where people leave keys for their friends, a common custom in New York. In our family, for example, when a friend wants to use our place while we are away for a weekend or everyone happens to be out during the day, or a visitor for whom we do not wish to wait up is spending the night, we tell such a friend that he can pick up the key at the delicatessen across the street. Joe Cornacchia, who keeps the delicatessen, usually has a dozen or so keys at a time for handing out like this. He has a special drawer for them.

Now why do I, and many others, select Joe as a logical custodian for keys? Because we trust him, first, to be a responsible custodian, but equally important because we know that he combines a feeling of good will with a feeling of no personal responsibility about our private affairs. Joe considers it no concern of his whom we choose to permit in our places and why.

Around on the other side of our block, people leave their keys at a Spanish grocery. On the other side of Joe's block, people leave them at the candy store. Down a block they leave them at the coffee shop, and a few hundred feet around the corner from that, in a barber shop. Around one corner from two fashionable blocks of town houses and apartments in the Upper East Side, people leave their keys in a butcher shop and a bookshop; around another corner they leave them in a cleaner's and a drug store. In unfashionable East Harlem keys are left with at least one florist, in bakeries, in luncheonettes, in Spanish and Italian groceries.

The point, wherever they are left, is not the kind of ostensible service that the enterprise offers, but the kind of proprietor it has.

A service like this cannot be formalized. Identifications . . . questions . . . insurance against mishaps. The all-essential line between public service and privacy would be transgressed by institutionalization. Nobody in his right mind would leave his key in such a place. The service must be given as a favor by someone with an unshakable understanding of the difference between a person's key and a person's private life, or it cannot be given at all.

Or consider the line drawn by Mr. Jaffe at the candy store around our corner – a line so well understood by his customers and by other storekeepers too that they can spend their whole lives in its presence and never think about it consciously. One ordinary morning last winter, Mr. Jaffe, whose formal business name is Bernie, and his wife, whose formal business name is Ann, supervised the small children crossing at the corner on the way to P.S. 41, as Bernie always does because he sees the need; lent an umbrella to one customer and a dollar to another; took custody of two keys; took in some packages for people in the next building who were away; lectured two youngsters who asked for cigarettes; gave street directions; took custody of a watch to give the repair man across the street when he opened later; gave out information on the range of rents in the neighborhood to an apartment seeker; listened to a tale of domestic difficulty and offered reassurance; told some rowdies they could not come in unless they behaved and then defined (and got) good behavior; provided an incidental forum for half a dozen conversations among customers who dropped in for oddments; set aside certain newly arrived papers and magazines for regular customers who would depend on getting them; advised a mother who came for a birthday present not to get the ship-model kit because another child going to the same birthday party was giving that; and got a back copy (this was for me) of the previous day's newspaper out of the deliverer's surplus returns when he came by.

After considering this multiplicity of extra-merchandising services I asked Bernie, "Do you ever introduce your customers to each other?"

He looked startled at the idea, even dismayed. "No," he said thoughtfully.

"That would just not be advisable. Sometimes, if I know two customers who are in at the same time have an interest in common, I bring up the subject in conversation and let them carry it on from there if they want to. But oh no, I wouldn't introduce them."

When I told this to an acquaintance in a suburb, she promptly assumed that Mr. Jaffe felt that to make an introduction would be to step above his social class. Not at all. In our neighborhood, storekeepers like the Jaffes enjoy an excellent social status, that of businessmen. In income they are apt to be the peers of the general run of customers and in independence they are the superiors. Their advice, as men or women of common sense and experience, is sought and respected. They are well known as individuals, rather than unknown as class symbols. No; this is that almost unconsciously enforced, well-balanced line showing, the line between the city public world and the world of privacy.

This line can be maintained, without awkwardness to anyone, because of the great plenty of opportunities for public contact in the enterprises along the sidewalks, or on the sidewalks themselves as people move to and fro or deliberately loiter when they feel like it, and also because of the presence of many public hosts, so to speak, proprietors of meeting-places like Bernie's where one is free either to hang around or dash in and out, no strings attached.

Under this system, it is possible in a city street neighborhood to know all kinds of people without unwelcome entanglements, without boredom, necessity for excuses, explanations, fears of giving offense, embarrassments respecting impositions or commitments, and all such paraphernalia of obligations which can accompany less limited relationships. It is possible to be on excellent sidewalk terms with people who are very different from oneself, and even, as time passes, on familiar public terms with them. Such relationships can, and do, endure for many years, for decades; they could never have formed without that line, much less endured. They form precisely because they are by-the-way to people's normal public sorties.

"Togetherness" is a fittingly nauseating name for an old ideal in planning theory. This ideal is that if anything is shared among people, much should be shared. "Togetherness," apparently a spiritual resource of the new suburbs, works destructively in cities. The requirement that much shall be shared drives city people apart.

When an area of a city lacks a sidewalk life, the people of the place must enlarge their private lives if they are to have anything approaching equivalent contact with their neighbors. They must settle for some form of "togetherness," in which more is shared with one another than in the life of the sidewalks, or else they must settle for lack of contact. Inevitably the outcome is one or the other; it has to be; and either has distressing results.

In the case of the first outcome, where people do share much, they become exceedingly choosy as to who their neighbors are, or with whom they associate at all. They have to become so. A friend of mine, Penny Kostritsky, is unwittingly and unwillingly in this fix on a street in Baltimore. Her street of nothing but residences, embedded in an area of almost nothing but residences, has been experimentally equipped with a charming sidewalk park. The sidewalk has been widened and attractively paved, wheeled traffic discouraged from the narrow street roadbed, trees and flowers planted, and a piece of play sculpture is to go in. All these are splendid ideas so far as they go.

However, there are no stores. The mothers from nearby blocks who bring small children here, and come here to find some contact with others themselves, perforce go into the houses of acquaintances along the street to warm up in winter, to make telephone calls, to take their children in emergencies to the bathroom. Their hostesses offer them coffee, for there is no other place to get coffee, and naturally considerable social life of this kind has arisen around the park. Much is shared.

Mrs. Kostritsky, who lives in one of the conveniently located houses, and who has two small children, is in the thick of this narrow and accidental social life. "I have lost the advantage of living in the city," she says, "without getting the advantages of living in the suburbs." Still more distressing, when mothers of different income or color or educational background bring their children to the

street park, they and their children are rudely and pointedly ostracized. They fit awkwardly into the suburbanlike sharing of private lives that has grown in default of city sidewalk life. The park lacks benches purposely; the "togetherness" people ruled them out because they might be interpreted as an invitation to people who cannot fit in.

"If only we had a couple of stores on the street," Mrs. Kostritsky laments.

> "If only there were a grocery store or a drug store or a snack joint. Then the telephone calls and the warming up and the gathering could be done naturally in public, and then people would act more decent to each other because everybody would have a right to be here."

Much the same thing that happens in this sidewalk park without a city public life happens sometimes in middle-class projects and colonies, such as Chatham Village in Pittsburgh for example, a famous model of Garden City planning.

The houses here are grouped in colonies around shared interior lawns and play yards, and the whole development is equipped with other devices for close sharing, such as a residents' club which holds parties, dances, reunions, has ladies' activities like bridge and sewing parties, and holds dances and parties for the children. There is no public life here, in any city sense. There are differing degrees of extended private life.

Chatham Village's success as a "model" neighborhood where much is shared has required that the residents be similar to one another in their standards, interests and backgrounds. In the main they are middle-class professionals and their families.[1] It has also required that residents set themselves distinctly apart from the different people in the surrounding city; these are in the main also middle class, but lower middle class, and this is too different for the degree of chumminess that neighborliness in Chatham Village entails.

The inevitable insularity (and homogeneity) of Chatham Village has practical consequences. As one illustration, the junior high school serving the area has problems, as all schools do. Chatham Village is large enough to dominate the elementary school to which its children go, and therefore to work at helping solve this school's problems. To deal with the junior high, however, Chatham Village's people must cooperate with entirely different neighborhoods. But there is no public acquaintanceship, no foundation of casual public trust, no cross-connections with the necessary people – and no practice or ease in applying the most ordinary techniques of city public life at lowly levels. Feeling helpless, as indeed they are, some Chatham Village families move away when their children reach junior high age; others contrive to send them to private high schools. Ironically, just such neighborhood islands as Chatham Village are encouraged in orthodox planning on the specific grounds that cities need the talents and stabilizing influence of the middle class. Presumably these qualities are to seep out by osmosis.

People who do not fit happily into such colonies eventually get out, and in time managements become sophisticated in knowing who among applicants will fit in. Along with basic similarities of standards, values and backgrounds, the arrangement seems to demand a formidable amount of forbearance and tact.

City residential planning that depends, for contact among neighbors, on personal sharing of this sort, and that cultivates it, often does work well socially, if rather narrowly, *for self-selected upper-middle-class people*. It solves easy problems for an easy kind of population. So far as I have been able to discover, it fails to work, however, even on its own terms, *with any other kind of population*.

The more common outcome in cities, where people are faced with the choice of sharing much or nothing, is nothing. In city areas that lack a natural and casual public life, it is common for residents to isolate themselves from each other to a fantastic degree. If mere contact with your neighbors threatens to entangle you in their private lives, or entangle them in yours, and if you cannot be so careful who your neighbors are as self-selected upper-middle-class people can be, the logical solution is absolutely to avoid friendliness or casual offers of help. Better to stay thoroughly distant. As a practical result, the ordinary public jobs – like keeping children in hand – for which people must take a little personal initiative, or those for which they must band together in limited common purposes, go undone. The abysses this opens up can be almost unbelievable.

For example, in one New York City project which is designed – like all orthodox residential city

planning – for sharing much or nothing, a remarkably outgoing woman prided herself that she had become acquainted, by making a deliberate effort, with the mothers of everyone of the ninety families in her building. She called on them. She buttonholed them at the door or in the hall. She struck up conversations if she sat beside them on a bench.

It so happened that her eight-year-old son, one day, got stuck in the elevator and was left there without help for more than two hours, although he screamed, cried and pounded. The next day the mother expressed her dismay to one of her ninety acquaintances. "Oh, was that *your* son?" said the other woman. "I didn't know whose boy he was. If I had realized he was *your* son I would have helped him."

This woman, who had not behaved in any such insanely calloused fashion on her old public street – to which she constantly returned, by the way, for public life – was afraid of a possible entanglement that might not be kept easily on a public plane.

Dozens of illustrations of this defense can be found wherever the choice is sharing much or nothing. A thorough and detailed report by Ellen Lurie, a social worker in East Harlem, on life in a low-income project there, has this to say:

It is . . . extremely important to recognize that for considerably complicated reasons, many adults either don't want to become involved in any friendship-relationships at all with their neighbors, or, if they do succumb to the need for some form of society, they strictly limit themselves to one or two friends, and no more. Over and over again, wives repeated their husband's warning:
"I'm not to get too friendly with anyone. My husband doesn't believe in it."
"People are too gossipy and they could get us in a lot of trouble."
"It's best to mind your own business."
One woman, Mrs. Abraham, always goes out the back door of the building because she doesn't want to interfere with the people standing around in the front. Another man, Mr. Colan . . . won't let his wife make any friends in the project, because he doesn't trust the people here. They have four children, ranging from 8 years to 14, but they are not allowed downstairs alone, because the parents are afraid someone

will hurt them.[2] What happens then is that all sorts of barriers to insure self-protection are being constructed by many families. To protect their children from a neighborhood they aren't sure of, they keep them upstairs in the apartment. To protect themselves, they make few, if any, friends. Some are afraid that friends will become angry or envious and make up a story to report to management, causing them great trouble. If the husband gets a bonus (which he decides not to report) and the wife buys new curtains, the visiting friends will see and might tell the management, who, in turn, investigates and issues a rent increase. Suspicion and fear of trouble often outweigh any need for neighborly advice and help. For these families the sense of privacy has already been extensively violated. The deepest secrets, all the family skeletons, are well known not only to management but often to other public agencies, such as the Welfare Department. To preserve any last remnants of privacy, they choose to avoid close relationships with others. This same phenomenon may be found to a much lesser degree in non-planned slum housing, for there too it is often necessary for other reasons to build up these forms of self-protection. But, it is surely true that this withdrawing from the society of others is much more extensive in planned housing. Even in England, this suspicion of the neighbors and the ensuing aloofness was found in studies of planned towns. Perhaps this pattern is nothing more than an elaborate group mechanism to protect and preserve inner dignity in the face of so many outside pressures to conform.

Along with nothingness, considerable "togetherness" can be found in such places, however. Mrs. Lurie reports on this type of relationship:

Often two women from two different buildings will meet in the laundry room, recognize each other; although they may never have spoken a single word to each other back on 99th Street, suddenly here they become "best friends." If one of these two already has a friend or two in her own building, the other is likely to be drawn into that circle and begins to make her friendships, not with women on her floor, but rather on her friend's floor.

These friendships do not go into an everwidening circle. There are certain definite welltraveled paths in the project, and after a while no new people are met.

Mrs. Lurie, who works at community organization in East Harlem, with remarkable success, has looked into the history of many past attempts at project tenant organization. She has told me that "togetherness," itself, is one of the factors that make this kind of organization so difficult. "These projects are not lacking in natural leaders," she says.

They contain people with real ability, wonderful people many of them, but the typical sequence is that in the course of organization leaders have found each other, gotten all involved in each others' social lives, and have ended up talking to nobody but each other. They have not found their followers. Everything tends to degenerate into ineffective cliques, as a natural course. There is no normal public life. Just the mechanics of people learning what is going on is so difficult. It all makes the simplest social gain extra hard for these people.

Residents of unplanned city residential areas that lack neighborhood commerce and sidewalk life seem sometimes to follow the same course as residents of public projects when faced with the choice of sharing much or nothing. Thus researchers hunting the secrets of the social structure in a dull gray-area district of Detroit came to the unexpected conclusion there was no social structure.

The social structure of sidewalk life hangs partly on what can be called self-appointed public characters. A public character is anyone who is in frequent contact with a wide circle of people and who is sufficiently interested to make himself a public character. A public character need have no special talents or wisdom to fulfill his function – although he often does. He just needs to be present, and there need to be enough of his counterparts. His main qualification is that he *is* public, that he talks to lots of different people. In this way, news travels that is of sidewalk interest.

Most public sidewalk characters are steadily stationed in public places. They are storekeepers or barkeepers or the like. These are the basic public characters. All other public characters of city sidewalks depend on them – if only indirectly because of the presence of sidewalk routes to such enterprises and their proprietors.

Settlement-house workers and pastors, two more formalized kinds of public characters, typically depend on the street grapevine news systems that have their ganglia in the stores. The director of a settlement on New York's Lower East Side, as an example, makes a regular round of stores. He learns from the cleaner who does his suits about the presence of dope pushers in the neighborhood. He learns from the grocer that the Dragons are working up to something and need attention. He learns from the candy store that two girls are agitating the Sportsmen toward a rumble. One of his most important information spots is an unused breadbox on Rivington Street. That is, it is not used for bread. It stands outside a grocery and is used for sitting on and lounging beside, between the settlement house, a candy store and a pool parlor. A message spoken there for any teen-ager within many blocks will reach his ears unerringly and surprisingly quickly, and the opposite flow along the grapevine similarly brings news quickly in to the breadbox.

Blake Hobbs, the head of the Union Settlement music school in East Harlem, notes that when he gets a first student from one block of the old busy street neighborhoods, he rapidly gets at least three or four more and sometimes almost every child on the block. But when he gets a child from the nearby projects – perhaps through the public school or a playground conversation he has initiated – he almost never gets another as a direct sequence. Word does not move around where public characters and sidewalk life are lacking.

Besides the anchored public characters of the sidewalk, and the well-recognized roving public characters, there are apt to be various more specialized public characters on a city sidewalk. In a curious way, some of these help establish an identity not only for themselves but for others. Describing the everyday life of a retired tenor at such sidewalk establishments as the restaurant and the *bocce* court, a San Francisco news story notes, "It is said of Meloni that because of his intensity, his dramatic manner and his lifelong interest in music, he transmits a feeling of vicarious importance to his many friends." Precisely.

One need not have either the artistry or the personality of such a man to become a specialized sidewalk character – but only a pertinent specialty of some sort. It is easy. I am a specialized public character of sorts along our street, owing of course to the fundamental presence of the basic, anchored public characters. The way I became one started with the fact that Greenwich Village, where I live, was waging an interminable and horrendous battle to save its main park from being bisected by a highway. During the course of battle I undertook, at the behest of a committee organizer away over on the other side of Greenwich Village, to deposit in stores on a few blocks of our street supplies of petition cards protesting the proposed roadway. Customers would sign the cards while in the stores, and from time to time I would make my pickups.[3] As a result of engaging in this messenger work, I have since become automatically the sidewalk public character on petition strategy. Before long, for instance, Mr. Fox at the liquor store was consulting me, as he wrapped up my bottle, on how we could get the city to remove a long abandoned and dangerous eyesore, a closed-up comfort station near his corner. If I would undertake to compose the petitions and find the effective way of presenting them to City Hall, he proposed, he and his partners would undertake to have them printed, circulated and picked up. Soon the stores round about had comfort station removal petitions. Our street by now has many public experts on petition tactics, including the children.

Not only do public characters spread the news and learn the news at retail, so to speak. They connect with each other and thus spread word wholesale, in effect.

A sidewalk life, so far as I can observe, arises out of no mysterious qualities or talents for it in this or that type of population. It arises only when the concrete, tangible facilities it requires are present. These happen to be the same facilities, in the same abundance and ubiquity, that are required for cultivating sidewalk safety. If they are absent, public sidewalk contacts are absent too.

The well-off have many ways of assuaging needs for which poorer people may depend much on sidewalk life – from hearing of jobs to being recognized by the headwaiter. But nevertheless, many of the rich or near-rich in cities appear to appreciate sidewalk life as much as anybody. At any rate, they pay enormous rents to move into areas with an exuberant and varied sidewalk life. They actually crowd out the middle class and the poor in lively areas like Yorkville or Greenwich Village in New York, or Telegraph Hill just off the North Beach streets of San Francisco. They capriciously desert, after only a few decades of fashion at most, the monotonous streets of "quiet residential areas" and leave them to the less fortunate. Talk to residents of Georgetown in the District of Columbia and by the second or third sentence at least you will begin to hear rhapsodies about the charming restaurants, "more good restaurants than in all the rest of the city put together," the uniqueness and friendliness of the stores, the pleasures of running into people when doing errands at the next corner – and nothing but pride over the fact that Georgetown has become a specialty shopping district for its whole metropolitan area. The city area, rich or poor or in between, harmed by an interesting sidewalk life and plentiful sidewalk contacts has yet to be found.

Efficiency of public sidewalk characters declines drastically if too much burden is put upon them. A store, for example, can reach a turnover in its contacts, or potential contacts, which is so large and so superficial that it is socially useless. An example of this can be seen at the candy and newspaper store owned by the housing cooperative of Corlears Hook on New York's Lower East Side. This planned project store replaces perhaps forty superficially similar stores which were wiped out (without compensation to their proprietors) on that project site and the adjoining sites. The place is a mill. Its clerks are so busy making change and screaming ineffectual imprecations at rowdies that they never hear anything except "I want that." This, or utter disinterest, is the usual atmosphere where shopping center planning or repressive zoning artificially contrives commercial monopolies for city neighborhoods. A store like this would fail economically if it had competition. Meantime, although monopoly insures the financial success planned for it, it fails the city socially.

Sidewalk public contact and sidewalk public safety, taken together, bear directly on our country's most serious social problem – segregation and racial discrimination.

I do not mean to imply that a city's planning and design, or its types of streets and street life, can automatically overcome segregation and discrimination. Too many other kinds of effort are also required to right these injustices.

But I do mean to say that to build and to rebuild big cities whose sidewalks are unsafe and whose people must settle for sharing much or nothing, *can* make it *much harder* for American cities to overcome discrimination no matter how much effort is expended.

Considering the amount of prejudice and fear that accompany discrimination and bolster it, overcoming residential discrimination is just that much harder if people feel unsafe on their sidewalks anyway. Overcoming residential discrimination comes hard where people have no means of keeping a civilized public life on a basically dignified public footing, and their private lives on a private footing.

To be sure, token model housing integration schemes here and there can be achieved in city areas handicapped by danger and by lack of public life – achieved by applying great effort and settling for abnormal (abnormal for cities) choosiness among new neighbors. This is an evasion of the size of the task and its urgency.

The tolerance, the room for great differences among neighbors – differences that often go far deeper than differences in color – which are possible and normal in intensely urban life, but which are so foreign to suburbs and pseudosuburbs, are possible and normal only when streets of great cities have built-in equipment allowing strangers to dwell in peace together on civilized but essentially dignified and reserved terms.

Lowly, unpurposeful and random as they may appear, sidewalk contacts are the small change from which a city's wealth of public life may grow.

Los Angeles is an extreme example of a metropolis with little public life, depending mainly instead on contacts of a more private social nature.

On one plane, for instance, an acquaintance there comments that although she has lived in the city for ten years and knows it contains Mexicans, she has never laid eyes on a Mexican or an item of Mexican culture, much less ever exchanged any words with a Mexican.

On another plane, Orson Welles has written that Hollywood is the only theatrical center in the world that has failed to develop a theatrical bistro.

And on still another plane, one of Los Angeles' most powerful businessmen comes upon a blank in public relationships which would be inconceivable in other cities of this size. This businessman, volunteering that the city is "culturally behind," as he put it, told me that he for one was at work to remedy this. He was heading a committee to raise funds for a first-rate art museum. Later in our conversation, after he had told me about the businessmen's club life of Los Angeles, a life with which he is involved as one of its leaders, I asked him how or where Hollywood people gathered in corresponding fashion. He was unable to answer this. He then added that he knew no one at all connected with the film industry, nor did he know anyone who did have such acquaintanceship. "I know that must sound strange," he reflected. "We are glad to have the film industry here, but those connected with it are just not people one would know socially."

Here again is "togetherness" or nothing. Consider this man's handicap in his attempts to get a metropolitan art museum established. He has no way of reaching with any ease, practice or trust some of his committee's potentially best prospects.

In its upper economic, political and cultural echelons, Los Angeles operates according to the same provincial premises of social insularity as the street with the sidewalk park in Baltimore or as Chatham Village in Pittsburgh. Such a metropolis lacks means for bringing together necessary ideas, necessary enthusiasms, necessary money. Los Angeles is embarked on a strange experiment: trying to run not just projects, not just gray areas, but a whole metropolis, by dint of "togetherness" or nothing. I think this is an inevitable outcome for great cities whose people lack city public life in ordinary living and working.

NOTES

1 One representative court, for example, contains as this is written four lawyers, two doctors, two engineers, a dentist, a salesman, a banker, a railroad executive, a planning executive.
2 This is very common in public projects in New York.
3 This, by the way, is an efficient device, accomplishing with a fraction of the effort what would be a mountainous task door to door. It also makes more public conversation and opinion than door-to-door visits.

"The Timeless Way"

from *The Timeless Way of Building* (1979)

Christopher Alexander

Editors' Introduction

Echoing the sentiments of other designers, planners and critics dissatisfied with the bankrupt results of twentieth-century city-making, Alexander proposes in *The Timeless Way of Building* a design method that recalls older, traditional ways of designing prior to the rise of professional expertise. In his practice and writings, he rejects the conventional design processes of the mainstream design professions, which are perceived to be overly reductionist and lacking the complexity that allows life, beauty and place-based harmonies to emerge. He and his colleagues at the Center for Environmental Structure at the University of California, Berkeley are highly critical of the profit-driven construction and development industries, which are understood here to be responsible for the tragedy of late-twentieth-century urban form and aesthetics.

This first chapter of *The Timeless Way of Building* reads similarly to the *Tao Te Ching*. Its mysterious and esoteric statements seemingly require meditative thought to gauge their meaning. At the beginning of the first chapter, Alexander calls for a return to a timeless way of building towns, structures and places where people can feel alive again: "There is one timeless way of building. It is thousands of years old, and the same today as it has always been." His thesis calls for a return to self-built places without the help of design professionals. These are the vernacular processes that produced simpler, more meaningful places that people have loved for eons, and consequently provided cultural differentiation across the globe. It is also a call for a more participatory design that is directed by the intuitive needs and desires of everyday life, rather than the abstract and over-intellectualized design practices influenced by scientific process, regulations, professional standards or academic theories. Just as important is Alexander's revalorization of the concept of beauty, which had long been discarded from intellectual design discourse.

As the first in a series of works that includes the well-known *A Pattern Language* (published prior to this text in 1977), *The Timeless Way of Building* serves as an introduction to Alexander's underlying theories and methods in the use of living patterns drawn from nature, everyday life, and traditional environments. Text in *The Timeless Way of Building* is written both for those who want to ponder the writing word-for-word and for those who have little time for this luxury. He has arranged each chapter by interspersing italicized headlines that summarize the text for readers who are time constrained to read it in full. Despite this, Alexander's writing will require deep thought to understand ideas such as *the quality without a name*, *egoless innocence*, and *the kernel of the way*.

Christopher Alexander is an architect, builder and Professor Emeritus at the University of California, Berkeley, where he began teaching in 1963. He gained early notoriety with his PhD dissertation, published as *Notes on the Synthesis of Form* (Cambridge, MA: Harvard University Press, 1964), as well as his attack on the sterility of modern planning in "A City is Not a Tree," *Architectural Forum* (vol. 122, no. 1, April 1965). In this work he suggests that the complexity of cities should be viewed as a multilayered latticework, rather than a branched diagram that separates and fragments functions and activities. He has long worked through the

Center for Environmental Structure that he founded at Berkeley, achieving a type of guru status among his followers. Together they have conducted design experiments, advised city leaders, and built hundreds of buildings in the Americas, Asia and Europe. Their design work is highly participatory and interactive, resulting in innovative building technologies and work that seemingly emerges from the ideas of their clients. His writings have been highly influential in reviving the craft of building, illuminating the value of vernacular environments, and transforming the status quo practices of architecture and town planning.

Other books in the Center for Environmental Structure Series published in New York by Oxford University Press are: *The Oregon Experiment* (1975), *A Pattern Language*, written with Sara Ishikawa and Murray Silverstein *et al.* (1977), *The Linz Café* (1981), *The Production of Houses*, written with Howard Davis, Julio Martinez, and Donald Cormer (1985), and *A New Theory of Urban Design*, written with Hajo Neis, Artemis Anninou, and Ingrid King (1985). The success of *A Pattern Language* has spawned a movement dedicated to the theories and work of its author, which can be found at http://www.patternlanguage.com, which is one of Alexander's websites. Other important texts by Alexander include the ambitious four-volume text *The Nature of Order: An Essay on the Art of Building and the Nature of the Universe* (Berkeley, CA: Centre for Environmental Structure, 2003): Volume 1, *The Phenomenon of Life*; Volume 2, *The Process of Creating Life*; Volume 3, *A Vision of a Living World*; Volume 4, *The Luminous Ground*. Additional material on Alexander's life and work can be found in his biography by Stephen Grabow, *Christopher Alexander: The Evolution of a New Paradigm in Architecture* (Boston, MA: Oriel Press, 1983). Two documentaries have been produced on his work: *Places for the Soul: The Architecture of Christopher Alexander* (1990) and *Christopher Alexander and Contemporary Architecture* (1993).

It is a process which brings order out of nothing but ourselves; it cannot be attained, but it will happen of its own accord, if we will only let it.

There is one timeless way of building. It is thousands of years old, and the same today as it has always been. The great traditional buildings of the past, the villages and tents and temples in which man feels at home, have always been made by people who were very close to the center of this way. It is not possible to make great buildings, or great towns, beautiful places, places where you feel yourself, places where you feel alive, except by following this way. And, as you will see, this way will lead anyone who looks for it to buildings which are themselves as ancient in their form, as the trees and hills, and as our faces are.

It is a process through which the order of a building or a town grows out directly from the inner nature of the people, and the animals, and plants, and matter which are in it.

It is a process which allows the life inside a person, or a family, or a town, to flourish, openly, in freedom, so vividly that it gives birth, of its own accord, to the natural order which is needed to sustain this life.

It is so powerful and fundamental that with its help you can make any building in the world as beautiful as any place that you have ever seen.

Once you understand this way, you will be able to make your room alive; you will be able to design a house together with your family; a garden for your children; places where you can work; beautiful terraces where you can sit and dream.

It is so powerful, that with its help hundreds of people together can create a town, which is alive and vibrant, peaceful and relaxed, a town as beautiful as any town in history.

Without the help of architects or planners, if you are working in the timeless way, a town will grow under your hands, as steady as the flowers in your garden.

And there is no other way in which a building or a town which lives can possibly be made.

This does not mean that all ways of making buildings are identical. It means that at the core of all successful acts of building and at the core of all successful processes of growth, even though there are a million different versions of these acts and processes, there is one fundamental invariant feature, which is responsible for their success. Although this way has taken on a thousand different forms at different times, in different places, still, there is an unavoidable, invariant core to all of them.

[. . .]

They are alive. They have that sleepy, awkward grace which comes from perfect ease. And, the Alhambra, some tiny gothic church, an old New England house, an Alpine hill village, an ancient Zen temple, a seat by a mountain stream, a courtyard filled with blue and yellow tiles among the earth. What is it they have in common? They are beautiful, ordered, harmonious – yes, all these things. But especially, and what strikes to the heart, they live.

Each one of us wants to be able to bring a building or part of a town to life like this.

It is a fundamental human instinct, as much a part of our desire as the desire for children. It is, quite simply, the desire to make a part of nature, to complete a world which is already made of mountains, streams, snowdrops, and stones, with something made by us, as much a part of nature, and a part of our immediate surroundings.

Each one of us has, somewhere in his heart, the dream to make a living world, a universe.

Those of us who have been trained as architects have this desire perhaps at the very center of our lives: that one day, somewhere, somehow, we shall build one building which is wonderful, beautiful, breathtaking, a place where people can walk and dream for centuries.

In some form, every person has some version of this dream: whoever you are, you may have the dream of one day building a most beautiful house for your family, a garden, a fountain, a fishpond, a big room with soft light, flowers outside and the smell of new grass.

In some less clear fashion, anyone who is concerned with towns has this same dream, perhaps, for an entire town.

And there is a way that a building or a town can actually be brought to life like this.

There is a definable sequence of activities which are at the heart of all acts of building, and it is possible to specify, precisely, under what conditions these activities will generate a building which is alive. All this can be made so explicit that anyone can do it.

And just so, the process by which a group of independent people make part of a town alive can equally be made precise. Again, there is a definable sequence of activities, more complex in this case, which are at the heart of all collective building processes, and it is possible to specify exactly when these processes will bring things to life. And, once again, these processes can be made so explicit, and so clear, that any group of people can make use of them.

This one way of building has always existed.

It is behind the building of traditional villages in Africa, and India, and Japan. It was behind the building of the great religious buildings: the mosques of Islam, the monasteries of the middle ages, and the temples of Japan. It was behind the building of the simple benches, and cloisters and arcades of English country towns; of the mountain huts of Norway and Austria; the roof tiles on the walls of castles and palaces; the bridges of the Italian middle ages; the cathedral of Pisa.

In an unconscious form, this way has been behind almost all ways of building for thousands of years.

But it has become possible to identify it, only now, by going to a level of analysis which is deep enough to show what is invariant in all the different versions of this way.

This hinges on a form of representation, which reveals all possible construction processes, as versions of one deeper process.

First, we have a way of looking at the ultimate constituents of the environment: the ultimate "things" which a building or a town is made of. Every building, every town, is made of certain entities which I call patterns: and once we understand buildings in terms of their patterns, we have a way of looking at them, which makes all buildings, all parts of a town similar, all members of the same class of physical structures.

Second, we have a way of understanding the generative processes which give rise to these patterns: in short, the source from which the ultimate constituents of building come. These patterns always come from certain combinatory processes, which are different in the specific patterns which they generate, but always similar in their overall structure, and in the way they work. They are essentially like languages. And again, in terms of these pattern languages, all the different ways of building, although different in detail, become similar in general outline.

At this level of analysis, we can compare many different building processes.

Then, once we see their differences clearly, it becomes possible to define the difference between those processes which make buildings live, and those which make them dead.

And it turns out that, invariant, behind all processes which allow us to make buildings live, there is a single common process.

This single process is operational and precise. It is not merely a vague idea, or a class of processes which we can understand: it is concrete enough and specific enough, so that it functions practically. It gives us the power to make towns and buildings live, as concretely as a match gives us the power to make a flame. It is a method or a discipline, which teaches us precisely what we have to do to make our buildings live.

But though this method is precise, it cannot be used mechanically.

The fact is, that even when we have seen deep into the processes by which it is possible to make a building or a town alive, in the end, it turns out that this knowledge only brings us back to that part of ourselves which is forgotten.

Although the process is precise, and can be defined in exact scientific terms, finally it becomes valuable, not so much because it shows us things which we don't know, but instead, because it shows us what we know already, only daren't admit because it seems so childish, and so primitive.

Indeed it turns out, in the end, that what this method does is simply free us from all method.

The more we learn to use this method, the more we find that what it does is not so much to teach us processes we did not know before, but rather opens up a process in us, which was part of us already.

We find out that we already know how to make buildings live, but that the power has been frozen in us: that we have it, but are afraid to use it: that we are crippled by our fears; and crippled by the methods and the images which we use to overcome these fears.

And what happens finally, is that we learn to overcome our fears, and reach that portion of ourselves which knows exactly how to make a building live, instinctively. But we learn too, that this capacity in us is not accessible, until we first go through the discipline which teaches us to let go of our fears.

And that is why the timeless way is, in the end, a timeless one.

It is not an external method, which can be imposed on things. It is instead a process which lies deep in us: and only needs to be released.

The power to make buildings beautiful lies in each of us already.

It is a core so simple, and so deep, that we are born with it. This is no metaphor. I mean it literally. Imagine the greatest possible beauty and harmony in the world – the most beautiful place that you have ever seen or dreamt of. You have the power to create it, at this very moment, just as you are.

And this power we have is so firmly rooted and coherent in every one of us that once it is liberated, it will allow us, by our individual, unconnected acts, to make a town, without the slightest need for plans, because, like every living process, it is a process which builds order out of nothing.

But as things are, we have so far beset ourselves with rules, and concepts, and ideas of what must be done to make a building or a town alive, that we have become afraid of what will happen naturally, and convinced that we must work within a "system" and with "methods" since without them our surroundings will come tumbling down in chaos.

We are afraid, perhaps, that without images and methods, chaos will break loose; worse still, that unless we use images of some kind, ourselves, our own creation will itself be chaos. And why are we afraid of that? Is it because people will laugh at us, if we make chaos? Or is it, perhaps, that we are most afraid of all that if we do make chaos, when we hope to create art, we will ourselves be chaos, hollow, nothing?

This is why it is so easy for others to play on our fears. They can persuade us that we must have more method, and more system, because we are afraid of our own chaos. Without method and more method, we are afraid the chaos which is in us will reveal itself. And yet these methods only make things worse.

The thoughts and fears which feed these methods are illusions.

It is the fears which these illusions have created in us, that make places which are dead and lifeless and artificial. And – greatest irony of all – it is the very methods we invent to free us from our fears which are themselves the chains whose grip on us creates our difficulties.

For the fact is, that this seeming, chaos which is in us is a rich, rolling, swelling, dying, lilting, singing, laughing, shouting, crying, sleeping order. If we will only let this order guide our acts of building, the buildings that we make, the towns we help to make, will be the forests and the meadows of the human heart.

To purge ourselves of these illusions, to become free of all the artificial images of order which distort the nature that is in us, we must first learn a discipline which teaches us the true relationship between ourselves and our surroundings.

Then, once this discipline has done its work, and pricked the bubbles of illusion which we cling to now, we will be ready to give up the discipline, and act as nature does.

This is the timeless way of building: learning the discipline – and shedding it.

"Toward an Urban Design Manifesto"

from *Journal of the American Planning Association* (1987)

Allan B. Jacobs and Donald Appleyard

Editors' Introduction

In the early 1980s, Allan B. Jacobs and Donald Appleyard (1928–1982), following a seminar on urban physical form in the Department of City and Regional Planning at the University of California, Berkeley that included strong criticism of the Le Corbusier led CIAM design manifesto, were urged by their students to write a design manifesto that articulated a counter-position. They took the challenge and wrote "Toward an Urban Design Manifesto." Initially rejected for publication by the *Journal of the American Planning Association* on the grounds that it was without scholarly merit because of its experiential methodology, Jacobs was later invited to have it included in a special urban design oriented issue of the same journal.

The manifesto – informed by Jacobs' extensive professional experience, Appleyard's research on street livability, and their shared feeling that CIAM and the Garden City Movement both represented overly strong design reactions to the physical decay and social inequities of industrial cities – was a validation of cities and urban life. At the same time, it was intended as a critique of the city planning profession for its growing lack of attention to urban physical form in favor of social planning. The argument was that social, economic, and cultural factors are necessarily influenced by physical form factors, not in a deterministic way but rather in terms of possibilities and probabilities, and so to neglect the physical is to neglect an essential part of planning.

The manifesto identifies problems for modern urban design – poor living environments, giantism and loss of control, large-scale privatization and loss of public life, centrifugal fragmentation, destruction of valued places, placelessness, injustice, and rootless professionalism – and then sets out central values for urban life. Goals that serve individuals and small social groups include authenticity and meaning, livability, identity and control, and access to opportunity, imagination, and joy. Those that serve larger social goals include community and public life, urban self-reliance, and an environment for all. From these goals, Jacobs and Appleyard theorize the essential qualities of city grain that must be present to achieve a good urban environment: livable streets and neighborhoods, a minimum residential density and intensity of use, integration of activities, buildings that define public space, and many different buildings and spaces with complex arrangements and relationships.

"Toward an Urban Design Manifesto" remains an important touchstone for present-day urban design practitioners, not least because it cautions against dehumanization, retreat into formalism, and over-dependence on standards that might or might not achieve desirable ends.

Allan B. Jacobs served as San Francisco's Planning Director from 1967 to 1975, where he and a group of talented young planners prepared a citywide urban design plan, which to this day is held to be an exemplary model of such a plan. Donald Appleyard worked as a consultant to the plan, researching the quality of life along neighborhood streets, which resulted in his well-known and highly influential street livability study. Jacobs and Appleyard became colleagues at the University of California, Berkeley, where they established an urban design concentration and collaborated until Appleyard's untimely death in 1982. Jacobs remains a professor and in the early 1990s started the Master of Urban Design program with colleagues from the architecture and landscape architecture programs.

Other writings by Appleyard include *The View from the Road*, co-authored with Kevin Lynch and John Myer (Cambridge, MA: MIT Press, 1963) and *Livable Streets*, co-authored with M. Sue Gerson and Mark Lintell (Berkeley, CA: University of California Press, 1981). Other writings by Jacobs include *Making City Planning Work* (Chicago, IL: American Society of Planning Officials, 1978), which recounts his professional experience in San Francisco through a series of case studies, *Looking at Cities* (Cambridge, MA: Harvard University Press, 1985), which argues that much knowledge about cities can be gained by looking closely at them and reading the clues, *Great Streets* (Cambridge, MA: MIT Press, 1993), which identifies the qualities of the best streets and presents hand-drawn scaled plans and sections of representative examples worldwide, and *The Boulevard Book: History, Evolution, Design of Multiway Boulevards*, co-authored with Elizabeth Macdonald and Yodan Rofé (Cambridge, MA: MIT Press, 2002), which reconsiders boulevards as a useful street type for modern cities.

Researchers and practitioners within the urban design field continue to develop normative theories of good city form. Of particular note is the public effort recently undertaken by the New Zealand Ministry for the Environment, documented in two reports, available for download from the ministry's website (http://www.mfe.govt.nz/). *The Value of Urban Design* surveys the urban design literature and research and identifies the economic, social, and environmental benefits of urban design. *The New Zealand Urban Design Protocol*, similar to Jacob's and Appleyard's manifesto, identifies goals for urban life and essential design qualities.

We think it's time for a new urban design manifesto. Almost 50 years have passed since Le Corbusier and the International Congress of Modern Architecture (CIAM) produced the Charter of Athens, and it is more than 20 years since the first Urban Design Conference, still in the CIAM tradition, was held (at Harvard in 1957). Since then the precepts of CIAM have been attacked by sociologists, planners, Jane Jacobs, and more recently by architects themselves. But it is still a strong influence, and we will take it as our starting point. Make no mistake: the charter was, simply, a manifesto – a public declaration that spelled out the ills of industrial cities as they existed in the 1930s and laid down physical requirements necessary to establish healthy, humane, and beautiful urban environments for people. It could not help but deal with social, economic, and political phenomena, but its basic subject matter was the physical design of cities. Its authors were (mostly) socially concerned architects, determined that their art and craft be responsive to social realities as well as to improving the lot of man. It would be a mistake to write them off as simply elitist designers and physical determinists.

So the charter decried the medium-size (up to six storys) high-density buildings with high land coverage that were associated so closely with slums. Similarly, buildings that faced streets were found to be detrimental to healthy living. These seemingly limitless horizontal expansion of urban areas devoured the countryside, and suburbs were viewed as symbols of terrible waste. Solutions could be found in the demolition of unsanitary housing, the provision of green areas in every residential district, and new high-rise, high-density buildings set in open space. Housing was to be removed from its traditional relationship facing streets, and the whole circulation system was to be revised to meet the needs of emerging mechanization (the automobile). Work areas should be close to but separate from residential areas. To achieve the new city, large land holdings, preferably owned by the public, should replace multiple small parcels (so that projects could be properly designed and developed).

Now thousands of housing estates and redevelopment projects in socialist and capitalist countries the world over, whether built on previously undeveloped land or developed as replacements for old urban areas, attest to the acceptance of the charter's dictums. The design notions it embraced have become part of a world design language, not just the intellectual property of an enlightened few, even though the principles have been devalued in many developments.

Of course, the Charter of Athens has not been the only major urban philosophy of this century to influence the development of urban areas. Ebenezer Howard, too, was responding to the ills of the nineteenth-century industrial city, and the Garden City movement has been at least as powerful as the Charter of Athens. New towns policies, where they exist, are rooted in Howard's thought. But you don't have to look to new towns to see the influence of Howard, Olmsted, Wright, and Stein. The superblock notion, if nothing else, pervades large housing projects around the world, in central cities as well as suburbs. The notion of buildings in a park is as common to garden city designs as it is to charter-inspired development. Indeed, the two movements have a great deal in common: superblocks, separate paths for people and cars, interior common spaces, housing divorced from streets, and central ownership of land. The garden city-inspired communities place greater emphasis on private outdoor space. The most significant difference, at least as they have evolved, is in density and building type: the garden city people preferred to accommodate people in row houses, garden apartments, and maisonettes, while Corbusier and the CIAM designers went for high-rise buildings and, inevitably, people living in flats and at significantly higher densities.

We are less than enthralled with what either the Charter of Athens or the Garden City movement has produced in the way of urban environments. The emphasis of CIAM was on buildings and what goes on within buildings that happen to sit in space, not on the public life that takes place constantly in public spaces. The orientation is often inward. Buildings tend to be islands, big or small. They could be placed anywhere. From the outside perspective, the building, like the work of art it was intended to be, sits where it can be seen and admired in full. And because it is large it is best seen from a distance (at a scale consistent with a moving auto). Diversity, spontaneity, and surprise are absent, at least for the person on foot. On the other hand, we find little joy or magic or spirit in the charter cities. They are not urban, to us, except according to some definition one might find in a census. Most garden cities, safe and healthy and even gracious as they may be, remind us more of suburbs than of cities.

But they weren't trying to be cities. The emphasis has always been on "garden" as much as or more than on "city."

Both movements represent overly strong design reactions to the physical decay and social inequities of industrial cities. In responding so strongly, albeit understandably, to crowded, lightless, airless, "utilitiless," congested buildings and cities that housed so many people, the utopians did not inquire what was good about those places, either socially or physically. Did not those physical environments reflect (and maybe even foster) values that were likely to be meaningful to people individually and collectively, such as publicness and community? Without knowing it, maybe these strong reactions to urban ills ended up by throwing the baby out with the bathwater.

In the meantime we have had a lot of experience with city building and rebuilding. New spokespeople with new urban visions have emerged. As more CIAM-style buildings were built people became more disenchanted. Many began to look through picturesque lenses back to the old preindustrial cities. From a concentration on the city as a kind of sculpture garden, the townscape movement, led by the *Architectural Review*, emphasized "urban experience." This phenomenological view of the city was espoused by Rasmussen, Kepes, and ultimately Kevin Lynch and Jane Jacobs. It identified a whole new vocabulary of urban form – one that depended on the sights, sounds, feels, and smells of the city, its materials and textures, floor surfaces, facades, style, signs, lights, seating, trees, sun, and shade all potential amenities for the attentive observer and user. This has permanently humanized the vocabulary of urban design, and we enthusiastically subscribe to most of its tenets, though some in the townscape movement ignored the social meanings and implications of what they were doing.

The 1960s saw the birth of community design and an active concern for the social groups affected, usually negatively, by urban design. Designers were the "soft cops," and many professionals left the design field for social or planning vocations, finding the physical environment to have no redeeming social value. But at the beginning of the 1980s the mood in the design professions is conservative. There is a withdrawal from social engagement back to formalism. Supported by semiology and other abstract themes, much of architecture has become a dilettantish and narcissistic pursuit, a chic component of the high art consumer culture, increasingly remote from most people's everyday lives, finding its ultimate manifestation in the art gallery and the art book. City planning is too immersed in the administration and survival of housing, environmental, and energy programs and in responding to budget cuts and community demands to have any clear sense of direction with regard to city form.

While all these professional ideologies have been working themselves out, massive economic, technological, and social changes have taken place in our cities. The scale of capitalism has continued to increase, as has the scale of bureaucracy, and the automobile has virtually destroyed cities as they once were.

In formulating a new manifesto, we react against other phenomena than did the leaders of CIAM 50 years ago. The automobile cities of California and the Southwest present utterly different problems from those of nineteenth-century European cities, as do the CIAM-influenced housing developments around European, Latin American, and Russian cities and the rash of squatter settlements around the fast-growing cities of the Third World. What are these problems?

PROBLEMS FOR MODERN URBAN DESIGN

Poor living environments

While housing conditions in most advanced countries have improved in terms of such fundamentals as light, air, and space, the surroundings of homes are still frequently dangerous, polluted, noisy, anonymous wastelands. Travel around such cities has become more and more fatiguing and stressful.

Giantism and loss of control

The urban environment is increasingly in the hands of the large-scale developers and public agencies. The elements of the city grow inexorably in size, massive transportation systems are segregated for single travel modes, and vast districts and complexes are created that make people feel irrelevant.

People, therefore, have less sense of control over their homes, neighborhoods, and cities than when they lived in slower-growing locally based communities. Such giantism can be found as readily in the housing projects of socialist cities as in the office buildings and commercial developments of capitalist cities.

Large-scale privatization and the loss of public life

Cities, especially American cities, have become privatized, partly because of the consumer society's emphasis on the individual and the private sector, creating Galbraith's "private affluence and public squalor," but escalated greatly by the spread of the automobile. Crime in the streets is both a cause and a consequence of this trend, which has resulted in a new form of city: one of closed, defended islands with blank and windowless facades surrounded by wastelands of parking lots and fast-moving traffic. As public transit systems have declined, the number of places in American cities where people of different social groups actually meet each other has dwindled. The public environment of many American cities has become an empty desert, leaving public life dependent for it survival solely on planned formal occasions, mostly in protected internal locations.

Centrifugal fragmentation

Advanced industrial societies took work out of the home, and then out of the neighborhood, while the automobile and the growing scale of commerce have taken shopping out of the local community. Fear has led social groups to flee from each other into homogeneous social enclaves. Communities themselves have become lower in density and increasingly homogeneous. Thus the city has spread out and separated to form extensive monocultures and specialized destinations reachable often only by long journeys – a fragile and extravagant urban system dependent on cheap, available gasoline, and an effective contributor to the isolation of social groups from each other.

Destruction of valued places

The quest for profit and prestige and the relentless exploitation of places that attract the public have led to the destruction of much of our heritage, of historic places that no longer turn a profit, of natural amenities that become overused. In many cases, as in San Francisco, the very value of the place threatens its destruction as hungry tourists and entrepreneurs flock to see and profit from it.

Placelessness

Cities are becoming meaningless places beyond their citizens' grasp. We no longer know the origins of the world around us. We rarely know where the materials and products come from, who owns what, who is behind what, what was intended. We live in cities where things happen without warning and without our participation. It is an alien world for most people. It is little surprise that most withdraw from community involvement to enjoy their own private and limited worlds.

Injustice

Cities are symbols of inequality. In most cities the discrepancy between the environments of the rich and the environments of the poor is striking. In many instances the environments of the rich, by occupying and dominating the prevailing patterns of transportation and access, make the environments of the poor relatively worse. This discrepancy may be less visible in the low-density modern city, where the display of affluence is more hidden than in the old city; but the discrepancy remains.

Rootless professionalism

Finally, design professionals today are often part of the problem. In too many cases, we design for places and people we do not know and grant them very little power or acknowledgment. Too many professionals are more part of a universal professional culture than part of the local cultures for whom we produce our plans and products. We carry our "bag of tricks" around the world and bring them

out wherever we land. This floating professional culture has only the most superficial conception of particular place. Rootless, it is more susceptible to changes in professional fashion and theory than to local events. There is too little inquiry, too much proposing. Quick surveys are made, instant solutions devised, and the rest of the time is spent persuading the clients. Limits on time and budgets drive us on, but so do lack of understanding and the placeless culture. Moreover, we designers are often unconscious of our own roots, which influence our preferences in hidden ways.

At the same time, the planning profession's retreat into trendism, under the positivist influence of social science, has left it virtually unable to resist the social pressures of capitalist economy and consumer sovereignty. Planners have lost their beliefs. Although we believe citizen participation is essential to urban planning, the professionals also must have a sense of what we believe is right, even though we may be vetoed.

GOALS FOR URBAN LIFE

We propose, therefore, a number of goals that we deem essential for the future of a good urban environment: livability; identity and control; access to opportunity, imagination, and joy; authenticity and meaning; open communities and public life; self-reliance; and justice.

Livability

A city should be a place where everyone can live in relative comfort. Most people want a kind of sanctuary for their living environment, a place where they can bring up children, have privacy, sleep, eat, relax, and restore themselves. This means a well-managed environment relatively devoid of nuisance, overcrowding, noise, danger, air pollution, dirt, trash, and other unwelcome intrusions.

Identity and control

People should feel that some part of the environment belongs to them, individually and collectively, some part for which they care and are ever responsible, whether they own it or not. The urban environment should be an environment that encourages people to express themselves, to become involved, to decide what they want and act on it. Like a seminar where everybody has something to contribute to communal discussion, the urban environment should encourage participation. Urbanites may not always want this. Many like the anonymity of the city, but we are not convinced that freedom of anonymity is a desirable freedom. It would be much better if people were sure enough of themselves to stand up and be counted. Environments should therefore be designed for those who use them or are affected by them, rather than for those who own them. This should reduce alienation and anonymity (even if people want them); it should increase people's sense of identity and rootedness and encourage more care and responsibility for the physical environment of cities.

Respect for the existing environment, both nature and city, is one fundamental difference we have with the CIAM movement. Urban design has too often assumed that new is better than old. But the new is justified only if it is better than what exists. Conservation encourages identity and control and, usually, a better sense of community, since old environments are more usually part of a common heritage.

Access to opportunity, imagination, and joy

People should find the city a place where they can break from traditional molds, extend their experience, meet new people, learn other viewpoints, have fun. At a functional level people should have access to alternative housing and job choices; at another level, they should find the city an enlightening cultural experience. A city should have magical places where fantasy is possible, a counter to and an escape from the mundaneness of everyday work and living. Architects and planners take cities and themselves too seriously; the result too often is deadliness and boredom, no imagination, no humor, alienating places. But people need an escape from the seriousness and meaning of the everyday. The city has always been a place of excitement; it is theater, a stage upon which citizens can display themselves and see others. It has magic, or should have, and that depends on a certain

sensuous, hedonistic mood, on signs, on night lights, on fantasy, color, and other imagery. There can be parts of the city where belief can be suspended, just as in the experience of fiction. It may be that such places have to be framed so that people know how to act. Until now such fantasy and experiment have been attempted mostly by commercial facilities, at rather low levels of quality and aspiration, seldom deeply experimental. One should not have to travel as far as the Himalayas or the South Sea Islands to stretch one's experience. Such challenges could be nearer home. There should be a place for community utopias; for historic, natural, and anthropological evocations of the modern city, for encounters with the truly exotic.

Authenticity and meaning

People should be able to understand their city (or other people's cities), its basic layout, public functions, and institutions; they should be aware of its opportunities. An authentic city is one where the origins of things and places are clear. All this means an urban environment should reveal its significant meanings; it should not be dominated only by one type of group, the powerful; neither should publicly important places be hidden. The city should symbolize the moral issues of society and educate its citizens to an awareness of them.

That does not mean everything has to be laid out as on a supermarket shelf. A city should present itself as a readable story, in an engaging and, if necessary, provocative way, for people are indifferent to the obvious, overwhelmed by complexity. A city's offerings should be revealed or they will be missed. This can affect the forms of the city, its signage, and other public information and education programs.

Livability, identity, authenticity, and opportunity are characteristics of the urban environment that should serve the individual and small social unit, but the city has to serve some higher social goals as well. It is these we especially wish to emphasize here.

Community and public life

Cities should encourage participation of their citizens in community and public life. In the face of giantism and fragmentation, public life, especially life in public places, has been seriously eroded. The neighborhood movement, by bringing thousands, probably millions of people out of their closed private lives into active participation in their local communities, has begun to counter that trend, but this movement has had its limitations. It can be purely defensive, parochial, and self-serving. A city should be more than a warring collection of interest groups, classes, and neighborhoods; it should breed a commitment to a larger whole, to tolerance, justice, law, and democracy. The structure of the city should invite and encourage public life, not only through its institutions, but directly and symbolically through its public spaces. The public environment, unlike the neighborhood, by definition should be open to all members of the community. It is where people of different kinds meet. No one should be excluded unless they threaten the balance of life.

Urban self-reliance

Increasingly cities will have to become more self-sustaining in their uses of energy and other scarce resources. "Soft energy paths" in particular not only will reduce dependence and exploitation across regions and countries but also will help re-establish a stronger sense of local and regional identity, authenticity, and meaning.

An environment for all

Good environments should be accessible to all. Every citizen is entitled to some minimal level of environmental livability and minimal levels of identity, control, and opportunity. Good urban design must be for the poor as well as the rich. Indeed, it is more needed by the poor.

We look toward a society that is truly pluralistic, one where power is more evenly distributed among social groups than it is today in virtually any country, but where the different values and cultures of interest – and place-based groups are acknowledged and negotiated in a just public arena.

These goals for the urban environment are both individual and collective, and as such they are frequently in conflict. The more a city promises for the individual, the less it seems to have a public

life; the more the city is built for public entities, the less the individual seems to count. The good urban environment is one that somehow balances these goals, allowing individual and group identity while maintaining a public concern, encouraging pleasure while maintaining responsibility, remaining open to outsiders while sustaining a strong sense of localism.

AN URBAN FABRIC FOR AN URBAN LIFE

We have some ideas, at least, for how the fabric or texture of cities might be conserved or created to encourage a livable urban environment. We emphasize the structural qualities of the good urban environment – qualities we hope will be successful in creating urban experiences that are consonant with our goals.

Do not misread this. We are not describing all the qualities of a city. We are not dealing with major transportation systems, open space, the natural environment, the structure of the large-scale city, or even the structure of neighborhoods, but only the grain of the good city.

There are five physical characteristics that must be present if there is to be a positive response to the goals and values we believe are central to urban life. They must be designed, they must exist, as prerequisites of a sound urban environment. All five must be present, not just one or two. There are other physical characteristics that are important, but these five are essential: livable streets and neighborhoods; some minimum density of residential development as well as intensity of land use; an integration of activities – living, working, shopping – in some reasonable proximity to each other; a manmade environment, particularly buildings, that defines public space (as opposed to buildings that, for the most part, sit in space); and many, many separate, distinct buildings with complex arrangements and relationships (as opposed to few, large buildings).

Let us explain, keeping in mind that all five of the characteristics must be present. People, we have said, should be able to live in reasonable (though not excessive) safety, cleanliness, and security. That means livable streets and neighborhoods: with adequate sunlight, clean air, trees, vegetation, gardens, open space, pleasantly scaled and designed buildings; without offensive noise; with cleanliness and physical safety. Many of these characteristics can be designed into the physical fabric of the city.

The reader will say, "Well of course, but what does that mean?" Usually it has meant specific standards and requirements, such as sun angles, decibel levels, lane widths, and distances between buildings. Many researchers have been trying to define the qualities of a livable environment. It depends on a wide array of attributes, some structural, some quite small details. There is no single right answer. We applaud these efforts and have participated in them ourselves. Nevertheless, desires for livability and individual comfort by themselves have led to fragmentation of the city. Livability standards, whether for urban or for suburban developments, have often been excessive.

Our approach to the details of this inclusive physical characteristic would center on the words "reasonable, though not excessive . . ." Too often, for example, the requirement of adequate sunlight has resulted in buildings and people inordinately far from each other, beyond what demonstrable need for light would dictate. Safety concerns have been the justifications for ever wider streets and wide, sweeping curves rather than narrow ways and sharp corners. Buildings are removed from streets because of noise considerations when there might be other ways to deal with this concern. So although livable streets and neighborhoods are a primary requirement for any good urban fabric — whether for existing, denser cities or for new development — the quest for livable neighborhoods, if pursued obsessively, can destroy the urban qualities we seek to achieve.

A *minimum density* is needed. By density we mean the number of people (sometimes expressed in terms of housing units) living on an area of land, or the number of people using an area of land.

Cities are not farms. A city is people living and working and doing the things they do in relatively close proximity to each other.

We are impressed with the importance of density as a perceived phenomenon and therefore relative to the beholder and agree that, for many purposes, perceived density is more important than an "objective" measurement of people per unit of land. We agree, too, that physical phenomena can be manipulated so as to render perceptions of greater or lesser density. Nevertheless, a narrow,

winding street, with a lot of signs and a small enclosed open space at the end, with no people, does not make a city. Cities are more than stage sets. Some minimum number of people living and using a given area of land is required if there is to be human exchange, public life and action, diversity and community.

Density of people alone will account for the presence or absence of certain uses and services we find important to urban life. We suspect, for example, that the number and diversity of small stores and services — for instance, groceries, bars, bakeries, laundries and cleaners, coffee shops, secondhand stores, and the like — to be found in a city or area is in part a function of density. That is, that such businesses are more likely to exist, and in greater variety, in an area, where people live in greater proximity to each other ("higher" density). The viability of mass transit, we know, depends partly on the density of residential areas and partly on the size and intensity of activity at commercial and service destinations. And more use of transit, in turn, reduces parking demands and permits increases in density. There must be a critical mass of people, and they must spend a lot of their time in reasonably close proximity to each other, including when they are at home, if there is to be an urban life. The goal of local control and community identity is associated with density as well. The notion of an optimum density is elusive and is easily confused with the health and livability of urban areas, with lifestyles, with housing types, with the size of area being considered (the building site or the neighborhood or the city), and with the economics of development. A density that might be best for child rearing might be less than adequate to support public transit. Most recently, energy efficiency has emerged as a concern associated with density, the notion being that conservation will demand more compact living arrangements.

Our conclusion, based largely on our experience and on the literature, is that a minimum net density (people or living units divided by the size of the building site, excluding public streets) of about 15 dwelling units (30–60 people) per acre of land is necessary to support city life. By way of illustration, that is the density produced with generous town houses (or row houses). It would permit parcel sizes up to 25 feet wide by about 115 feet deep. But other building types and lot sizes also would produce that density. Some areas could be developed with lower densities, but not very many. We don't think you get cities at 6 dwellings to the acre, let alone on half-acre lots. On the other hand, it is possible to go as high as 48 dwelling units per acre (96 to 192 people) for a very large part of the city and still provide for a spacious and gracious urban life. Much of San Francisco, for example, is developed with three-story buildings (one unit per floor) above a parking story, on parcels that measure 25 feet by 100 or 125 feet. At those densities, with that kind of housing, there can be private or shared gardens for most people, no common hallways are required, and people can have direct access to the ground. Public streets and walks adequate to handle pedestrian and vehicular traffic generated by these densities can be accommodated in rights-of-way that are 50 feet wide or less. Higher densities, for parts of the city, to suit particular needs and lifestyles, would be both possible and desirable. We are not sure what the upper limits would be but suspect that as the numbers get much higher than 200 people per net residential acre, for larger parts of the city, the concessions to less desirable living environments mount rapidly.

Beyond residential density, there must be a minimum intensity of people using an area for it to be urban, as we are defining that word. We aren't sure what the numbers are or even how best to measure this kind of intensity. We are speaking here, particularly, of the public or "meeting" areas of our city. We are confident that our lowest residential densities will provide most meeting areas with life and human exchange, but are not sure if they will generate enough activity for the most intense central districts.

There must be an *integration of activities* – living, working, and shopping as well as public, spiritual, and recreational activities – reasonably near each other.

The best urban places have some mixtures of uses. The mixture responds to the values of publicness and diversity that encourage local community identity. Excitement, spirit, sense, stimulation, and exchange are more likely when there is a mixture of activities than when there is not. There are many examples that we all know. It is the mix, not just the density of people and uses, that brings life to an area, the life of people going about a full range of normal activities without having to get into an automobile.

We are not saying that every area of the city should have a full mix of all uses. That would be impossible. The ultimate in mixture would be for each building to have a range of uses from living, to working, to shopping, to recreation. We are not calling for a return to the medieval city. There is a lot to be said for the notion of "living sanctuaries," which consist almost wholly of housing. But we think these should be relatively small, of a few blocks, and they should be close and easily accessible (by foot) to areas where people meet to shop or work or recreate or do public business. And except for a few of the most intensely developed office blocks of a central business district or a heavy industrial area, the meeting areas should have housing within them. Stores should be mixed with offices. If we envision the urban landscape as a fabric, then it would be a salt-and-pepper fabric of many colors, each color for a separate use or a combination. Of course, some areas would be much more heavily one color than another, and some would be an even mix of colors. Some areas, if you squinted your eyes, or if you got so close as to see only a small part of the fabric, would read as one color, a red or a brown or a green. But by and large there would be few if any distinct patterns, where one color stopped and another started. It would not be patchwork quilt, or an even-colored fabric. The fabric would be mixed.

In an urban environment, *buildings* (and other objects that people place in the environment) *should be arranged in such a way as to define and even enclose public space, rather than sit in space.* It is not enough to have high densities and an integration of activities to have cities. A tall enough building with enough people living (or even working) in it, sited on a large parcel, can easily produce the densities we have talked about and can have internally mixed uses, like most "mixed use" projects. But that building and its neighbors will be unrelated objects sitting in space if they are far enough apart, and the mixed uses might be only privately available. In large measure that is what the Charter of Athens, the garden cities, and standard suburban development produce.

Buildings close to each other along a street, regardless of whether the street is straight, or curved, or angled, tend to define space if the street is not too wide in relation to the buildings. The same is true of a plaza or a square. As the spaces

PART THREE

Place Theories in Urban Design

Plate 3 Café culture at Bryant Park, New York City: Ray Oldenburg's concept of the "third place" (settings where an informal social life can thrive) is recognized most easily in sidewalk cafés and coffee houses where people can spend time with friends and meet new people. (Photo: E. Macdonald)

INTRODUCTION TO PART THREE

Dissatisfied with the increasing homogeneity and soullessness of mid-twentieth-century urban spaces, a number of authors began investigating issues of space and place as a means of correction. In Part Two we examined how theorists approached these issues through normatively based theories to better guide design and physical outcomes. The authors in this part approach the same manifest problems by trying to understand the relationship of personal experience to environmental settings. While this body of literature explores the relationship between physical settings and human subjects, the focus here is primarily on the physicality of the tangible world and our ability to provide and design places. Research in place studies tends to be both empirical and phenomenological in nature, largely incorporating observational and environmental behavior methods. Researchers in this tradition utilize an eclectic array of theories in defining notions of place, but seemingly they all stem from this pervasive unhappiness with non-place-based forms and difficulties in reading modern urbanism. In many ways it is a highly reactive and critical literature, often looking backwards to traditional environments where places were better differentiated and place-based meanings were more easily understood.

This part of the reader begins with a chapter from Edward Relph's book *Place and Placelessness*, where he defines and differentiates the concept of place from the placeless geographies of "endless similarity." To Relph, places are "directly experienced phenomena of the lived-world and hence are full with meanings, with real objects, and with ongoing activities." In this reading he warns about the power of placelessness and its potential consumption of our place-based world. In the next reading, Christian Norberg-Schulz introduces us to phenomenological understandings of place and the Roman concept of *genius loci*, or the guardian spirit of place. Here, the concept of place is examined through our ability to connect with the physical character of geographic settings, thereby bestowing the place with identity and meaning. For Norberg-Schulz, like Martin Heidegger, place is inextricably connected to existential realities and helps us define our personal identities relative to those places we inhabit. The last reading in Part Three concerns the need for places within our communities. Ray Oldenburg's chapter from *The Great Good Place* suggests we need and utilize "third places" to supplement our home and work lives with meaningful places of leisure and informal social relations. Here the concept of place is derived through the socio-spatial opportunities allowed by cafés, bookstores, pubs, and other hangouts. It is a notion of place that turns on society's ability to provide places for people to come together freely and voluntarily to make contact and enjoy public life.

The readings presented here are theoretical in nature, rather than dealing directly with urban design or design guidance. Approaches to design and place-making can be found in Part Four. Other theoretical authors within the place studies tradition, who have not been included here, deserve mention as well. Henri Lefebvre's classic text *The Production of Space*, translated by David Nicholson-Smith (Oxford: Blackwell, 1991), provides a critical unitary theory of space that integrates the physical space of the everyday lifeworld, the mental space of abstractions and the relational realm of social space. In a similar critical manner to the writings above, he suggests that physical everyday places are disappearing in favor of the abstract spaces of capitalism, which is a social product subject to the mode of production. In like manner, David Harvey examines the capitalist production of space and analyzes how command over the forces of production relates to power in urban space: *Consciousness and the Urban Experience* (Baltimore, MD: Johns Hopkins Press, 1985) and *The Condition of Postmodernity* (Oxford: Blackwell, 1989). Authors looking at society's role in the provision of urban space include Mark Gottdiener, *The Social Production of Urban Space* (Austin, TX: University of Texas Press, 1985); Ali Madanipour, *Design of Urban Space: An Inquiry into a Socio-Spatial Process* (Chichester, UK: John Wiley, 1996). A much loved and widely read perspective on domestic space, spatial psychology, and daydreaming can be found in Gaston Bachelard, *The Poetics of Space* (Boston, MA: Beacon Press, reprint edition, 1994).

"Prospects for Places"
from *Place and Placelessness* (1976)

Edward Relph

Editors' Introduction

A leading early voice to identify and analyze the growing sense of placelessness that was occurring through-out the world by the latter half of the twentieth century was geographer Edward Relph. In his now classic book *Place and Placelessness*, Relph uses critical observation to make connections between the visible land-scape, everyday life experiences, and the abstract social and economic processes that contribute to their transformation. Written in straightforward language and grounded in experience of actual places, Relph's ideas were easily accessible to physical planners and urban designers, providing an intellectual base on which place-making proposals could rest.

Within "Prospects for Places," Relph describes the main features of place and placelessness. He identifies meaningful experience, a sense of belonging, human scale, fit with local physical and cultural contexts, and local significance as the important qualities of place. Placelessness, on the other hand, is associated with an overriding concern for efficiency, mass culture, and anonymous, exchangeable environments. Going beyond critical description to provide useful direction for planning and urban design professionals, he suggests how authentic place-making could be achieved in modern times. Dismissing either unstructured laissez-faire chaos or rigid bureaucratic prescription, he argues instead for a self-conscious planned diversity that allows people to make their own places, rooted in local contexts and filled with local meaning.

Relph teaches geography at the University of Toronto. His other writings that focus on landscape and place include *Rational Landscapes and Humanistic Geography* (Totowa, NJ: Barnes & Noble, 1981) and *The Modern Urban Landscape* (London: Croom Helm, 1987; Baltimore, MD: Johns Hopkins University Press, 1987). Other writings that deal with placelessness include James Kuntsler, *The Geography of Nowhere: The Rise and Decline of America's Man-Made Landscape* (New York: Simon & Schuster, 1993); Joel Garreau, *Edge City: Life on the New Frontier* (New York: Doubleday, 1991); Michael Sorkin, *Variations on a Theme Park: The American City and the End of Public Space* (New York: Hill & Wang, 1992); and Sharon Zukin, *Landscapes of Power: From Detroit to Disney World* (Berkeley, CA: University of California Press, 1991).

There are at least two experienced geographies: there is a geography of places, characterised by variety and meaning, and there is a placeless geography, a labyrinth of endless similarities. The current scale of the destruction and replacement of the distinctive places of the world suggests that place-less geography is increasingly the more forceful of these, even though a considerable diversity of places persists. It is not immediately apparent whether this persistence is the remnant of an old place-making tradition and is shortly to disappear beneath a tide of uniformity, or whether there exist ongoing and developing sources of diversity that can be encouraged. In other words the prospects for geography of places are uncertain, but one possibility is the inevitable spread of placelessness, and an alternative possibility is the transcending of placelessness through the formulation and application of an approach for the design of a lived-world of significant places. [Here] . . . these possibilities are considered in the context of summaries of the main features of place and placelessness.

PLACE

Places are fusions of human and natural order and are the significant centres of our immediate experiences of the world. They are defined less by unique locations, landscape, and communities than by the focusing of experiences and intentions onto particular settings. Places are not abstractions or concepts, but are directly experienced phenomena of the lived-world and hence are full with meanings, with real objects, and with ongoing activities. They are important sources of individual and communal identity, and are often profound centres of human existence to which people have deep emotional and psychological ties. Indeed our relationships with places are just as necessary, varied, and sometimes perhaps just as unpleasant, as our relationships with other people.

Experience of place can range in scale from part of a room to an entire continent, but at all scales places are whole entities, syntheses of natural and man-made objects, activities and functions, and meanings given by intentions. Out of these components the identity of a particular place is moulded, but they do not define this identity – it is the special quality of insideness and the experience of being inside that sets places apart in space. Insideness may relate to and be reflected in a physical form, such as the walls of a medieval town, or it may be expressed in rituals and repeated activities that maintain the peculiar properties of a place. But above all it is related to the intensity of experience of a place. Alan Gussow (1971?, p. 27) has written of this: "The catalyst that converts any physical location – any environment if you will – into a place, is the process of experiencing deeply. A place is a piece of the whole environment that has been claimed by feelings."

It is possible to distinguish several levels of experience of the insideness of places, and it is perhaps these that tell us most about the nature of the phenomenon of place. At the deepest levels there is an unselfconscious, perhaps even subconscious, association with place. It is home, where your roots are, a centre of safety and security, a field of care and concern, a point of orientation. Such insideness is individual but also intersubjective, a personal experience with which many people can sympathise; it is the essence of a sense of place. And it is perhaps presymbolic and universal insofar as it is an aspect of profound place experience anywhere, yet is not associated with the culturally defined meanings of specific places. This is, in fact, existential insideness – the unselfconscious and authentic experience of place as central to existence. The next level of experience is also authentic and unselfconscious, but it is cultural and communal rather than individual: it involves a deep and unreflective participation in the symbols of a place for what they are. It is associated particularly with the sacred experience of involvement in holy places, and with the secular experience of being known in and knowing the named and significant places of a home region. At a shallower level of insideness there is an authentic sense of place that is selfconscious, and which involves a deliberate attempt to appreciate fully the significance of places without the adoption of narrow intellectual or social conventions and fashions. This is the experience of a sensitive and open-minded outsider seeking to grasp places for what they are to those who dwell in them and for what they mean to him. It is an attitude of particular importance in terms of the possibilities it offers to contemporary and authentic place-making. In contrast is the superficial level of insideness, which involves simply being in a place without attending in any sensitive way to its qualities or significances. Though each of us must experience many of the places we visit like this, since concern with our activities takes precedence and it becomes impossible to concentrate on the place itself, when this is the only form of experience of place it denotes a real failure to 'see' or to be involved in places. For those swayed by the easy charms of mass culture or the cool attractions of technique this does seem to be the primary, perhaps the only, way of experiencing environments; and consequently they feel no care or commitment for places: they are geographically alienated.

The various levels of insideness are manifest in the creation of distinctive types of places. The deep levels of existential insideness are apparent in the unselfconscious making of places which are human in their scale and organisation, which fit both their physical and cultural contexts and hence are as varied as those contexts, and which are filled with significances for those who live in them. Authentic and selfconscious insideness offers a similar, though less completely involved, possibility for expressing man's humanity in places. In both instances "the making of places is", as Rapoport (1972, p. 3-3-10) writes, "the ordering of the world", for it differentiates the world into qualitatively distinct centres and gives a structure that both reflects and guides experiences. This is not so with incidental insideness, for such non-commitment opens the way for the development of environments ordered by conceptual principles or mass fashions rather than by patterns of direct experience. In short, uncommitted insideness is the basis for placelessness.

PLACELESSNESS

Placelessness describes both an environment without significant places and the underlying attitude, which does not acknowledge significance in places. It reaches back into the deepest levels of place, cutting roots, eroding symbols, replacing diversity with uniformity and experiential order with conceptual order. At its most profound it consists of a pervasive and perhaps irreversible alienation from places as the homes of men: "He who has no home now will not build one anymore", Rilke declared, and this was echoed by Heidegger – "Homelessness is becoming a world fate" (both cited in Pappenheim, 1959, p. 33). At less deep levels placelessness is the adoption of the attitude described by Harvey Cox (1968, p. 424) as an "abstract geometric view of place, denuded of its human meaning", and it is manifest in landscapes that can be aptly described by Stephen Kurtz' specific account (1973, p. 23) of Howard Johnson's restaurants: "Nothing calls attention to itself; it is all remarkably unremarkable . . . You have seen it, heard it, experienced it all before, and yet . . . you have seen and experienced nothing . . ."

As a selfconsciously adopted posture placelessness is particularly apparent in *technique*, the over-riding concern with efficiency as an end in itself. In *technique* places can be treated as the interchangeable, replaceable locations of things, as indeed they are by multinational corporations, powerful central governments, and uninvolved planners. As an unselfconscious attitude placelessness is particularly associated with mass culture – the adoption of fashions and ideas about landscapes and places that are coined by a few 'experts' and disseminated to the people through the mass media. The products of these two attitudes are combined in uniform, sterile, other-directed, and kitschy places – places which have few significances and symbols, only more or less gaudy signs and things performing functions with greater or lesser efficiency. The overall result is the undermining of the importance of place for both individuals and cultures, and the casual replacement of the diverse and significant places of the world with anonymous spaces and exchangeable environments.

THE INEVITABILITY OF PLACELESSNESS?

The places that we have known belong now only to the little world of space on which we map them for our own convenience. None of them was ever more than a thin slice held between the contiguous impressions that composed our life at that time; remembrance for a particular form is but regret for a particular moment, and houses, roads, avenues, are as fugitive, alas, as the years.

Thus Marcel Proust (1970, p. 288) expressed with nostalgia the insignificance of places for modern man. No more is there the "sense of continuity with place" which Harvey Cox (1968, p. 423) believes is so necessary for people's sense of reality and so essential for their identity; the meanings of places have become as ephemeral as their physical forms. Cox judges this as "one of the most deplorable characteristics of our time", but deplore it, condemn it, criticise it as we might, there often appears to be little that can be done to prevent the diminishing of significant relations with places.

The prospect of inevitable placelessness is supported by Jacques Ellul's view of *technique*, one of the main forces behind the developing placeless geography. He writes (1964, p. 436): "The attitude of scientists, at any rate, is clear. Technique exists because it is technique. The golden age will be because it will be. Any other answer is superfluous." In other words *technique* has a drive of its own that is universal, we can no longer think in terms other than those of *technique* because it is the only language we know, and the only possibility is that placelessness will come to dominate. If we regret the disappearance of significant places this is only sentimentality and we should at least acknowledge the benefits of the new geography. As George Grant (1969, p. 138) expresses it:

> It might be said that the older systems of meaning have been replaced by a new one. The enchantment of our souls by myth, philosophy or revelation has been replaced by a more immediate meaning – the building of free and equal men by the overcoming of chance.

But in what sense freedom and in what sense equality? To master chance in human and non-human nature requires the most efficient use of *technique* that is possible, and that in turn requires the perfection of science and powerful central government. Louch (1966, p. 239) has declared: "Totalitarianism is too weak a word and too inefficient an instrument to describe the perfect scientific society." Alexis de Tocqueville (1945, vol. II, p. 337) wrote: "The will of man is not shattered but softened, bent and guided – such centralised power does not destroy, but compresses, enervates, extinguishes and stupefies a people."

If Tocqueville, Grant, and Ellul are correct, and in the landscape of industrial cultures there is massive evidence to support them, then opposition to *technique* and to central authorities – two of the primary sources of placelessness – seems either futile or impossible. We may protest it, deplore it, propose alternatives to it, but the fundamental basis for our experience of the landscapes we live in is increasingly becoming the attitude of placelessness.

DESIGNING A LIVED-WORLD OF PLACES

But such pessimism and fatalism are not yet justified. There may indeed come a time when placelessness is inevitable because it is the only geography we know, but so long as there are what Grant (1969, p. 139) calls "intimations of authentic deprival", then the possibility of some different way of thinking and acting must remain. David Brower (in Gussow, 1971?, p. 15) is in fact quite specific about what must be done: "The best weapon against the unending deprivation that would be the consequence of . . . unending demand is a revival of man's sense of place." How this is to be achieved he does not make clear, but it is certain that loss of attachment to places and the decline of the ability to make places authentically do constitute real deprivations, and that the redevelopment of such attachments and abilities is essential if we are to create environments that do not have to be ignored or endured. Furthermore, there appears to be a possibility of doing this outside the context of *technique*, for sense of place is in its essence both prescientific and intersubjective.

The possibilities for maintaining and reviving man's sense of place do not lie in the preservation of old places – that would be museumisation; nor can they lie in a selfconscious return to the traditional ways of placemaking – that would require the regaining of a lost state of innocence. Instead, placelessness must be transcended. "That human activity should become more dispersed is inevitable", Georges Matoré (1966, p. 6) has written, "but to compensate let the occupied, lived-in space acquire more cohesion, become as rich as possible, and grow large with the experience of living." Similarly Harvey Cox (1968, p. 424) has argued that beyond the stage of homogeneous space, in which every place is interchangeable with every other place, lies a stage of human space in which "space is for man and places are understood as giving pace, variety and orientation to man". This will not come about automatically but through deliberate effort and the development of 'secularisation', an attitude which corresponds closely to selfconscious authenticity. Secularisation "dislodges ancient oppressions and overturns stultifying conventions. It turns man's social and cultural life over to him, demanding a constant expenditure of vision and competence" (Cox, 1965, p. 86). While the danger always remains of this being short-circuited by new orthodoxies that will result in placelessness, secularisation provides a very real basis for optimism about places so long as we can live up to the responsibilities it demands. Cox continues: "A

(Heidegger, 1971, p. 181).[12] It is therefore not only important that our environment has a spatial structure which facilitates orientation, but that it consists of concrete objects of identification. *Human identity presupposes the identity of place.* Identification and orientation are primary aspects of man's being-in-the-world. Whereas identification is the basis for man's sense of *belonging*, orientation is the function which enables him to be that *homo viator* which is part of his nature. It is characteristic for modern man that for a long time he gave the role as a wanderer pride of place. He wanted to be "free" and conquer the world. Today we start to realize that true freedom presupposes belonging, and that "dwelling" means belonging to a concrete place.

The word to "dwell" has several connotations which confirm and illuminate our thesis. Firstly it ought to be mentioned that "dwell" is derived from the Old Norse *dvelja*, which meant to linger or remain. Analogously, Heidegger related the German "wohnen" to "bleiben" and "sich aufhalten" (Heidegger, 1971, pp. 146ff). Furthermore he points out that the Gothic *wunian* meant to "be at peace," "to remain in peace." The German word for "peace," *Friede*, means to be free, that is, protected from harm and danger. This protection is achieved by means of an *Umfriedung* or enclosure. *Friede* is also related to *zufrieden* (content), *Freund* (friend) and the Gothic *frijön* (love). Heidegger uses these linguistic relationships to show that *dwelling means to be at peace in a protected place.* We should also mention that the German word for dwelling *Wohnung*, derives from *das Gewohnte*, which means what is known or habitual. "Habit" and "habitat" show an analogous relationship. In other words, man knows what has become accessible to him through dwelling. We here return to the *Übereinstimmung* or correspondence between man and his environment, and arrive at the very root of the problem of "gathering." To gather means that the everyday life-world has become "gewohnt" or "habitual." But gathering is a concrete phenomenon, and thus leads us to the final connotation of "dwelling." Again it is Heidegger who has uncovered a fundamental relationship. Thus he points out that the Old English and High German word for "building," *buan*, meant to dwell, and that it is intimately related to the verb *to be*.

What then does *ich bin* mean? The old word *bauen*, to which the *bin* belongs, answers: *ich bin, du bist*, mean: I dwell, you dwell. The way in which you are and I am, the manner in which we humans *are* on earth, is *buan*, dwelling.

Heidegger, 1971, p. 147

We may conclude that dwelling means to gather the world as a concrete building or "thing," and that the archetypal act of building is the *Umfriedung* or enclosure. Trakl's poetic intuition of the inside–outside relationship thus gets its confirmation, and we understand that our concept of *concretization* denotes the essence of dwelling (Norberg-Schulz, 1963, pp. 61ff, 68).

Man dwells when he is able to concretize the world in buildings and things. As we have mentioned above, "concretization" is the function of the work of art, as opposed to the "abstraction" of science (Norberg-Schulz, 1963, pp. 168ff). Works of art concretize what remains "between" the pure objects of science. Our everyday life-world *consists of* such "intermediary" objects, and we understand that the fundamental function of art is to gather the contradictions and complexities of the life-world. Being an *imago mundi*, the work of art helps man to dwell. Hölderlin was right when he said:

Full of merit, yet poetically, man
Dwells on this earth.

This means: man's merits do not count much if he is unable to dwell *poetically*, that is, to dwell in the true sense of the word. Thus Heidegger says: "Poetry does not fly above and surmount the earth in order to escape it and hover over it. Poetry is what first brings man onto the earth, making him belong to it, and thus brings him into dwelling" (Heidegger, 1971, p. 218). Only poetry in all its forms (also as the "art of living") makes human existence meaningful, and *meaning* is the fundamental human need.

Architecture belongs to poetry, and its purpose is to help man to dwell. But architecture is a difficult art. To make practical towns and buildings is not enough. Architecture comes into being when a "total environment is made visible," to quote the definition of Susanne Langer (1953). In general, this means to concretize the *genius loci*. We have seen that this is done by means of buildings which gather the properties of the place and bring them close to man. The basic act of architecture is therefore to understand the "vocation" of the place. In this way we protect the earth and become ourselves part of a comprehensive totality. What is here advocated is not some kind of "environmental determinism." We only recognize the fact that man *is* an integral part of the environment, and that it can only lead to human alienation and environmental disruption if he forgets that. To belong to a place means to have an existential foothold, in a concrete every-day sense. When God said to Adam: "You shall be a fugitive and a wanderer on the Earth,"[13] he put man in front of his most basic problem: to cross the threshold and regain the lost place.

NOTES

1   The concept "everyday life-world" was intro-duced by Husserl in *The Crisis of European Sciences and Transcendental Phenomenology* (1936).

2   Heidegger, "Bauen Wohnen Denken"; Bollnow, "Mensch und Raum"; Merleau-Ponty, "Phe-nomenology of Perception"; Bachelard, "Poetics of Space"; also L. Kruse, *Räumliche Umwelt* (Berlin: 1974).

3   Ein Winterabend

   Wenn der Schnee ans Fenster fällt,
   Lang die Abendglocke läutet,
   Vielen ist der Tisch bereitet
   Und das Haus ist wohlbestellt.
       Mancher auf der Wanderschaft
   Kommt ans Tor auf dunklen Pfaden.
   Golden blüht der Baum der Gnaden
   Aus der Erde kühlem Saft.
       Wanderer tritt still herein;
   Schmerz versteinerte die Schwelle.
   Da erglänzt in reiner Helle
   Auf dem Tische Brot und Wein.

4   Heidegger points out the relationship between the words *gegen* (against, opposite) and *Gegend* (environment, locality).

5   This has been done by some writers, such as K. Graf von Dürckheim, E. Straus, and O.F. Bollnow.

6   We may compare with Alberti's distinction between "beauty" and "ornament."

7   For the concept of "capacity" see Norberg-Schulz, *Intentions* (1963).

8   See M.M. Webber, *Explorations into Urban Structure* (1963), who talks about "non-place urban realm."

9   Norberg-Schulz, *Intentions* (1963), where the concepts "cognitive orientation" and "cathetic orientation" are used.

10   For a detailed discussion, see Norberg-Schulz, *Existence* (1971).

11   Seltsam, im Nebel zu wandern! Einsam ist jeder Busch und Stein, kein Baum sieht den anderen, jeder ist allein.

12   Heidegger, "We are the be-thinged," the con-ditioned ones.

13   *Genesis*, chapter 4, verse 12.

REFERENCES

Appleton, J. (1975) *The Experience of Landscape*. London.

Bollnow, O.F. (1956) *Das Wesen der Stimmungen*. Frankfurt am Main.

Durrell, L. (1969) *Spirit of Place*. London.

Frey, D. (1949) *Grundlegung zu einer vergleichenden Kunstwissenschaft*. Vienna and Innsbruck.

Giedion, S. (1964) *The Eternal Present: The Beginnings of Architecture*. London.

Goethe, J.W. von (1786) *Italienische Reise*, 8, October.

Heidegger, M. (1954) Die Frage nach der Technik. In *Vorträge und Aufsätze*. Pfullingen.

Heidegger, M. (1957) *Hebel der Hausfreund*. Pfullingen.

Heidegger, M. (1971) *Poetry, Language, Thought*, ed. A. Hofstadter. New York.

Husserl, E. (1936) *The Crisis of European Sciences and Transcendental Phenomenology*. Evanston, IL.

Langer, S. (1953) *Feeling and Form: A Theory of Art*. New York.

Lynch, K. (1960) *The Image of the City*. Cambridge, MA.

Norberg-Schulz, C. (1963) *Intentions in Architecture*. Oslo and London.

Norberg-Schulz, C. (1971) *Existence, Space and Architecture*. London and New York.

Norberg-Schulz, C. (1975) *Meaning in Western Archi-tecture*. London and New York.

Paulys (n.d.) *Realencyclopedie der Klassischen Alter-tumwissenschaft*, VII.

Portoghesi, P. (1975) *Le inibizioni dell'architettura moderna*. Bari, Italy.

Rapoport, A. (1975) Australian Aborigines and the definition of place. In P. Oliver (ed.), *Shelter, Sign and Symbol*. London

Richardson, W.J. (1974) *Heidegger: Through Phenom-enology to Thought*. The Hague.

Rilke, R.M. (1972) *The Duino Elegies*. New York.

Venturi, R. (1967) *Complexity and Contradiction in Architecture*. New York.

Webber, M.M. (1963) *Explorations into Urban Structure*. Philadelphia.

"The Problem of Place in America"

from *The Great Good Place* (1989)

Ray Oldenburg

Editors' Introduction

As the previous two readings show, issues of place in modern society have become important because of declines in both design quality and resulting activity levels in urban space. In this first chapter of *The Great Good Place*, Ray Oldenburg suggests that urban decline is associated with historical post-war trends of suburbanization, urban renewal, increasing residential mobility, growing auto dependency, freeway expansion, and single-use zoning – all of which have contributed to the disappearance of informal gathering spaces. Newer suburban subdivisions and neighborhoods have failed at providing spaces for community life for their inhabitants largely due to the increasing isolation of family life and the extreme individualization championed by American society. In physical settings dependent on single-occupancy auto trips to the strip mall and the zoned illegality of neighborhood retail uses, there is little opportunity for chance meetings on sidewalks, in corner bars or at a local café within walking distance from home. In households where one's worklife takes up so much time, and where television has become the primary source of nightly entertainment, it becomes no surprise that people have little time for community-oriented activities such as bowling, bocce, and billiards. In a related manner, the increasing prevalence of obesity, chronic disease, and high stress levels can also be attributed to unwalkable suburban form, the increase in auto dependency, and lack of places to relax and blow off steam.

In light of this systematic loss of social space, Ray Oldenburg posits that a possible solution to "the problem of place in America" might be the championing of the "third place." He defines third places as those informal public gathering spaces where people can come together on neutral ground, free of charge, to develop friendships, enjoy conversation, voluntarily interact and enjoy being part of a larger spatial community. Oldenburg suggests third places are essential ingredients to a well-functioning democracy, for developing social cohesion, endowing a sense of identity and providing psychological support outside of home (the first place) and the work setting (the second place). They are the pubs, coffee houses, general stores, bookshops, post offices, laundromats, beauty salons, community centers, bowling alleys, and other public social spaces (including streets, sidewalks, and parks) where people can come together to enjoy each other's company and conversation. Third places might be thought of as regular local hangouts (without the negative connotation of a dive) or like the French *rendezvous* (without the romantic connotation). Because of their accessibility and inclusiveness, third places promote social equality and are considered social levelers, places where little distinction is made on the basis of demographic, economic, social or cultural differences. These places tend to allow people of different backgrounds to get to know one another in settings that are socially expansive and non-threatening.

Although he focuses primarily on programmatic land use elements, urban designers can take inspiration in Oldenburg's valorization of the public realm as a locus for social life. While not addressing specific design issues, those interested in improving streets, parks, and other public meeting grounds will be forced to consider the physical elements that make these places function for active human occupation, including: streetscaping, seating, lighting, climate protection and other amenities that help to make places comfortable and useable.

Third places were much more prevalent in earlier times, especially in the denser, urban villages of traditional pedestrian-oriented towns and cities. Historically they provided compensatory social space away from home and work. In modern American society, however, they became rare with the rise of suburbia and zoning. With renewed interest for in-town living, as well as the rise of creative class interests in place-based urban lifestyles, we are beginning to see a rebound in the number of cafés, pubs, and local-serving retail nodes in closer proximity to housing. Counter-productive to the concept of the third place, however, are recent efforts at the privatization of public space, the growing prevalence of private security and policing, and heightened surveillance activities in light of real and perceived threats of crime and terrorism. These new concerns can be witnessed in the rise of gated communities, theme parks, malls, office parks, entertainment centers, mega-projects, and new towns, places where behavior is monitored and other personal freedoms are often limited.

Ray Oldenburg is Professor Emeritus of Sociology at the University of West Florida in Pensacola, and works as a consultant to cities and community-based advocacy groups. He also edited a companion piece to *The Great Good Place* titled *Celebrating the Third Place: Inspiring Stories from the "Great Good Places" at the Heart of our Communities* (New York: Marlowe/Avalon, 2001). For other books on third spaces and informal social meeting grounds see: Bernard Rudofsky, *Streets for People: A Primer for Americans* (Garden City, NY: Doubleday, 1969); Claude Fisher, *To Dwell Among Friends* (Chicago, IL: University of Chicago Press, 1982); Anne Vernez Moudon (ed.), *Public Streets for Public Use* (New York: Van Nostrand Reinhold, 1987); Jan Gehl, *Life Between Buildings: Using Public Space* (New York: Van Nostrand Reinhold, 1987); and Christian Mikunda, *Brand Lands, Hot Spots and Cool Spaces: Welcome to the Third Place and the Total Marketing Experience* (London: Kogan Page, 2004).

Books on the decline of social life and public space, and increasing privatization include: David Riesman, Nathan Glazer, and Reuel Denney, *The Lonely Crowd* (New Haven, CT: Yale University Press, 1950); Vance Packard, *A Nation of Strangers* (New York: Pocket Books, 1972); Martin Pawley, *The Private Future: Causes and Consequences of Community Collapse in the West* (London: Pan, 1973); Richard Sennett, *The Fall of Public Man* (New York: Alfred A. Knopf, 1977); David Popenoe, *Public Pleasure, Private Plight* (New Brunswick, NJ: Transaction, 1984); Philip Slater, *The Pursuit of Loneliness: American Culture at the Breaking Point* (Boston, MA: Beacon Press, 20th anniversary edition, 1990); Michael Sorkin (ed.), *Variations on a Theme Park: The New American City and the End of Public Space* (New York: Noonday Press, 1992); Robert D. Putnam, *Bowling Alone: The Collapse and Revival of America Community* (New York: Simon & Schuster, 2001) and *Better Together: Restoring the American Community* (New York: Simon & Schuster, 2003); Don Mitchell, *The Right to the City: Social Justice and the Fight for Public Space* (New York: Guilford Press, 2003); and Margaret Kohn, *Brave New Neighborhoods: The Privatization of Public Space* (London: Routledge, 2005).

▪ ▪ ▪ ▪ ▪ ▪

A number of recent American writings indicate that the nostalgia for the small town need not be construed as directed toward the town itself: it is rather a "quest for community" (as Robert Nisbet puts it) – a nostalgia for a compassable and integral living unit. The critical question is not whether the small town can be rehabilitated in the image of its earlier strength and growth – for clearly it cannot – but whether American life will be able to evolve any other integral community to replace it. This is what I call the problem of place in America, and unless it is somehow resolved, American life will become more jangled and fragmented than it is, and American personality will continue to be unquiet and unfulfilled.

MAX LERNER. *America as a Civilization*, 1957

The ensuing years have confirmed Lerner's diagnosis. The problem of place in America has not been resolved and life *has* become more jangled and fragmented. No new form of integral community has been found; the small town has yet to greet its replacement. And Americans are not a contented people.

What may have seemed like the new form of community – the automobile suburb – multiplied rapidly after World War II. Thirteen million plus returning veterans qualified for single-family dwellings requiring no down payments in the new developments. In building and equipping these millions of new private domains, American industry found a major alternative to military production and companionate marriages appeared to have found ideal nesting places. But we did not live happily ever after.

Life in the subdivision may have satisfied the combat veteran's longing for a safe, orderly, and quiet haven, but it rarely offered the sense of place and belonging that had rooted his parents and grandparents. Houses alone do not a community make, and the typical subdivision proved hostile to the emergence of any structure or space utilization beyond the uniform houses and streets that characterized it.

Like all-residential city blocks, observed one student of the American condition, the suburb is "merely a base from which the individual reaches out to the scattered components of social existence."[1] Though proclaimed as offering the best of both rural and urban life, the automobile suburb had the effect of fragmenting the individual's world. As one observer wrote: "A man works in one place, sleeps in another, shops somewhere else, finds pleasure or companionship where he can, and cares about none of these places."

The typical suburban home is easy to leave behind as its occupants move to another. What people cherish most in them can be taken along in the move. There are no sad farewells at the local taverns or the corner store because there are no local taverns or corner stores. Indeed, there is often more encouragement to leave a given subdivision than to stay in it, for neither the homes nor the neighborhoods are equipped to see families or individuals through the cycle of life. Each is designed for families of particular sizes, incomes, and ages. There is little sense of place and even less opportunity to put down roots.

Transplanted Europeans are acutely aware of the lack of a community life in our residential areas. We recently talked with an outgoing lady who had lived in many countries and was used to adapting to local ways. The problem of place in America had become her problem as well:

> After four years here, I still feel more of a foreigner than in any other place in the world I have been. People here are proud to live in a 'good' area, but to us these so-called desirable areas are like prisons. There is no contact between the various households, we rarely see the neighbors and certainly do not know any of them. In Luxembourg, however, we would frequently stroll down to one of the local cafés in the evening, and there pass a very congenial few hours in the company of the local fireman, dentist, bank employee or whoever happened to be there at the time. There is no pleasure to be had in driving to a sleazy, dark bar where one keeps strictly to one's self and becomes fearful if approached by some drunk.

Sounding the same note, Kenneth Harris has commented on one of the things British people miss most in the United States. It is some reasonable approximation of the village inn or local pub; our neighborhoods do not have it. Harris comments:

> The American does not walk around to the local two or three times a week with his wife or with his son, to have his pint, chat with the neighbors, and then walk home. He does not take out the dog last thing every night, and break his journey with a quick one at the Crown.[2]

The contrast in cultures is keenly felt by those who enjoy a dual residence in Europe and America. Victor Gruen and his wife have a large place in Los Angeles and a small one in Vienna. He finds that: "In Los Angeles we are hesitant to leave our sheltered home in order to visit friends or to participate in cultural or entertainment events because every such outing involves a major investment of time and nervous strain in driving long distances."[3] But, he says, the European experience is much different:

> In Vienna, we are persuaded to go out often because we are within easy walking distance of two concert halls, the opera, a number of theatres, and a variety of restaurants, cafés, and shops. Seeing old friends does not have to be a

prearranged affair as in Los Angeles, and more often than not, one bumps into them on the street or in a café.

The Gruens have a hundred times more residential space in America but give the impression that they don't enjoy it half as much as their little corner of Vienna.

But one needn't call upon foreign visitors to point up the shortcomings of the suburban experiment. As a setting for marriage and family life, it has given those institutions a bad name. By the 1960s, a picture had emerged of the suburban housewife as "bored, isolated, and preoccupied with material things."[4] The suburban wife without a car to escape in epitomized the experience of being alone in America.[5] Those who could afford it compensated for the loneliness, isolation, and lack of community with the "frantic scheduling syndrome" as described by a counselor in the northeastern region of the United States:

> The loneliness I'm most familiar with in my job is that of wives and mothers of small children who are dumped in the suburbs and whose husbands are commuters . . . I see a lot of generalized loneliness, but I think that in well-to-do communities they cover it up with a wealth of frantic activity. That's the reason tennis has gotten so big. They all go out and play tennis.[6]

A majority of the former stay-at-home wives are now in the labor force. As both father and mother gain some semblance of a community life via their daily escapes from the subdivision, children are even more cut off from ties with adults. Home offers less and the neighborhood offers nothing for the typical suburban adolescent. The situation in the early seventies as described by Richard Sennett is worsening:

> In the past ten years, many middle-class children have tried to break out of the communities, the schools and the homes that their parents have spent so much of their own lives creating. If any one feeling can be said to run through the diverse groups and life-styles of the youth movements, it is a feeling that these middle-class communities of the parents were like pens, like cages keeping the youth from being free and alive. The source of the feeling lies in the perception that while these middle-class environments are secure and orderly regimes, people suffocate there for lack of the new, the unexpected, the diverse in their lives.[7]

The adolescent houseguest, I would suggest, is probably the best and quickest test of the vitality of a neighborhood; the visiting teenager in the subdivision soon acts like an animal in a cage. He or she paces, looks unhappy and uncomfortable, and by the second day is putting heavy pressure on the parents to leave. There is no place to which they can escape and join their own kind. There is nothing for them to do on their own. There is nothing in the surroundings but the houses of strangers and nobody on the streets. Adults make a more successful adjustment, largely because they demand less. But few at any age find vitality in the housing developments. David Riesman, an esteemed elder statesman among social scientists, once attempted to describe the import of suburbia upon most of those who live there. "There would seem," he wrote, "to be an aimlessness, a pervasive low-keyed unpleasure."[8] The word he seemed averse to using is *boring*. A teenager would not have had to struggle for the right phrasing.

Their failure to solve the problem of place in America and to provide a community life for their inhabitants has not effectively discouraged the growth of the postwar suburbs. To the contrary, there have emerged new generations of suburban development in which there is even less life outside the houses than before. Why does failure succeed? Dolores Hayden supplies part of the answer when she observes that Americans have substituted the vision of the ideal home for that of the ideal city.[9] The purchase of the even larger home on the even larger lot in the even more lifeless neighborhood is not so much a matter of joining community as retreating from it. Encouraged by a continuing decline in the civilities and amenities of the public or shared environment, people invest more hopes in their private acreage. They proceed as though a house can substitute for a community if only it is spacious enough, entertaining enough, comfortable enough, splendid enough – and suitably isolated from that common horde that politicians still refer to as our "fellow Americans."

Observers disagree about the reasons for the growing estrangement between the family and the city in American society.[10] Richard Sennett, whose research spans several generations, argues that as soon as an American family became middle class and could afford to do something about its fear of the outside world and its confusions, it drew in upon itself, and "in America, unlike France or Germany, the urban middle-class shunned public forms of social life like cafés and banquet halls."[11] Philippe Ariès, who also knows his history, counters with the argument that modern urban development has killed the essential relationships that once made a city and, as a consequence, "the role of the family over-expanded like a hypertrophied cell" trying to take up the slack.[12]

In some countries, television broadcasting is suspended one night a week so that people will not abandon the habit of getting out of their homes and maintaining contact with one another. This tactic would probably not work in America. Sennett would argue that the middle-class family, given its assessment of the public domain, would stay at home anyway. Ariès would argue that most would stay home for want of places to get together with their friends and neighbors. As Richard Goodwin declared, "there is virtually no place where neighbors can anticipate unplanned meetings – no pub or corner store or park."[13] The bright spot in this dispute is that the same set of remedies would cure both the family and the city of major ills.

Meantime, new generations are encouraged to shun a community life in favor of a highly privatized one and to set personal aggrandizement above public good. The attitudes may be learned from parents but they are also learned in each generation's experiences. The modest housing developments, those un-exclusive suburbs from which middle-class people graduate as they grow older and more affluent, teach their residents that future hopes for a good life are pretty much confined to one's house and yard. Community life amid tract housing is a disappointing experience. The space within the development has been equipped and staged for isolated family living and little else. The processes by which potential friends might find one another and by which friendships not suited to the home might be nurtured outside it are severely thwarted by the limited features and facilities of the modern suburb.

The housing development's lack of informal social centers or informal public gathering places puts people too much at the mercy of their closest neighbors. The small town taught us that people's best friends and favorite companions rarely lived right next door to one another. Why should it be any different in the automobile suburbs? What are the odds, given that a hundred households are within easy walking distance, that one is most likely to hit it off with the people next door? Small! Yet, the closest neighbors are the ones with whom friendships are most likely to be attempted, for how does one even find out enough about someone a block and a half away to justify an introduction?

What opportunity is there for two men who both enjoy shooting, fishing, or flying to get together and gab if their families are not compatible? Where do people entertain and enjoy one another if, for whatever reason, they are not comfortable in one another's homes? Where do people have a chance to get to know one another casually and without commitment before deciding whether to involve other family members in their relationship? Tract housing offers no such places.

Getting together with neighbors in the development entails considerable hosting efforts, and it depends upon continuing good relationships between households and their members. In the usual course of things, these relationships are easily strained or ruptured. Having been lately formed and built on little, they are not easy to mend. Worse, some of the few good friends will move and are not easily replaced. In time, the overtures toward friendship, neighborliness, and a semblance of community hardly seem worth the effort.

IN THE ABSENCE OF AN INFORMAL PUBLIC LIFE

We have noted Sennett's observation that middle-class Americans are not like their French or German counterparts. Americans do not make daily visits to sidewalk cafés or banquet halls. We do not have that third realm of satisfaction and social cohesion beyond the portals of home and work that for others is an essential element of the good life. Our comings and goings are more restricted to the home and work settings, and those two spheres have become preemptive. Multitudes shuttle back and forth between the "womb" and the "rat race" in a

INTRODUCTION TO PART FOUR

The desire for more meaningful places has resulted in a wide variety of approaches to place reinforcement and place-making within the design community. Designers attempting to reinsert meaning into place utilize various elements of physical character to highlight local distinctions, such as environmental imagery, natural history, craft and cultural traditions, memory, history, formal aesthetics, and beauty. While the bulk of design practice remains oriented to functional, pragmatic, and economic concerns, a number of theorists and practitioners have sought a deeper design discourse that employs local and contextually-based imagery to create distinctive place identities. To defeat the globalized placelessness recognized in the previous readings, evoking a sense of place has become a primary concern in urban design. A number of authors describe this trend as a key signifier of late-twentieth-century Postmodernism. Key attributes of Postmodern design and urbanism include a desire for history, comfort, entertainment, and importantly, readable meaning. These design interests are supported both by common public desires for more meaningful places, as well as by key economic agents in contemporary society, such as real estate developers, chambers of commerce and city marketers. Not surprising, the growing tourism industry benefits from places that are distinct and imageable, encouraging likely tourist destinations to reinforce their historical and place-based identities. The results have been mixed, resulting both in places that authentically incorporate a sense of place, as well as places that utilize inauthentic and shallow forms of "theming" to evoke past histories and otherness. This movement to place-based design has resulted in new roles for the urban designer, including those of storyteller, midwife, public educator, local historian, urban repair specialist, and heretic (demanding change within the field itself).

Part Four starts with an early and influential piece of writing by Kevin Lynch that highlights the importance of visual imagery in making cities legible and memorable. Image studies such as Lynch's *The Image of the City* are investigations into environmental psychology and how people perceive their cities. This work is indicative of the growing influence that social science methods have had in shaping urban design research, as well as of the growing importance of public participation in urban design practice. This reading is one of the best known in the urban design canon, and describes Lynch's famous five elements of city imageability: paths, edges, districts, nodes, and landmarks. Concurrent with Lynch's work in the United States, Gordon Cullen's work in Europe focuses on picturesque and emotional qualities of city design, bodily experience, and memory. Cullen relies on the pictorial aspects of cities that people recognize as they move through space, which are then subject to later recall and memory. Cullen's contributions in this writing include concepts of "serial vision" and "townscape analysis." Both of these readings focus on perceptual approaches to place and the reinforcement of those physical qualities that are likely to influence orientation, memory and the public image of places.

In a different manner, the next three readings focus on various natural, regional, and historical aspects of the city to reinforce local sense of place. In Michael Hough's chapter "Principles for Regional Design," from *Out of Place*, he examines the role of identity creation through landscape design and the use of local plant materials, topography, and environmental values. He stresses designing with an economy of means and minimum interference in the processes of sustainable place-making. His is a natural process perspective that suggests "doing as little as possible" and "starting where it's easiest." Continuing this contextual approach, Douglas S. Kelbaugh advocates a Critical Regionalism that reinforces place qualities through the operationalization of place character in design. He describes several physical and process aspects to place that can be used for design inspiration without reducing design to cliché, camp, or inauthenticity, for example, sense of place, nature, history, craft, and a sense of limits. Dolores Hayden's approach focuses primarily on highlighting the place memories and urban public histories of various diverse, and often forgotten, groups in society: women, ethnic minorities, labor, and lower income groups. She advocates the preservation of everyday places and the creation of public art projects that chronicle the experiences and histories of these groups and their leaders. With these three readings, we see three very different approaches to place-making that reinforce contextual material that is already in evidence at the site.

The last two readings in Part Four are critical descriptions focusing on the practices of contemporary city-building and place-making at the end of the twentieth century. Nan Ellin's discussion of Postmodern urbanism brings to light and describes the practices of place-making and urban design in the period of the 1970s to the 1990s. She is highly critical of several aspects of Postmodern design (e.g. inauthentic fictionalizing, apolitical designers, capitalist profit-seeking, design narcissism, and the growing culture of fear in society), but also acknowledges the benefits that have accrued through the reintroduction of context, local history, and urbanity, as well as the revalorization of the public realm. In opposition to the previous readings that reinforce place qualities, Rem Koolhaas extols the virtues of modern design and homogenous urbanism found in "The Generic City" from his book *S, M, L, XL*. He is an existential realist who suggests that the placeless and generic contemporary city are both acceptable and reflective of modern society; in this way the most authentic representation of modern life that is currently available. Although recognizing its usefulness as "modernism's little helper," Koolhaas abhors the Postmodern reflex described by Ellin, as well as place strategies that highlight past histories or traditional urbanism. In these two readings we hear very different critical voices willing to find virtue in the places of the present.

"The Image of the Environment" and "The City Image and Its Elements"

from *The Image of the City* (1960)

Kevin Lynch

Editors' Introduction

This second reading by Kevin Lynch highlights a very different aspect of his research on the legibility and visual perception of cities. It is by far the best known of his writings and has had a profound influence on how designers perceive cities and urban form. His underlying idea in "Chapter I: The Image of the Environment" is that people understand and mentally process the form of cities through the recognition of key physical elements. By utilizing visual elements, Lynch argues that urban designers have a kit of tools for making more legible and psychologically satisfying places. Not only do these elements provide organizational clues and way-finding devices for people to orient themselves in space, but also they can help in engendering emotional security and a sense of place-based ownership that comes from one's ability to recognize familiar territory.

Lynch defines "imageability" as

> that quality in a physical object which gives it a high probability of evoking a strong image in any given observer. It is that shape, color, or arrangement which facilitates the making of vividly identified, powerfully structured, highly useful mental images of the environment.

Imageability to Lynch combines both the ability of the physical object to project a strong distinctive image, as well as the ability of the observer to mentally select, process, store, organize, and endow the image with meaning. In the selection from "Chapter III: The City Image and Its Elements," the author identifies five key elements that provide urban imageability: paths, edges, districts, nodes, and landmarks. In the conclusion to the book, he suggests ways in which designers can process this information to provide visual plans for reinforcing the form, physical controls and public image of cities.

The book is important not only for its findings on the visual form of cities, but also in highlighting Lynch's research methods in environmental perception. These methods allowed researchers to "get into the heads" of research subjects to better understand how they perceived their everyday environments. His methods included cognitive mapping, in-depth oral interviews, travel maps, direct observation, field reconnaissance walks, random pedestrian interviews, aerial and ground-level photography and synthesis maps. Data from extensive use of cognitive maps (mental maps of their city that people were asked to draw from memory) was easily compiled to provide synthetic illustrations of those elements that were most recognized or remembered. The same was done with data culled from oral interviews, which was then correlated across the data pulled from the cognitive mapping. From these different methods, Lynch's research team was able to triangulate similar findings from a relatively small sample of interviewees, although he later notes the biases in these small sample sets.

Although this work has not been particularly fruitful in effecting public policy on a broad scale, it has been important to urban plan making in some specific places, such as San Francisco (California), Ciudad Guyana (Venezuela), and Brookline (Massachusetts), as well as with research on childhood experience for UNESCO. Lynch's techniques have proven particularly valuable in environmental design research, in many types of urban design plan making, within SWOT analysis (Strengths, Weaknesses, Opportunities, and Threats), in understanding public images of the city for marketing purposes, and for knowledge in place memorability.

With its publication in 1960, Lynch's contribution belongs to the first generation of works in environmental-psychology and environmental behavior. This literature has burgeoned since then, including major influences on a generation of researchers such as Amos Rapoport, Clare Cooper Marcus, Oscar Newman, William H. Whyte, and Donald Appleyard – as well as planning and design departments at several universities. With regard to its lasting impacts, *The Image of the City* helped to highlight the importance of urban form-making at a time when city planners were looking to social science methods to replace what was perceived to be an underperforming physical planning tradition. And with respect to our current planning interest in public participation, it was influential in consulting substantive users and residents of the city, and bringing them back into the planning conversation at a time when decision-making and design relied primarily on expert and elite knowledge.

Kevin Lynch's biographical profile and primary works are included in his earlier reading in Part Two. In Tridib Banerjee and Michael Southworth's edited book of Lynch's shorter essays and articles, *City Sense and City Design: Writings and Projects of Kevin Lynch* (Cambridge, MA: MIT Press, 1990), one can find supplemental material by Lynch on environmental perception and the visual form of cities: "Environmental Perception: Research and Public Policy" (MIT Libraries' Institute Archives and Special Collections), "Reconsidering *The Image of the City*," in Lloyd Rodwin and Robert Hollister (eds.), *Cities in Mind* (New York: Plenum, 1984), "A Process of Community Visual Survey" (MIT Libraries' Institute Archives and Special Collections), and, "The Visual Shape of the Shapeless Metropolis" (MIT Libraries' Institute Archives and Special Collections).

Additional material on the visual image of cities and environmental psychology can be found in the following: Gyorgy Kepes, *Language of Vision* (Chicago, IL: P. Theobald, 1944) and *Sign, Image, Symbol* (New York: George Braziller, 1966); D. De Jonge, "Images of Urban Areas: Their Structure and Psychological Foundations," *Journal of the American Institute of Planners* (vol. 28, pp. 266–276, 1962); Harold M. Proshansky, William Ittleson, and Leanne Rivlin (eds.), *Environmental Psychology: Man and his Physical Setting* (New York: Holt, Rinehart & Winston, 1970); Roger M. Downs and David Stea (eds.), *Image and Environment: Cognitive Mapping and Spatial Behavior* (Chicago, IL: Aldine, 1973); S. Kaplan and R. Kaplan, *Cognition and Environment: Functioning in an Uncertain World* (New York: Praeger, 1982); M. Gottdiener and A. Lagopoulos, *The City and the Sign: An Introduction to Urban Semiotics* (New York: Columbia University Press, 1986); P.A. Bell, J.D. Fisher, A. Baum, and T.D. Greene, *Environmental Psychology*, 3rd edn (London: Holt, Rinehart & Winston, 1990); and Robert B. Bechtel and Arza Churchman (eds.), *Handbook of Environmental Psychology* (New York: John Wiley, 2002).

THE IMAGE OF THE ENVIRONMENT

Looking at cities can give a special pleasure, however commonplace the sight may be. Like a piece of architecture, the city is a construction in space, but one of vast scale, a thing perceived only in the course of long spans of time. City design is therefore a temporal art, but it can rarely use the controlled and limited sequences of other temporal arts like music. On different occasions and for different people, the sequences are reversed, interrupted, abandoned, cut across. It is seen in all lights and all weathers.

At every instant, there is more than the eye can see, more than the ear can hear, a setting or a view waiting to be explored. Nothing is experienced by itself, but always in relation to its surroundings, the sequences of events leading up to it, the memory of past experiences. Washington Street set in a farmer's field might look like the shopping street in the heart of Boston, and yet it would seem utterly different. Every citizen has had long associations not merely be simplified, but also extended and deepened. Such a city would be one that could be apprehended over time as a pattern of high continuity with many distinctive parts clearly interconnected. The perceptive and familiar observer could absorb new sensuous impacts without disruption of his basic image, and each new impact would touch upon many previous elements. He would be well oriented, and he could move easily. He would be highly aware of his environment. The city of Venice might be an example of such a highly imageable environment. In the United States, one is tempted to cite parts of Manhattan, San Francisco, Boston, or perhaps the lake front of Chicago.

These are characterizations that flow from our definitions. The concept of imageability does not necessarily connote something fixed, limited, precise, unified, or regularly ordered, although it may sometimes have these qualities. Nor does it mean apparent at a glance, obvious, patent, or plain. The total environment to be patterned is highly complex, while the obvious image is soon boring, and can point to only a few features of the living world.

The imageability of city form will be the center of the study to follow. There are other basic properties in a beautiful environment: meaning or expressiveness, sensuous delight, rhythm, stimulus, choice. Our concentration on imageability does not deny their importance. Our purpose is simply to consider the need for identity and structure in our perceptual world, and to illustrate the special relevance of this quality to the particular case of the complex, shifting urban environment.

Since image development is a two-way process between observer and observed, it is possible to strengthen the image either by symbolic devices, by the retraining of the perceiver, or by reshaping one's surroundings. You can provide the viewer with a symbolic diagram of how the world fits together: a map or a set of written instructions. As long as he can fit reality to the diagram, he has a clue to the relatedness of things. You can even install a machine for giving directions, as has recently been done in New York (*New York Times* 1957). While such devices are extremely useful for providing condensed data on interconnections, they are also precarious, since orientation fails if the device is lost, and the device itself must constantly be referred and fitted to reality. [. . .] Moreover, the complete experience of interconnection, the full depth of a vivid image, is lacking.

You may also train the observer. Brown remarks that a maze through which subjects were asked to move blindfolded seemed to them at first to be one unbroken problem. On repetition, parts of the pattern, particularly the beginning and end, became familiar and assumed the character of localities. Finally, when they could tread the maze without error, the whole system seemed to have become one locality (Brown 1932). DeSilva describes the case of a boy who seemed to have "automatic" directional orientation, but proved to have been trained from infancy (by a mother who could not distinguish right from left) to respond to "the east side of the porch" or "the south end of the dresser" (deSilva 1931).

Shipton's account of the reconnaissance for the ascent of Everest offers a dramatic case of such learning. Approaching Everest from a new direction, Shipton immediately recognized the main peaks and saddles that he knew from the north side. But the Sherpa guide accompanying him, to whom both sides were long familiar, had never realized that these were the same features, and he greeted the revelation with surprise and delight (Shipton 1952).

Kilpatrick describes the process of perceptual learning forced on an observer by new stimuli that no longer fit into previous images (Kilpatrick 1954). It begins with hypothetical forms that explain the new stimuli conceptually, while the illusion of the old forms persists. The personal experience of most of us will testify to this persistence of an illusory image long after its inadequacy is conceptually realized. We stare into the jungle and see only the sunlight on the green leaves, but a warning noise tells us that an animal is hidden there. The observer then learns to interpret the scene by singling out "give-away" clues and by reweighting previous signals. The camouflaged animal may now be picked up by the reflection of his eyes. Finally by repeated experience the entire pattern of perception is changed, and the observer need no longer consciously search for give-aways, or add new data to an old framework. He has achieved an image which will operate successfully in the new situation, seeming natural and right. Quite suddenly the hidden animal appears among the leaves, "as plain as day."

In the same way, we must learn to see the hidden forms in the vast sprawl of our cities. We are not accustomed to organizing and imaging an artificial environment on such a large scale; yet our activities are pushing us toward that end. Curt Sachs gives an example of a failure to make connections beyond a certain level (Sachs 1953). The voice and drumbeat of the North American Indian follow entirely different tempos, the two being perceived independently. Searching for a musical analogy of our own, he mentions our church services, where we do not think of coordinating the choir inside with the bells above.

In our vast metropolitan areas we do not connect the choir and the bells; like the Sherpa, we see only the sides of Everest and not the mountain. To extend and deepen our perception of the environment would be to continue a long biological and cultural development which has gone from the contact senses to the distant senses and from the distant senses to symbolic communications. Our thesis is that we are now able to develop our image of the environment by operation on the external physical shape as well as by an internal learning process. Indeed, the complexity of our environment now compels us to do so.

Primitive man was forced to improve his environmental image by adapting his perception to the given landscape. He could effect minor changes in his environment with cairns, beacons, or tree blazes, but substantial modifications for visual clarity or visual interconnection were confined to house sites or religious enclosures. Only powerful civilizations can begin to act on their total environment at a significant scale. The conscious remolding of the large-scale physical environment has been possible only recently, and so the problem of environmental imageability is a new one. Technically, we can now make completely new landscapes in a brief time, as in the Dutch polders. Here the designers are already at grips with the question of how to form the total scene so that it is easy for the human observer to identify its parts and to structure the whole (Granpré-Molière 1955).

We are rapidly building a new functional unit, the metropolitan region, but we have yet to grasp that this unit, too, should have its corresponding image. Suzanne Langer sets the problem in her capsule definition of architecture: "It is the total environment made visible" (Langer 1953).

THE CITY IMAGE AND ITS ELEMENTS

There seems to be a public image of any given city which is the overlap of many individual images. Or perhaps there is a series of public images, each held by some significant number of citizens. Such group images are necessary if an individual is to operate successfully within his environment and to cooperate with his fellows. Each individual picture is unique, with some content that is rarely or never communicated, yet it approximates the public image, which, in different environments, is more or less compelling, more or less embracing.

This analysis limits itself to the effects of physical, perceptible objects. There are other influences on imageability, such as the social meaning of an area, its function, its history, or even its name. These will be glossed over, since the objective here is to uncover the role of form itself. It is taken for granted that in actual design form should be used to reinforce meaning, and not to negate it.

The contents of the city images so far studied, which are referable to physical forms, can conveniently be classified into five types of elements: paths, edges, districts, nodes, and landmarks. Indeed, these elements may be of more general application, since they seem to reappear in many types of environmental images. These elements may be defined as follows:

1. *Paths.* Paths are the channels along which the observer customarily, occasionally, or potentially moves. They may be streets, walkways, transit lines, canals, railroads. For many people, these are the predominant elements in their image. People observe the city while moving through it, and along these paths the other environmental elements are arranged and related.

2. *Edges.* Edges are the linear elements not used or considered as paths by the observer. They are the boundaries between two phases, linear breaks in continuity: shores, railroad cuts, edges of development, walls. They are lateral references rather than coordinate axes. Such edges may be barriers, more or less penetrable, which close one region off from another; or they may be seams, lines

along which two regions are related and joined together. These edge elements, although probably not as dominant as paths, are for many people important organizing features, particularly in the role of holding together generalized areas, as in the outline of a city by water or wall.

3. *Districts*. Districts are the medium-to-large sections of the city, conceived of as having two-dimensional extent, which the observer mentally enters "inside of," and which are recognizable as having some common, identifying character. Always identifiable from the inside, they are also used for exterior reference if visible from the outside. Most people structure their city to some extent in this way, with individual differences as to whether paths or districts are the dominant elements. It seems to depend not only upon the individual but also upon the given city.

4. *Nodes*. Nodes are points, the strategic spots in a city into which an observer can enter, and which are the intensive foci to and from which he is traveling. They may be primarily junctions, places of a break in transportation, a crossing or convergence of paths, moments of shift from one structure to another. Or the nodes may be simply concentrations, which gain their importance from being the condensation of some use or physical character, as a street-corner hangout or an enclosed square. Some of these concentration nodes are the focus and epitome of a district, over which their influence radiates and of which they stand as a symbol. They may be called cores. Many nodes, of course, partake of the nature of both junctions and concentrations. The concept of node is related to the concept of path, since junctions are

typically the convergence of paths, events on the journey. It is similarly related to the concept of district, since cores are typically the intensive foci of districts, their polarizing center. In any event, some nodal points are to be found in almost every image, and in certain cases they may be the dominant feature.

5. *Landmarks*. Landmarks are another type of point-reference, but in this case the observer does not enter within them, they are external. They are usually a rather simply defined physical object: building, sign, store, or mountain. Their use involves the singling out of one element from a host of possibilities. Some landmarks are distant ones, typically seen from many angles and distances, over the tops of smaller elements, and used as radial references. They may be within the city or at such a distance that for all practical purposes they symbolize a constant direction. Such are isolated towers, golden domes, great hills. Even a mobile point, like the sun, whose motion is sufficiently slow and regular, may be employed. Other landmarks are primarily local, being visible only in restricted localities and from certain approaches. These are the innumerable signs, store fronts, trees, doorknobs, and other urban detail, which fill in the image of most observers. They are frequently used clues of identity and even of structure, and seem to be increasingly relied upon as a journey becomes more and more familiar.

The image of a given physical reality may occasionally shift its type with different circumstances of viewing. Thus an expressway may be a path for the driver, and edge for the pedestrian. Or a central area may be a district when a city is organized on a medium scale, and a node when the entire metropolitan area is considered. But the categories seem to have stability for a given observer when he is operating at a given level.

None of the element types isolated above exist in isolation in the real case. Districts are structured with nodes, defined by edges, penetrated by paths, and sprinkled with landmarks. Elements regularly overlap and pierce one another. If this analysis begins with the differentiation of the data into categories, it must end with their reintegration into the whole image. Our studies have furnished much information about the visual character of the element types. This will be discussed below. Only to a lesser extent, unfortunately, did the work make revelations about the interrelations between elements, or about image levels, image qualities, or the development of the image. These latter topics will be treated at the end of this chapter.

[. . .]

ELEMENT INTERRELATIONS

These elements are simply the raw material of the environmental image at the city scale. They must be patterned together to provide a satisfying form. The preceding discussions have gone as far as groups of similar elements (nets of paths, clusters of landmarks, mosaics of regions). The next logical step is to consider the interaction of pairs of unlike elements.

Such pairs may reinforce one another, resonate so that they enhance each other's power; or they may conflict and destroy themselves. A great landmark may dwarf and throw out of scale a small region at its base. Properly located, another landmark may fix and strengthen a core; placed off center, it may only mislead, as does the John Hancock Building in relation to Boston's Copley Square. A large street, with its ambiguous character of both edge and path, may penetrate and thus expose a region to view, while at the same time disrupting it. A landmark feature may be so alien to the character of a district as to dissolve the

regional continuity, or it may, on the other hand, stand in just the contrast that intensifies that continuity.

Districts in particular, which tend to be of larger size than the other elements, contain within themselves, and are thus related to, various paths, nodes, and landmarks. These other elements not only structure the region internally, they also intensify the identity of the whole by enriching and deepening its character. Beacon Hill in Boston is one example of this effect. In fact, the components of structure and identity (which are the parts of the image in which we are interested) seem to leapfrog as the observer moves up from level to level. The identity of a window may be structured into a pattern of windows, which is the cue for the identification of a building. The buildings themselves are interrelated so as to form an identifiable space, and so on.

Paths, which are dominant in many individual images, and which may be a principal resource in organization at the metropolitan scale, have intimate interrelations with other element types. Junction nodes occur automatically at major intersections and termini, and by their form should reinforce those critical moments in a journey. These nodes, in turn, are not only strengthened by the presence of landmarks (as is Copley Square) but provide a setting which almost guarantees attention for any such mark. The paths, again, are given identity and tempo not only by their own form, or by their nodal junctions, but by the regions they pass through, the edges they move along, and the landmarks distributed along their length.

All these elements operate together, in a context. It would be interesting to study the characteristics of various pairings: landmark-region, node-path, etc. Eventually, one should try to go beyond such pairings to consider total patterns.

Most observers seem to group their elements into intermediate organizations, which might be called complexes. The observer senses the complex as a whole whose parts are interdependent and are relatively fixed in relation to each other. Thus many Bostonians would be able to fit most of the major elements of the Back Bay, the Common, Beacon Hill, and the central shopping, into a single complex. This whole area, in the terms used by Brown (1932) in his experiments referred to earlier, has become one locality. For others, the size of their locality may be much smaller: the central shopping and the near edge of the Common alone, for example. Outside of this complex there are gaps of identity; the observer must run blind to the next whole, even if only momentarily. Although they are close together in physical reality, most people seem to feel only a vague link between Boston's office and financial district and the central shopping district on Washington Street. This peculiar remoteness was also exemplified in the puzzling gap between Scollay Square and Dock Square which are only a block apart. The psychological distance between two localities may be much greater, or more difficult to surmount, than mere physical separation seems to warrant.

Our preoccupation here with parts rather than wholes is a necessary feature of an investigation in a primitive stage. After successful differentiation and understanding of parts, a study can move on to consideration of a total system. There were indications that the image may be a continuous field, the disturbance of one element in some way affecting all others. Even the recognition of an object is as much dependent on context as on the form of the object itself. One major distortion, such as a twisting of the shape of the Common, seemed to be reflected throughout the image of Boston. The disturbance of large-scale construction affected more than its immediate environs. But such field effects have hardly been studied here.

THE SHIFTING IMAGE

Rather than a single comprehensive image for the entire environment, there seemed to be sets of images, which more or less overlapped and interrelated. They were typically arranged in a series of levels, roughly by the scale of area involved, so that the observer moved as necessary from an image at street level to levels of a neighborhood, a city, or a metropolitan region.

This arrangement by levels is a necessity in a large and complex environment. Yet it imposes an extra burden of organization on the observer, especially if there is little relation between levels. If a tall building is unmistakable in the city-wide panorama yet unrecognizable from its base, then a chance has been lost to pin together the images at two different levels of organization. The State

House on Beacon Hill, on the other hand, seems to pierce through several image levels. It holds a strategic place in the organization of the center.

Images may differ not only by the scale of area involved, but by viewpoint, time of day, or season. The image of Faneuil Hall as seen from the markets should be related to its image from a car on the Artery. Washington-Street-by-night should have some continuity, some element of invariance, with Washington-Street-by-day. In order to accomplish this continuity in the face of sensuous confusion, many observers drained their images of visual content, using abstractions such as "restaurant" or "second street." These will operate both day and night, driving or walking, rain or shine, albeit with some effort and loss.

The observer must also adjust his image to secular shifts in the physical reality around him. Los Angeles illustrated the practical and emotional strains induced as the image is confronted with constant physical changes. It would be important to know how to maintain continuity through these changes. Just as ties are needed between level and level of organization, so are continuities required which persist through a major change. This might be facilitated by the retention of an old tree, a path trace, or some regional character.

The sequence in which sketch maps were drawn seemed to indicate that the image develops, or grows, in different ways. This may perhaps have some relation to the way in which it first develops as an individual becomes familiar with his environment. Several types were apparent:

a. Quite frequently, images were developed along, and then outward from, familiar lines of movement. Thus a map might be drawn as branching out from a point of entrance, or beginning from some base line such as Massachusetts Avenue.

b. Other maps were begun by the construction of an enclosing outline, such as the Boston peninsula, which was then filled in toward the center.

c. Still others, particularly in Los Angeles, began by laying down a basic repeating pattern (the path gridiron) and then adding detail.

d. Somewhat fewer maps started as a set of adjacent regions, which were then detailed as to connections and interiors.

e. A few Boston examples developed from a familiar kernel, a dense familiar element on which everything was ultimately hung.

The image itself was not a precise, miniaturized model of reality, reduced in scale and consistently abstracted. As a purposive simplification, it was made by reducing, eliminating, or even adding elements to reality, by fusion and distortion, by relating and structuring the parts. It was sufficient, perhaps better, for its purpose if rearranged, distorted, "illogical." It resembled that famous cartoon of the New Yorker's view of the United States.

However distorted, there was a strong element of topological invariance with respect to reality. It was as if the map were drawn on an infinitely flexible rubber sheet; directions were twisted, distances stretched or compressed, large forms so changed from their accurate scale projection as to be at first unrecognizable. But the sequence was usually correct, the map was rarely torn and sewn back together in another order. This continuity is necessary if the image is to be of any value.

IMAGE QUALITY

Study of various individual images among the Bostonians revealed certain other distinctions between them. For example, images of an element differed between observers in terms of their relative density, i.e., the extent to which they were packed with detail. They might be relatively dense, as a picture of Newbury Street which identifies each building along its length, or relatively thin, when Newbury Street is characterized simply as a street bordered by old houses of mixed use.

Another distinction could be made between concrete, sensuously vivid images, and those which were highly abstract, generalized, and void of sensuous content. Thus the mental picture of a building might be vivid, involving its shape, color, texture, and detail, or be relatively abstract, the structure being identified as "a restaurant" or the "third building from the corner."

Vivid does not necessarily equate with dense, nor thin with abstract. An image might be both dense and abstract, as in the case of the taxicab dispatcher's knowledge of a city street, which related house numbers to uses along block after block,

yet could not describe those buildings in any concrete sense.

Images could be further distinguished according to their structural quality: the manner in which their parts were arranged and interrelated. There were four stages along a continuum of increasing structural precision:

a. The various elements were free; there was no structure or interrelation between parts. We found no pure cases of this type, but several images were definitely disjointed, with vast gaps and many unrelated elements. Here rational movement was impossible without outside help, unless a systematic coverage of the entire area were to be resorted to (which meant the building up of a new structure on the spot).

b. In others, the structure became positional; the parts were roughly related in terms of their general direction and perhaps even relative distance from each other, while still remaining disconnected. One subject in particular always related herself to a few elements, without knowing definite connections between them. Movement was accomplished by searching, by moving out in the correct general direction, while weaving back and forth to cover a band and having an estimate of distance to correct overshooting.

c. Most often, perhaps, the structure was flexible; parts were connected one to the other, but in a loose and flexible manner, as if by limp or elastic ties. The sequence of events was known, but the mental map might be quite distorted, and its distortion might shift at different moments. To quote one subject: "I like to think of a few focal points and how to get from one to another, and the rest I don't bother to learn." With a flexible structure, movement was easier, since it proceeded along known paths, through known sequences. Motion between pairs of elements not habitually connected, or along other than habitual paths, might still be very confusing, however.

d. As connections multiplied, the structure tended to become rigid; parts were firmly interconnected in all dimensions; and any distortions became built in. The possessor of such a map can move much more freely, and can interconnect new points at will. As the density of the image builds up, it begins to take on the characteristics of a total field, in which interaction is possible in any direction and at any distance.

These characteristics of structure might apply in different ways at different levels. For example, two city regions may each possess rigid internal structures, and both connect at some seam or node. But this connection may fail to interlock with the internal structures, so that the connection itself is simply flexible. This effect seemed to occur for many Bostonians at Scollay Square, for example.

Total structure may also be distinguished in a still different way. For some, their images were organized rather instantaneously, as a series of wholes and parts descending from the general to the particular. This organization had the quality of a static map. Connection was made by moving up to the necessary bridging generality, and back down to the desired particular. To go from City Hospital to the Old North Church, for example, one might first consider that the hospital was in the South End and that the South End was in central Boston, then locate the North End in Boston and the church within the North End. This type of image might be called hierarchical.

For others, the image was put together in a more dynamic way, parts being interconnected by a sequence over time (even if the time was quite brief), and pictured as though seen by a motion picture camera. It was more closely related to the actual experience of moving through the city. This might be called a continuous organization, employing unrolling interconnections instead of static hierarchies.

One might infer from this that the images of greatest value are those which most closely approach a strong total field: dense, rigid, and vivid; which make use of all element types and form characteristics without narrow concentration; and which can be put together either hierarchically or continuously, as occasion demands. We may find, of course, that such an image is rare or impossible, that there are strong individual or cultural types which cannot transcend their basic abilities. In this case, an environment should be geared to the appropriate cultural type, or shaped in many ways so as to satisfy the varying demands of the individuals who inhabit it.

We are continuously engaged in the attempt to organize our surroundings, to structure and identify them. Various environments are more or less amenable to such treatment. When reshaping cities it should be possible to give them a form which facilitates these organizing efforts rather than frustrates them.

REFERENCES

Angyal, A., "Über die Raumlage vorgestellter Oerter," *Archiv für die Gesamte Psychologie*, Vol. 78, 1930, pp. 47–94.

Binet, M.A., "Reverse Illusions of Orientation," *Psychological Rewiew*, Vol. I, No. 4, July 1894, pp. 337–350.

Brown, Warner, "Spatial Integrations in a Human Maze," *University of California Publications in Psychology*, Vol. V, No. 5, 1932, pp. 123–134.

Casamajor, Jean, "Le Mystérieux Sens de l'Espace," *Revue Scientifique*, Vol. 65, No. 18, 1927, pp. 554–565.

Claparède, Edouard, "L'Orientation Lointaine," *Nouveau Traité de Psychologie*, Tome VIII, Fasc. 3, Paris, Presses Universitaires de France, 1943.

Fischer, M.H., "Die Orientierung im Raume bei Wirbeltieren und beim Meschen," in *Handbuch der Normalen und Pathologischen Physiologie*, Berlin, J. Springer, 1931, pp. 909–1022.

Granpé-Molière, M.J., "Landscape of the N.E. Polder," translated from *Forum*, Vol. 10:1-2, 1955.

Griffin, Donald R. "Sensory Physiology and the Orientation of Animals," *American Scientist*, April 1953, pp. 209–244.

Jaccard, Pierre, *Le Sens de la direction et l'orientation lointaine chez l'homme*, Paris, Payot, 1932.

Kilpatrick, Franklin P., "Recent Experiments in Perception," *New York Academy of Science, Transaction*, No. 8, Vol. 16. June 1954, pp. 420–425.

Langer, Susanne, *Feeling and Form: A Theory of Art*, New York, Scribner, 1953.

New York Times, April 30, 1957, article of the "Directomat."

Rabaud, Etienne, *L'Orientation Lointaine et la Reconnaissance des Lieux*, Paris, ALcan, 1927.

Ryan, T.A. and M.S., "Geographical Orientation," *American Journal of Psychology*, Vol. 53, 1940, pp. 204–215.

Sachs, Curt, *Rhythm and Tempo*, New York, Norton, 1953.

Sandström, Carl Ivan, *Orientation on the Present Space*, Stockholm, Almqvist and Wiksell, 1951.

Shipton, Eric Earle, *The Mount Everest Reconnaissance Expedition*, London, Hodder and Stoughton, 1952.

deSilva, H.R., "A Case of a Boy Possessing an Automatic Directional Orientation," *Science*, Vol. 73, No. 1893, April 10, 1931, pp. 393–394.

Stern, Paul, "On the Problem of Artistic Form," *Logos*, Vol. V, 1914–15, pp. 165–172.

Trowbridge, C.C., "On Fundamental Methods of Orientation and Imaginary Maps," *Science*, Vol. 38, No. 990, Dec. 9, 1913, pp. 888–897.

Witkin, H.A., "Orientation in Space," *Research Reviews*, Office of Naval Research, December 1949.

"Introduction to The Concise Townscape"

from *The Concise Townscape* (1961)

Gordon Cullen

Editors' Introduction

Like Lynch, Gordon Cullen (1914–1994) was interested in how people perceive urban environments through their sense of sight, but his emphasis was on emotional impacts rather than legibility. In his seminal book *Townscape*, later republished as *The Concise Townscape*, he defined urban design as The Art of Relationship. The goal was to manipulate groups of buildings and physical town elements so as to achieve visual impact and drama.

Cullen understood that people apprehend urban environments through kinesthetic experience as they move through them in everyday life, and felt that this fundamental body–environment relationship was a basis for design: cities should be designed from the point of view of the moving person. From this, he developed the concept of Serial Vision, which theorizes that urban scenes are experienced as a series of revelations, as current views juxtapose with emerging views. Tensions related to an observer's position in the environment – here versus there, enclosure versus exposure, constraint versus relief, etc. – could be purposefully and artfully designed for.

Cullen was also concerned with sense of place, which he theorized through the concept of This and That. He argued that to achieve a unique sense of place, individual townscape elements should be designed as part of a whole. Taking a cue from the qualities of pre-modern towns, he advocated that a sense of wholeness would be best achieved by allowing diversity within an agreed-upon common visual framework, rather than through complete visual conformity and regularity.

Within *The Concise Townscape*, Cullen illustrated his theoretical ideas with freehand ink drawings as well as photographs. Some of the most memorable drawings include a set that shows the plan of a medieval hill town with the path of a walk through it indicated, and related sketches of the sequential views seen along the path. Like Camillo Sitte before him, Cullen focused on picturesque visual effects, but he analyzed and illustrated these effects largely through perspective views rather than projected plans as Sitte had done, making his ideas perhaps more accessible.

As well as theory, Cullen laid out practical ideas for how to accomplish desired design objectives. After *Townscape*, Cullen developed a method for applying townscape ideas to urban places. The conceptual underpinning of the method involved the idea of two interlinked chains: an integrated chain of human activity, and a spatial chain of physical elements. He developed matrixes of human factors and physical factors that designers could use to map a design problem. Cullen's principles for creating a unique sense of place can be derived from his proposal for the new (never built) town of Maryculter, in Scotland: fit development to the site, provide a center, provide distinctive housing areas, create distinct edges and boundaries, provide a network of recognizable landmarks, use topography and planting to create drama, and provide a series of sequential enclosures and climaxes to create a memorable unfolding sense of drama.

Cullen's ideas had an enormous influence on a generation of British urban designers, although his work has been criticized as backward-looking because of its focus on picturesque aesthetic qualities. Other resources on the kinesthetic experience of urban space are Peter Bosselmann, *Representation of Places: Reality and Realism in City Design* (Berkeley, CA: University of California Press, 1998) and Donald Appleyard, Kevin Lynch, and John R. Myer, *The View from the Road* (Cambridge, MA: MIT Press, 1965).

INTRODUCTION TO THE 1959 EDITION

There are advantages to be gained from the gathering together of people to form a town. A single family living in the country can scarcely hope to drop into a theatre, have a meal out or browse in a library, whereas the same family living in a town can enjoy these amenities. The little money that one family can afford is multiplied by thousands and so a collective amenity is made possible. A city is more than the sum of its inhabitants. It has the power to generate a surplus of amenity, which is one reason why people like to live in communities rather than in isolation.

Now turn to the visual impact which a city has on those who live in it or visit it. I wish to show that an argument parallel to the one put forward above holds good for buildings: bring people together and they create a collective surplus of enjoyment; bring buildings together and collectively they can give visual pleasure which none can give separately.

One building standing alone in the countryside is experienced as a work of architecture, but bring half a dozen buildings together and an art other than architecture is made possible. Several things begin to happen in the group which would be impossible for the isolated building. We may walk through and past the buildings, and as a corner is turned an unsuspected building is suddenly revealed. We may be surprised, even astonished (a reaction generated by the composition of the group and not by the individual building). Again, suppose that the buildings have been put together in a group so that one can get inside the group, then the space created between the buildings is seen to have a life of its own over and above the buildings which create it and one's reaction is to say 'I am inside IT' or 'I am entering IT'. Note also that in this group of half a dozen buildings there may be one which through reason of function does not conform. It may be a bank, a temple or a church amongst houses.

Suppose that we are just looking at the temple by itself, it would stand in front of us and all its qualities, size, colour and intricacy, would be evident. But put the temple back amongst the small houses and immediately its size is made more real and more obvious by the comparison between the two scales. Instead of being a big temple it TOWERS. The difference in meaning between bigness and towering is the measure of the relationship.

In fact there is an *art of relationship* just as there is an art of architecture. Its purpose is to take all the elements that go to create the environment: buildings, trees, nature, water, traffic, advertisements and so on, and to weave them together in such a way that drama is released. For a city is a dramatic event in the environment. Look at the research that is put into making a city work: demographers, sociologists, engineers, traffic experts; all co-operating to form the myriad factors into a workable, viable and healthy organization. It is a tremendous human undertaking.

And yet . . . if at the end of it all the city appears dull, uninteresting and soulless, then it is not fulfilling itself. It has failed. The fire has been laid but nobody has put a match to it.

Firstly we have to rid ourselves of the thought that the excitement and drama that we seek can be born automatically out of the scientific research and solutions arrived at by the technical man (or the technical half of the brain). We naturally accept these solutions, but are not entirely bound by them. In fact we cannot be entirely bound by them because the scientific solution is based on the best that can be made of the average: of averages of human behaviour, averages of weather, factors of safety and so on. And these averages do not give an inevitable result for any particular problem. They are, so to speak, wandering facts which may synchronize or, just as likely, may conflict with each other. The upshot is that a town could take one of several patterns and still operate with success, equal success. Here then we discover a pliability in the scientific solution and it is precisely in the *manipulation of this pliability* that the art of relationship is made possible. As will be seen, the aim is not to dictate the shape of the town or environment, but is a modest one: simply to *manipulate within the tolerances.*

This means that we can get no further help from the scientific attitude and that we must therefore turn to other values and other standards.

We turn to the *faculty of sight,* for it is almost entirely through vision that the environment is apprehended. If someone knocks at your door and you open it to let him in, it sometimes happens that a gust of wind comes in too, sweeping round the room, blowing the curtains and making a great fuss. Vision is somewhat the same; we often get more than we bargained for. Glance at the clock to see the time and you see the wallpaper, the clock's carved brown mahogany frame, the fly crawling over the glass and the delicate rapier-like pointers. Cézanne might have made a painting of it. In fact, of course, vision is not only useful but it evokes our memories and experiences, those responsive emotions inside us which have the power to disturb the mind when aroused. It is this unlooked-for surplus that we are dealing with, for clearly if the environment is going to produce an emotional reaction, with or without our volition, it is up to us to try to understand the three ways in which this happens.

1. Concerning OPTICS. Let us suppose that we are walking through a town: here is a straight road off which is a courtyard, at the far side of which another street leads out and bends slightly before reaching a monument. Not very unusual. We take this path and our first view is that of the street. Upon turning into the courtyard the new view is revealed instantaneously at the point of turning, and this view remains with us whilst we walk across the courtyard. Leaving the courtyard we enter the further street. Again a new view is suddenly revealed although we are travelling at a uniform speed. Finally as the road bends the monument swings into view. The significance of all this is that although the pedestrian walks through the town at a uniform speed, the scenery of towns is often revealed in a series of jerks or revelations. This we call SERIAL VISION [Figure 1].

Examine what this means. Our original aim is to manipulate the elements of the town so that

an impact on the emotions is achieved. A long straight road has little impact because the initial view is soon digested and becomes monotonous. The human mind reacts to a contrast, to the difference between things, and when two pictures (the street and the courtyard) are in the mind at the same time, a vivid contrast is felt and the town becomes visible in a deeper sense. It comes alive through the drama of juxtaposition. Unless this happens the town will slip past us featureless and inert.

There is a further observation to be made concerning Serial Vision. Although from a scientific or commercial point of view the town may be a unity, from our optical viewpoint we have split it into two elements: the *existing view* and the *emerging view*. In the normal way this is an accidental chain of events and whatever significance may arise out of the linking of views will be fortuitous. Suppose, however, that we take over this linking as a branch of the art of relationship; then we are finding a tool with which human imagination can begin to mould the city into a coherent drama. The process of manipulation has begun to turn the blind facts into a taut emotional situation.

2. Concerning PLACE. This second point is concerned with our reactions to the position of our body in its environment. This is as simple as it appears to be. It means, for instance, that when you go into a room you utter to yourself the unspoken words 'I am outside IT, I am entering IT, I am in the middle of IT'. At this level of consciousness we are dealing with a range of experience stemming from the major impacts of exposure and enclosure (which if taken to their morbid extremes result in the symptoms of agoraphobia and claustrophobia). Place a man on the edge of a 500-ft. cliff and he will have a very lively sense of position, put him at the end of a deep cave and he will react to the fact of enclosure.

Since it is an instinctive and continuous habit of the body to relate itself to the environment, this sense of position cannot be ignored; it becomes a factor in the design of the environment (just as an additional source of light must be reckoned with by a photographer, however annoying it may be). I would go further and say that it should be exploited.

Here is an example. Suppose you are visiting one of the hill towns in the south of France. You climb laboriously up the winding road and eventually

Figure 1 Serial Vision: to walk from one end of the plan to another, at a uniform pace, will provide a sequence of revelations which are suggested in the serial drawings, reading from left to right. Each arrow on the plan represents a drawing. The even progress of travel is illuminated by a series of sudden contrasts and so an impact is made on the eye, bringing the plan to life.

find yourself in a tiny village street at the summit. You feel thirsty and go to a nearby restaurant, your drink is served to you on a veranda and as you go out to it you find to your exhilaration or horror that the veranda is cantilevered out over a thousand-foot drop. By this device of the containment (street) and the revelation (cantilever) the fact of height is dramatized and made real.

In a town we do not normally have such a dramatic situation to manipulate but the principle still holds good. There is, for instance, a typical emotional reaction to being below the general ground level and there is another resulting from being above it. There is a reaction to being hemmed in as in a tunnel and another to the wideness of the square. If, therefore, we design our towns from the point of view of the moving person (pedestrian or car-borne) it is easy to see how the whole city becomes a plastic experience, a journey through pressures and vacuums, a sequence of exposures and enclosures, of constraint and relief.

Arising out of this sense of identity or sympathy with the environment, this feeling of a person in street or square that he is in IT or entering IT or leaving IT, we discover that no sooner do we postulate a HERE than automatically we must create a THERE, for you cannot have one without the other. Some of the greatest townscape effects are created by a skilful relationship between the two, and I will name an example in India, where this introduction is being written: the approach from the Central Vista to the Rashtrapathi Bhawan,[1] in New Delhi. There is an open-ended courtyard composed of the two Secretariat buildings and, at the end, the Rashtrapathi Bhawan. All this is raised above normal ground level and the approach is by a ramp. At the top of the ramp and in front of the axis building is a tall screen of railings. This is the setting. Travelling through it from the Central Vista we see the two Secretariats in full, but the Rashtrapathi Bhawan is partially hidden by the ramp; only its upper part is visible. This effect of truncation serves to isolate and make remote. The building is withheld. We are Here and it is There. As we climb the ramp the Rashtrapathi Bhawan is gradually revealed, the mystery culminates in fulfilment as it becomes immediate to us, standing on the same floor. But at this point the railing, the wrought iron screen, is inserted; which again creates a form of Here and There by means of the screened vista. A brilliant, if painfully conceived, sequence.[2]

3. Concerning CONTENT. In this last category we turn to an examination of the fabric of towns: colour, texture, scale, style, character, personality and uniqueness. Accepting the fact that most towns are of old foundation, their fabric will show evidence of differing periods in its architectural styles and also in the various accidents of layout. Many towns do so display this mixture of styles, materials and scales.

Yet there exists at the back of our minds a feeling that could we only start again we would get rid of this hotchpotch and make all new and fine and perfect. We would create an orderly scene with straight roads and with buildings that conformed in height and style. Given a free hand that is what we might do . . . create symmetry, balance, perfection and conformity. After all, that is the popular conception of the purpose of town planning.

But what is this conformity? Let us approach it by a simile. Let us suppose a party in a private house, where are gathered together half a dozen people who are strangers to each other. The early part of the evening is passed in polite conversation on general subjects such as the weather and the current news. Cigarettes are passed and lights offered punctiliously. In fact it is all an exhibition of manners, of how one ought to behave. It is also very boring. This is conformity. However, later on the ice begins to break and out of the straightjacket of orthodox manners and conformity real human beings begin to emerge. It is found that Miss X's sharp but good-natured wit is just the right foil to Major Y's somewhat simple exuberance. And so on. It begins to be fun. Conformity gives way to the agreement to differ within a recognized tolerance of behaviour.

Conformity, from the point of view of the planner, is difficult to avoid but to avoid it deliberately, by creating artificial diversions, is surely worse than the original boredom. Here, for instance, is a programme to rehouse 5,000 people. They are all treated the same, they get the same kind of house. How *can* one differentiate? Yet if we start from a much wider point of view we will see that tropical housing differs from temperate zone housing, that buildings in a brick country differ from buildings in a stone country, that religion and social manners vary the buildings. And as the field of observation narrows, so our sensitivity to the local gods must grow sharper. There is too much insensitivity in the building of towns, too much reliance on the tank and the armoured car where the telescopic rifle is wanted.

Within a commonly accepted framework – one that produces lucidity and not anarchy – we can manipulate the nuances of scale and style, of texture and colour and of character and individuality, juxtaposing them in order to create collective benefits. In fact the environment thus resolves itself into not conformity but the interplay of This and That.

It is a matter of observation that in a successful contrast of colours not only do we experience the harmony released but, equally, the colours become more truly themselves. In a large landscape by Corot, I forget its name, a landscape of sombre greens, almost a monochrome, there is a small figure in red. It is probably the reddest thing I have ever seen.

Statistics are abstracts: when they are plucked out of the completeness of life and converted into plans and the plans into buildings they will be lifeless. The result will be a three-dimensional diagram in which people are asked to live. In trying to colonize such a wasteland, to translate it from an environment for walking stomachs into a home for human beings, the difficulty lay in finding the point of application, in finding the gateway into the castle. We discovered three gateways, that of motion, that of position and that of content. By the exercise of vision it became apparent that motion was not one simple, measurable progression useful in planning, it was in fact two things, the Existing and the Revealed view. We discovered that the human being is constantly aware of his position in the environment, that he feels the need for a sense of place and that this sense of identity is coupled with an awareness of elsewhere. Conformity killed, whereas the agreement to differ gave life. In this way the void of statistics, of the diagram city, has been split into two parts, whether they be those of Serial Vision, Here and There or This and That. All that remains is to join them together into a new pattern created by the warmth and power and vitality of human imagination so that we build the home of man.

That is the theory of the game, the background. In fact the most difficult part lies ahead, the Art of Playing. As in any other game there are recognized gambits and moves built up from experience and precedent. In the pages that follow an attempt is made to chart these moves under the three main heads as a series of cases.

INTRODUCTION TO THE 1971 EDITION

In writing an introduction to this edition of *Townscape* I find little to alter in the attitude expressed in the original introduction written ten years ago.

It has been said that a new edition of *Townscape* should rely on modern work for its examples instead of these being culled from the past. This has not been done for two reasons.

First the task of finding the sharp little needles in the vast haystack of post-war building would be quite uneconomical. This leads to the second point, why should it be so difficult? Because, in my view, the original message of *Townscape* has not been delivered effectively.

We have witnessed a superficial civic style of decoration using bollards and cobbles, we have seen traffic-free pedestrian precincts and we have noted the rise of conservation.

But none of these is germane to townscape. The sadness of the situation is that the superficials have become the currency but the spirit, the Environment Game Itself, is still locked away in its little red and gilt box.

The position may indeed have deteriorated over the last ten years for reasons which are set out below.

Man meets environment: unfamiliarity, shock, ugliness and boredom according to what kind of man you are. The problem is not new but is this generation getting more than its fair share? Yes. Reason? The reason in my view is the speed of change which has disrupted the normal communication between planner and planee. The list is familiar enough: more people, more houses, more amenities, faster communications and unfamiliar building methods.

The speed of change prevents the environment organisers from settling down and learning by experience how to humanise the raw material thrown at them. In consequence the environment is ill-digested. London is suffering from indigestion. The gastric juices, as represented by planners, have not been able to break down all the vast chunks of hastily swallowed stodge into emotional nutriment. We may be able to do many things our grandparents could not do but we cannot digest any faster. The process, be it in stomach or brain, is part of our human bondage. And so we have to make organisational changes in order that human scale can be brought into effective contact with the forces of development.

The first change is to popularise the art of environment on the principle that the game improves with the amount of popular emotion invested and this is the crux of the situation. The stumbling block here is that in the popular mind administrative planning is dull, technical and forbidding whilst good planning is conceived as a wide, straight street with bushy-topped trees on either side, full stop. On the contrary! The way the environment is put together is potentially one of our most exciting and widespread pleasure sources. It is no use complaining of ugliness without realising that the shoes that pinch are really a pair of ten-league boots.

How to explain? Example: the nearest to hand at the time of writing is Sées cathedral near Alençon. The Gothic builders were fascinated by the problem of weight, how to support the culmination of their structures, the vault, and guide its weight safely down to earth. In this building weight has been divided into two parts. The walls are supported by sturdy cylindrical columns: the vault itself, the pride of the endeavour, appears to be supported on fantastically attenuated applied columns which act almost as lightning conductors of gravity between heaven and the solid earth. The walls are held up by man, the vault is clearly held up by angels. 'I understand weight, I am strong', 'I have overcome weight, I am ethereal'. 'We both spring from the same earth together, we need each other'. Through the centuries they commune together in serenity.

As soon as the game or dialogue is understood the whole place begins to shake hands with you. It bursts all through the dull business of who did what and when and who did it first. We know who did it, it was a chap with a twinkle in his eye.

This is the Environment Game and it is going on all round us. You will see that I am not discussing absolute values such as beauty, perfection, art with a big A, or morals. I am trying to describe an environment that chats away happily, plain folk talking together. Apart from a handful of noble exceptions our world is being filled with system-built dumb blondes and a scatter of Irish confetti. Only when the dialogue commences will people stop to listen.

Until such happy day arrives when people in the street throw their caps in the air at the sight of a planner (the volume of sardonic laughter is the measure of your deprivation) as they now do for footballers and pop singers, a holding operation in two parts will be necessary.

First, streaming the environment. It is difficult to fight for a general principle, easier to protect the particular. By breaking down the environment into its constituent parts the ecologist can fight for his national parks, local authority for its green belts, antiquarians for conservation areas and so on. This is already happening.

Second, the time scaling of these streams. Change, of itself, is often resented even if it can be seen to be a change for the better. Continuity is a desirable characteristic of cities. Consequently while planning consent in a development stream might be automatic one may have to expect a built-in delay often or even twenty years in an important conservation area. This is not necessarily to improve the design but simply to slow down the process. This also is happening, if grudgingly, in the case of Piccadilly Circus.

But the main endeavour is for the environment makers to reach their public, not democratically but emotionally. As the great Max Miller once remarked across the footlights on a dull evening, "I know you're out there, I can hear you breathing."

NOTES

1 The President's Residence, lately Viceregal Lodge.
2 It was the cause of bitterness between Lutyens and Baker.

"Principles for Regional Design"

from *Out of Place: Restoring Identity to the Regional Landscape* (1990)

Michael Hough

Editors' Introduction

The idea that regional identity is a fundamental component of sense of place has been gaining prominence over the last several decades, hand in hand with the increasing advance of landscape homogenization. As the profession of landscape architecture has steadily grown to encompass ecological concerns as well as aesthetic concerns, arguments are being made that regional identity is not only important for place-making, but also critical for sustainability because it is associated with bio-diversity.

Michael Hough has been a leading voice in associating regional landscape identity with sense of place and ecological responsibility. In *Out of Place: Restoring Identity to the Regional Landscape*, he asks why modern landscapes tend to look alike despite their different regional settings and determines that the causes include globalization, the rise of consumerism, the loss of rootedness and rise of transience that has come with the information society, and the rise of bureaucratic standardization. Having established the main contributing factors to placelessness, he outlines a regionally based approach to place-making in the final chapter of his book.

In "Principles for Regional Design," reprinted here, Hough, critical of most environmental design theory because of the emphasis on ideal forms and complete design, offers instead a design philosophy based on notions of restraint, minimal intervention, and respect for what is local. Instead of rigid design guidelines, he sets out principles for action, which include learning about places through direct experience of them, maintaining a sense of history, promoting environmental education, doing as little as possible, and starting where it's easiest. While all the principles are well articulated and useful, the latter two are particularly important for urban designers to pay attention to because they encourage modest design attitudes of a kind not generally prevalent within the design fields.

Hough's focus on regional eco-systems harkens back to ideas developed by Patrick Geddes over one hundred years ago. Geddes emphasized regional diversity and advocated extensive ecological surveys prior to undertaking any design or planning work. His innovative Valley Section, which cut a transect through a metropolitan region, identified the intricate ties between central cities and their rural hinterlands. A good discussion of Geddes' ideas can be found in Walter Stephen *et al.*'s *Think Global Act Local: The Life and Legacy of Patrick Geddes* (Edinburgh: Luath Press, 2004). Hough's ideas also relate to landscape architect Ian McHarg's work from the 1960s and 1970s. Concerned with ecological planning – though not necessarily sense of place – and representing an early attempt to integrate emerging green design concepts at the regional level, McHarg developed a method for analyzing landscapes through mapped overlays of soil, hydrology, slope, geology, viewshed, and vegetation. The idea was that this process would then indicate the best areas for development, and help designers avoid the least desirable locations. A thorough articulation of McHarg's method can be found in his *Design with Nature* (New York: Natural History Press, 1969).

Michael Hough is a Professor in the Faculty of Environmental Studies at York University and a principal and founding partner in the Landscape Architecture firm of Hough Woodland Naylor Dance Leinster Limited in Toronto, a firm widely recognized for its work in ecological design. His other writings include *City Form and Natural Processes* (London: Croom Helm, 1984), which is considered a classic of the environmental design field. Writings by others on regional landscapes include Paul Gobster's *Restoring Nature: Perspectives from the Social Sciences and Humanities* (Washington, DC: Island Press, 2000) and Robert L. Thayer Jr.'s *LifePlace: Bioregional Thought and Practice* (Berkeley: University of California Press, 2003).

▧ ▧ ▧ ▪ ▧ ▧

What role does design play in the development of a contemporary regional landscape? A historical perspective suggests that the differences between one place and another have arisen, not from efforts to create long-range visions and grand designs, but from vernacular responses to the practical problems of everyday life. Indeed, it can be argued that purposeful design has done more to generate placelessness than to promote a sense of place. The new forces shaping the landscape are no longer small and local in scope but are great in scale and consequence. The technological and economic impact of these forces on the environment has never before had such profound potential for the destruction of life systems. As a discipline dedicated to fitting man to the land and to giving it form, contemporary design is faced with solving problems that have traditionally not been a part of the agenda in the creation of vernacular places.

In the past, there were limits to what one was able to do and the extent to which one could modify the natural environment. The constraints of environment and society created an undisputed sense of being rooted to the place, but they were, nonetheless, limitations to be overcome, not inherent motivations to be at one with nature. In today's landscape the heterogeneity of the past is giving way to a more homogeneous, information-based society. In design terms, therefore, it becomes as much romantic nonsense to force the old regional differences upon this new landscape as it is to expect people to give up cars, washing machines, and television in the interest of a better environment. We are locked into our times and ways of doing things.

Yet, there is a dilemma for designers in the new and evolving landscape. The determinants that shaped the settlements and countryside of pre-industrial society and that gave rise to the physical forms which we now admire are now no longer those of environmental limitation but of choice.

Creating a sense of place involves a conscious decision to do so. At the same time, the need to invest in the protection of nature has never been so urgent. The connections between regional identity and the sustainability of the land are essential and fundamental. A valid design philosophy, therefore, is tied to ecological values and principles; to the notions of environmental and social health; to the essential bond of people to nature, and to the biological sustainability of life itself. This is the new necessity that will counter-balance and bring some sanity to a world whose goals are focused on helping us "live in a society of abundance and leisure."[1] Yet values that espouse a truly sustainable future will only emerge when it is perceived that there are no alternatives. It is possible that over time the fragility of earth's life systems will create an imperative for survival on which a new ethic can flourish. The international agreement to protect the earth's ozone layer, signed in 1987, may be one indication of this trend. And it is only on this basis that regionalism can become an imperative – a fundamental platform for understanding and shaping the future landscape.

In the preceding chapters I examined the various factors affecting regional identity in order to establish a framework for a design philosophy for the contemporary landscape. This chapter suggests the principles that seem most appropriate to this objective.

KNOWING THE PLACE

Recognizing how people use different places to fulfill the practical needs of living is one of the building blocks on which a distinctive sense of place can be enhanced in the urban landscape. Regional identity is connected with the peculiar characteristics of a location that tell us something about its physical and social environment. It is what a place has when it somehow belongs to its location and nowhere else. It has to do, therefore, with two fundamental criteria: first, with the natural processes of the region or locality – what nature has put there; second, with social processes – what people have put there. It has to do with the way people adapt to their living environment; how they change it to suit their needs in the process of living; how they make it their own. In effect, regional identity is the collective reaction of people to the environment over time.

At the turn of the century Patrick Geddes taught that before attempting to change a place, one must seek out its essential character on foot in order to understand its patterns of movement, its social dynamics, history and traditions, its environmental possibilities. He commented on the way planners dictated form and solutions to problems with little reference to the reality. In his design studies for Madura in the Madras Presidency he wrote:

> One of the poor quarters is at present threatened with "relief from congestion" and we are shown a rough plan in which the usual gridiron of new thoroughfares is hacked through its old-world village life . . . the sanitary improvements begin by destroying an excellent house for the sole purpose of inclining the present lane from the position slightly oblique to the edge of the drawing board to one strictly parallel to it.[2]

In effect, he was saying that modifications to city plans, and for that matter modifications to any landscape, are based on thought processes that begin and end with paper, not the environmental and social realities of the place.

Underlying every urban or urbanizing environment that has developed an image of increasing sameness are unique natural or cultural attributes waiting to be revealed. A place's identity is rarely completely destroyed. There, are always elements of the original landscape that remain, sometimes deeply buried beneath the new. Landform, remnant native plant communities, an old hedge, a barn, old paving stones speak to natural and cultural origins and changing uses. The task is to build an identity based on these remnants.

The hidden elements of a place affect our senses, albeit unconsciously. Tony Hiss describes this in his analysis of experiencing places:

Small, unnoticed changes in level play a larger organizing role in our activities than we suspect: in Manhattan, the right-angle street grid, which keeps people's eyes focused straight ahead, and the uniform paving of streets and sidewalks, together with the solid blocks of buildings on both sides, tend to keep New Yorkers from noticing the natural contours – or what's left of the natural contours – beneath their feet. The nineteenth-century Manhattan developers who covered midtown fields and meadows with brownstones did such a good job of lopping off the tops of hills and filling in valleys that a hundred years or so later . . . no one really knows what the original topography was . . . Nevertheless, almost every block has some rise or dip to it, and these hints of elevation do help people define certain districts.[3]

Several other examples of natural and cultural attributes will illustrate how these affect our sense of a place's identity.

Identity through the landscape

The deep, densely wooded ravines of Toronto that were cut from tableland by streams following the last ice age are part of a major system of rivers draining south to Lake Ontario. Twenty-one meters below the flat, urbanizing plateau of the growing city, they formed a unique system of remnant southern hardwood forest and streams, a habitat for animals and birds within an urban area, a place where the original forest and the natural history of the land could still be experienced. It was where sounds of traffic were no longer heard, where smells and tactile feelings were enhanced by the utter contrast of enclosing woods, suddenly experienced as one reached the valley floor from the level of the street. As the city expanded, many ravines were obliterated by encroaching development, and by the 1950s, 840 acres (340 hectares) of the 1,900 acres (770 hectares) of original ravine land in the city had been given over to houses, factories, and roads.[4] The unique character of Toronto's landscape – the city's structure and identity – rapidly gave way to featureless urban growth. Others were left alone, not as a consequence of planning but because they were simply a nuisance,

Association (vol. 58, no. 1, pp. 49–59, 1992), Ann Forsyth, *Constructing Suburbs: Competing Voices in a Debate over Urban Growth* (London: Routledge, 1999), and Gail Lee Dubrow and Jennifer B. Goodman (eds), *Restoring Women's History through Historic Preservation* (Baltimore, MD: Johns Hopkins University Press, 2003).

■ ■ ■ ■ ■ ■ ■

A look at the fields of public history, architectural preservation, environmental activism, and public art suggests that in the 1990s there is a growing desire to engage urban landscape history as a unifying framework for urban preservation. Many practitioners in these fields are dissatisfied with the old narrative of city building as "conquest." There is broad interest in ethnic history and women's history as part of interpretive projects of all kinds, and a growing sympathy for cultural landscapes in preference to isolated monuments. There is a concern for public processes. But there is not often a sense of how, in practice, the public presentation of historic urban landscapes might become more than the sum of the parts.

Different kinds of organizations may find it difficult to work together on large urban themes. Often, groups simply ignore the other areas of activity. In the worst case, they criticize each other's points of view: social historians are baited as overconcerned with class, race, and gender; architectural preservationists are attacked as being in the grip of real estate developers promoting gentrification; environmentalists are lampooned as idealists defending untouched nature and unimportant species while human needs go unattended; commemorative public art is debated as ugly or irrelevant to social needs. There needs to be, and there can be, a more coherent way of conceptualizing and planning the work each group is able to contribute to the presence of the past in the city. Cultural landscape history can strengthen the links between previously disparate areas of practice that draw on public memory. And conscious effort to draw out public memory suggests new processes for developing projects.

"The relationship between history and memory is peculiarly and perhaps uniquely fractured in contemporary American life," writes public historian Michael Frisch. His colleagues, Jack Tchen and Michael Wallace, observe "historical amnesia" and "historical" culture. Urbanist M. Christine Boyer

writes of architectural history manipulated for commercial purposes. Geographer David Lowenthal wryly calls the past a "foreign country." Citizens surveyed about history will often speak disparagingly of memorized dates, great men, "boring stuff from school" disconnected from their own lives, families, neighborhoods, and work. And certainly there are many people for whom the past is something they want to escape. Yet every year tens of millions of Americans travel to visit historic sites and museums (including some of doubtful quality), as well as historically oriented theme parks and "theme towns."[1] If Americans were to find their own social history preserved in the public landscapes of their own neighborhoods and cities, then connection to the past might be very different.

PLACE MEMORY

"We all come to know each other by asking for accounts, by giving accounts, and by believing or disbelieving stories about each other's pasts and identities," writes Paul Connerton in *How Societies Remember*.[2] Social memory relies on storytelling, but what specialists call place memory can be used to help trigger social memory through the urban landscape. "Place memory" is philosopher Edward S. Casey's formulation:

> It is the stabilizing persistence of place as a container of experiences that contributes so powerfully to its intrinsic memorability. An alert and alive memory connects spontaneously with place, finding in it features that favor and parallel its own activities. We might even say that memory is naturally place-oriented or at least place-supported.[3]

Place memory encapsulates the human ability to connect with both the built and natural environments that are entwined in the cultural landscape. It is the key to the power of historic places to help citizens define their public pasts: places trigger memories for insiders, who have shared a common past, and at the same time places often can represent shared pasts to outsiders who might be interested in knowing about them in the present.

Place memory is so strong that many different cultures have used "memory palaces" – sequences of imaginary spaces within an imaginary landscape or building or series of buildings – as mnemonic devices.[4] Many cultures have also attempted to embed public memory in narrative elements of buildings, from imperial monuments in Augustan Rome to doctrinal sculptural programs for Gothic cathedrals. The importance of ordinary buildings for public memory has largely been ignored, although, like monumental architecture, common urban places like union halls, schools, and residences have the power to evoke visual, social memory.

A strategy to foster urban public history should certainly exploit place memory as well as social memory. (For example, place memory might include personal memory of one's arrival in the city and emotional attachments there, cognitive memory of its street names and street layout, and body memory of routine journeys to home and work.) According to Connerton, cognitive memory is understood to be "encoded" according to semantic, verbal, and visual codes, and seems especially place-oriented because images are "much better retained than abstract items because such concrete items undergo a double encoding in terms of visual coding as well as verbal expression."[5]

Because the urban landscape stimulates visual memory, it is an important but underutilized resource for public history. While many museum curators concerned with artifacts have long understood the strength of visual memory, social historians often have not had much visual training and are not always well equipped to evaluate environmental memory's component of visual evidence.[6] For example, one well-known social historian on the Baltimore Neighborhood Heritage Project, Linda Shopes, complained in an account of her experiences that oral history interviewers were always hearing about places, eliciting "descriptions of the area in years gone by." She was looking for abstractions such as "the workings of local institutions or the local political machine, the conditions and social relations of work, immigration and the

process of assimilation or non-assimilation into American life."[7] Yet stories about places could convey all these themes, and memories of places would probably trigger more stories. More sense of the possibilities of place memory is conveyed by historian Paul Buhle in his recent plea that labor historians turn their attention to photographic collections, as a way to document vanishing working-class neighborhoods developed in the era of industrial capitalism. This could be the beginning of documenting a three-dimensional urban landscape history with a strong social component.[8]

Body memory is also difficult to convey as part of books or exhibits. It connects into places because the shared experience of dwellings, public spaces, and workplaces, and the paths traveled between home and work, give body memory its social component, modified by the postures of gender, race, and class. The experience of physical labor is also part of body memory. In a dusty vineyard, a crowded sweatshop, or an oil field, people acquire the characteristic postures of certain occupations – picking grapes, sewing dresses, pumping gas. In the sphere of domestic work, one thinks of suckling a baby, sweeping, or kneading bread.[9] Thus, Casey argues that body memory "moves us directly into place, whose very immobility contributes to its distinct potency in matters of memory." He suggests that "what is remembered is well grounded if it is remembered as being in a particular place – a place that may well take precedence over the time of its occurrence."[10]

URBAN PUBLIC HISTORY

The field of public history embraces many different kinds of efforts to bring history to the public, from blockbuster museum exhibitions to documentary films to community-based projects. Within the field there are many different views of its content and audience, but for urban landscape history, community-based public history is a natural ally. The great strength of this approach to public history is its desire for a "shared authority" (Michael Frisch's phrase) or a "dialogic" history (Jack Tchen's term) that gives power to communities to define their own collective pasts.[11] This approach is based on the understanding that the history of workers, women, ethnic groups, and the poor requires broad source materials, including oral histories, because often people, rather than professors, are the best authorities on their own pasts.[12] In the search for new materials, including oral histories, many professionally trained historians have seen how communities gained from defining their own economic and social histories. Hundreds of projects now pursue this approach across the United States. Some also have an interest in community empowerment, and connect work in public history to other kinds of community organizing.

[. . .]

ARCHITECTURAL PRESERVATION

Architectural preservation is usually less concerned with accountability and more expensive than community-based public history, but it does assert its visual presence in the spaces of the city. Since the mid-nineteenth century in the United States, most preservation groups have directed their efforts toward saving historic buildings as a unifying focus for national pride and patriotism in a nation of immigrants, or as examples of stylistic excellence in architecture. Preservation at the national level (the National Park Service) and state level thus tends to the creation of museums, and as historian John Bodnar has argued, these often come complete with patriotic exhortations about the glories of the national past.[13] Preservation at the local level, in most cities and towns, tends to the adaptive reuse of historic structures by local real estate developers, with little public access or interpretation, and often involves gentrification and displacement for low-income residents. There are some notable efforts to integrate the preservation of vernacular buildings with local economic development, as in the National Trust for Historic Preservation's Main Street Program, or to preserve working people's neighborhoods without gentrification, as in the Mt. Auburn neighborhood of Cincinnati, but it is difficult work.

The state of preservation today is uneven, and, like public history, many groups struggle without adequate funds. Almost two decades after the Gans-Huxtable debate, New York does recognize more of its social history, with a National Park Service Museum of Immigration at Ellis Island and a Lower East Side Tenement Museum at 97

Orchard Street. But the old focus on great buildings dies hard. As Antoinette J. Lee, historian with the National Park Service, has observed, "disagreement between preservation agencies that prize historical and structural integrity on the one hand and historians interested in vernacular and ethnic history on the other, will likely continue for years to come."[14] Only about five percent of national, state, and local landmark designations reflect women's history, and an even tinier proportion deal with so-called "minority" history.[15] How to preserve is as much debated as what to preserve. Women and people of color need to be making policy. And most social history landmarks cannot be turned into commercial real estate to pay for their physical preservation, nor can they all function as income-producing museums. Yet two projects at the national level suggest how important the sense of place can be in supporting African American or women's history with tangible public forms. These projects show how museums of national importance can extend the resources of an urban neighborhood, although they do not seem to have as much local process behind them as the previous examples.

[. . .]

ENVIRONMENTAL PROTECTION AND LANDSCAPE PRESERVATION

Most American cities and towns host a variety of environmental organizations whose mission is to defend natural landscapes against destruction through carelessness, neglect, toxic substances, or development. There are dozens of different types of environmental groups, some with participatory processes that permit local groups to assess their own sense of what the most important places are to defend, and others that set national agendas (to campaign for the preservation of wetlands as habitations for wildlife, for example). In addition, most cities and states, as well as the federal government, have agencies concerned with regulating treatment of the environment.

While there is much political and scientific activity concerned with protection of the natural environment, there is also a broad cultural and historical debate taking place about the extent to which "nature" and "culture" are intertwined. Urban landscape history shows that simply trying

FOUR

to protect untouched parts of nature has a very limited possibility of success against the constant production of urbanized space. Understanding this interplay suggests that making common cause with other groups concerned with the presence of the past may be a very useful strategy for environmentalists.

Working together with architectural preservationists, landscape architecture historians and landscape architects are increasingly concerned about methods for defining and protecting historic cultural landscapes, although these concerns often seem to focus more on designed landscapes and rural areas than urban landscapes.[16] More promising is the work of some large environmental organizations in protecting rural landscapes with their human activities of gaining a livelihood intact.[17] Most promising of all is work being undertaken in urban regions to protect a variety of natural landscapes interwoven with urban development. At the same time some landscape architects are active in the heart of the city, developing community gardens and educational projects to reach the poorest residents in the most devastated urban areas.[18] Cultural landscape history could become a part of all such ventures to connect efforts to nurture green spaces with a broader understanding of the urban past.

[. . .]

PUBLIC ART AND PUBLIC MEMORY

Public historians may conduct interviews, construct territorial histories, chart workers' landscapes, research political histories of housing types or neighborhoods, and document vernacular arts traditions. Many artists, especially those interested in working with urban communities in public places, are attempting similar kinds of tasks, reaching out to history and geography to study the meanings of historic urban landscapes as a way of making more resonant public work.

This couldn't be more different from the conjunction of public history and public art achieved in the nineteenth-century statues that adorn American parks. "It buttered no parsnips that it was raining on some statues of older men" runs the opening of John Ashberry's poem "A Call for Papers," summoning our distaste for things stuffy and irrelevant.[19]

"Statuomania" preoccupied the capitals of imperial power – London and Washington – at the end of the nineteenth century, as parks and other public places filled up with statues of political leaders and military figures on horseback. It also spread to many small towns, and in every case a political consensus was assumed to support the presence of these statues in public places, although they rarely represented a full range of citizens and often had clichés about white men's conquests as their implicit or explicit narratives, the legacy that artist Judith Baca terms "the cannon in the park."[20] It usually excluded heroic representations of women or members of diverse groups. (As one disgruntled female official remarked in New York after a survey of the civic statuary in Central Park, the only representative of her sex to be found was Mother Goose!)

Today there are many new ways to be an artist: sculptors and painters, muralists and printmakers are joined by environmental artists, performance artists, book artists, and new media artists. For all of them, the key to acquiring an audience is making meaning for people in resonant and original ways. Along with new media come new definitions of public. The older definition of public art is art that is accessible to the public because it is permanently sited in public places. (It is not in galleries or museums or private offices or private homes, but in the streets, the parks, the public realm.) But many artists would agree that a better definition is art that has public content. James Clark says, "Public art is artwork that depends on its context; it is an amalgamation of events – the physical appearance of a site, its history, the socio-economic dimensions of the community, and the artist's intervention."[21] Or as Lucy Lippard, critic and author of the notable book on multicultural art, *Mixed Blessings*, puts it,

> Public art is accessible art of any kind that cares about/challenges/involves and consults the audience for or with whom it is made, respecting community and environment; the other stuff is still private art, no matter how big or exposed or intrusive or hyped it may be.[22]

While these debates continue, two decades of provocative work on urban sites has been done across the country, and artist Suzanne Lacy suggests there is now a convergence of many different artists on a common direction, dealing with identity through connecting art to the history of places, and moving away from a feeling of marginalization toward a sense of centrality in the city.[23] Encouraged by public art planners and curators like Richard Andrews in Seattle, Ronald Fleming and Renata Von Tscharner in Cambridge, Pamela Worden of Urban Arts in Boston, or Mary Jane Jacob in Charleston, artists are undertaking projects that involve complex processes of community engagement, as well as works that claim public places in new ways.[24] Worden has organized the public art on the Orange Line, nine stations leading through the Roxbury and Jamaica Plain areas into downtown Boston. In addition to permanent artworks in the stations, she commissioned works of prose and poetry to be sited in public and organized local history, and photography workshops, Jacob, an independent curator in Chicago, has grouped artists with an interest in history in a series of temporary installations for Charleston, *Places with a Past*.[25]

[. . .]

PUBLIC PROCESSES AND URBAN PRESERVATION

Public history, architectural preservation, environmental protection, and public art can take on a special evocative role in helping to define a city's history if, and only if, they are complemented by a strong community process that establishes the context of social memory. That is not to say that there are simple guidelines for a good public process. This is an emerging area of interdisciplinary work, and in all of these related fields some practitioners are looking for ways to merge their knowledge and concerns with those of residents. There are bound to be conflicts between the outsiders and the insiders, as well as between various individuals, whether members of a project team or residents. Artists, historians, citizens, and planners who come to this kind of work need both historical and spatial imagination to learn to work together to identify and interpret people's history in the urban landscape.

All of the participants in such a process transcend their traditional roles. For the historian, this means leaving the security of the library to listen to the community's evaluation of its own history and the ambiguities this implies. It means working in media – from pamphlets to stone walls – that offer less control and a less predictable audience than academic journals or university presses do. It also means exchanging the well-established roles of academic life for the uncertainties of collaboration with others who may take history for granted as the raw material for their own creativity, rather than a creative work in itself.

For the artist or designer seeking a broader audience in the urban landscape than a single patron or a gallery or museum can provide, it means being willing to engage historical and political material. The kind of public art that truly contributes to a sense of place needs to start with a new kind of relationship to the people whose history is being represented. This means that the artist is involved in an art-making process very different from conventional conceptions of art as the progression of an idiosyncratic, personal style; this kind of process exchanges the established system of shows in galleries or museums for the uncertainties of collaboration with others and community review.

For the public art curator, environmental planner, or urban designer, it means being willing to work for the community, in incremental ways, rather than trying to control grand plans and strategies from the top down. It means understanding that citizens and their grass roots organizations are a source of meaning as legitimate as elected officials and the real estate lobby that often provides the bulk of campaign funding in city politics. It means working alongside – rather than within – the established system of getting things done by phone calls from powerful people in City Hall or the local museum. Funding is always difficult, and the planner on such projects often becomes a nonprofit developer.

For the community member or local resident, it means being willing to engage in a lengthy process of developing priorities for a place, and working through their meanings with a group. This demands patience as insiders educate outsiders and people of different ages, genders, and political views try to agree on what is meaningful and what is creative. Like any other local issue, from parking to street widenings, protection of the public landscape can be exasperating, but it is never unimportant.

While interdisciplinary, community-based projects are not always easy to accomplish, they are not necessarily enormously expensive. They require a labor of love from everyone involved, transcending old roles and expectations, but these are not million-dollar projects. They compare favorably with the funding required to mount a major museum exhibit on city history, or to produce a documentary film for television. And unlike either of these media, public installations in the city refresh the memories of citizens who are passing. A large and diverse audience for urban history exists today in American cities – people who will never go to history museums, attend public humanities programs, or read scholarly journals. Entrepreneurial public historians may be able to reach them occasionally in community centers, churches, or union halls. Successful installations in public places, in all parts of the city, may reach even more people, and, if these are permanent installations, may reach them on a continual basis.

While a single, preserved historic place may trigger potent memories, networks of such places begin to reconnect social memory on an urban scale. Networks of related places, organized in a thematic way, exploit the potential of reaching urban audiences more fully and with more complex histories. There is "a politics to place construction," David Harvey has observed, including "material, representational and symbolic activities which find their hallmark in the way in which individuals invest in places and thereby empower themselves collectively."[26] People invest places with social and cultural meaning, and urban landscape history can provide a frame-work for connecting those meanings into contemporary urban life.

NOTES

1   John Kuo Wei Tchen, "Historical Amnesia and Collective Reclamation: Building Identity with Chinese Laundry Workers in the United States," in *Vers des Sociétés Pluriculturelles: Etudes Comparatives et Situation en France* (Paris Éditions de l'Orstrom, 1987), 242–250; Michael Frisch, "The Memory of History," in Susan Porter Benson, Stephen Brier, and Roy Rosenzweig, eds, *Presenting the Past: Essays on History and the Public* (Philadelphia: Temple University Press, 1986); Micheal Wallace, "Visiting the Past: History Museums in the United States," in Benson, Brier, and Rosenzweig, eds, *Presenting the Past*, 160; M. Christine Boyer, *The City of Collective Memory: Its Historical Imagery and Architectural Entertainments* (Cambridge, Mass.: MIT Press, 1994), 367–476; David Lowenthal, *The Past is a Foreign Country* (Oxford and New York: Oxford University Press, 1985), xvi. For other examples of recent work on this topic, see Michael Kammen, *Mystic Chords of Memory: The Transformation of Traditions in American Culture* (New York: Knopf, 1991); John Bodnar, *Remaking America* (Princeton: Princeton University Press, 1992); Lois Silverman, ed., *A Bibliography on History-Making* (Washington, D.C.: American Associations of Museums, 1989); Peter Fowler, *The Past in Contemporary Society* (London: Routledge, 1992); Martha Norkunas, *The Politics of Public Memory: Tourism, History, and Ethnicity in Monterey, California* (Albany: State University of New York, 1993); David Thelen, *Memory and American History* (Bloomington, Ind.: Indiana University Press, 1990); Melissa Keane, "Asking Questions about the Past," *Mosaic: The Newsletter of the Center on History Making in America*, 1 (Spring/Summer 1992), 9.

2   Paul Connerton, *How Societies Remember* (Cambridge and New York: Cambridge University Press, 1989), 36. He cites Maurice Halbwachs, *Les Cadres sociaux de mémoire* (Paris, 1925), and *La mémoire collective* (Paris: Presses Universitaires des France, 1950). See Maurice Halbwachs, *The Collective Memory*, tr. Francis J. Ditter, Jr., and Vida Yazdi Ditter (New York: Harper and Row, 1980). On the importance of social memory, also see Lowenthal, *The Past is a Foreign Country*, xix; Boyer, *The City of Collective Memory* (especially chapters 1 and 4); Edmund Blair Bolles, *Remembering and Forgetting* (New York: Walker and Company, 1988); George Lipsitz, *Time Passages: Collective Memory and American Popular Culture* (Minneapolis: University of Minnesota Press, 1989); Pierre Nora, "Between Memory and History: les Lieux de Mémoire," *Representations* 26 (Spring 1989), 7–24.

and gated communities, and the growing importance of the home. Privatization also takes a corporate face in the controlled and policed spaces of the shopping mall, theme park, office park, and new town development. Accompanying these is the decline of public space in the city, attributable to both fear in the city (associated with perceptions of crime, the ghetto, poverty and the "other") and many cities' neglect and poor maintenance of these spaces. Here Postmodern Urbanism has become a language of security, which includes the privatization and control suggested above, but also the use of comfortable neo-traditional trappings.

Ellin's critique of *Form Following Finesse* is the most difficult to grasp, dealing with elite concerns for aesthetics, semiotics, and political neutrality. Postmodern Urbanism is viewed here as a narcissistic undertaking of architects engaging in "archi-speak" among themselves, producing work for the sake of image and fame, and a preoccupation with aesthetics rather than solving social problems. Ellin points out the economically regressive nature of some postmodern projects, in light of their use of progressive planning and design theories that, indeed, make the city less affordable and less accessible to moderate income residents. With respect to the social critiques and idealism apparent in Modernism, Postmodern Urbanism abandons most discussions of politics, critical social theory, or political economy.

The final critique of *Form Following Finance* suggests that because of the apolitical stance of many designers, Postmodern Urbanism exacerbates existing urban inequalities and reinforces corporate capitalist agendas. Because of its populist nature, postmodern design may in fact be promoting enhanced opportunities for consumption and profit-making. The adaptive reuse of historic buildings, the rise of the festival marketplace, the growth of themed resorts, and the prevalent post-industrial redevelopment formula (cineplexes, food courts, entertainment, bookstores, stadia, malls, and museums) – all suggest an increasing commercialization of urban development and the importance of market forces in Postmodern Urbanism.

Nan Ellin is an Associate Professor at the School of Architecture and Landscape Architecture at Arizona State University. In addition to *Postmodern Urbanism* she is the author of *Integral Urbanism* (London: Routledge, 2006) and the editor of *Architecture of Fear* (New York: Princeton Architectural Press, 1997). She holds a PhD from Columbia University and was a Fulbright Scholar in France, where she studied European New Urbanism.

Theoretical works on the concepts, history and critiques of Postmodernism include these works: Jean-François Lyotard, *The Postmodern Condition* (Manchester: Manchester University Press, 1984); Frederic Jameson, *Postmodernism or the Cultural Logic of Late Capitalism* (Durham, NC: Duke University Press, 1984); David Harvey, *The Condition of Postmodernity: An Enquiry into the Origins of Cultural Change* (Oxford: Blackwell, 1989); Edward W. Soja, *Postmodern Geographies: The Reassertion of Space in Critical Social Theory* (London: Verso, 1989); Mike Featherstone, *Undoing Culture: Globalization, Postmodernism and Modernity* (London: Sage, 1995); Michael J. Dear, *The Postmodern Urban Condition* (Oxford: Blackwell, 2000).

Classic texts on postmodern design, urbanism and planning include: Charles Jencks, *The Language of Post-modern Architecture* (New York: Rizzoli, 1977); Robert Venturi, Denise Scott Brown, and Steven Izenour, *Learning from Las Vegas: The Forgotten Symbolism of Architectural Form* (Cambridge, MA: MIT Press, 1977); Tom Wolfe, *From Bauhaus to Our House* (New York: Farrar Strauss Giroux, 1980); Michael J. Dear, "Postmodernism and Planning," *Environment and Planning D: Society and Space* (vol. 4, pp. 367–384, 1986); Sharon Zukin, "The Postmodern Debate over Urban Form," *Theory, Culture, Society* (vol. 5, nos. 2–3, pp. 431–446, 1988); Michael J. Dear, "The Premature Demise of Postmodern Urbanism," *Cultural Anthropology* (vol. 6, no. 4, pp. 538–552, 1991); Michael Sorkin (ed.), *Variations on a Theme Park: The New American City and the End of Public Space* (New York: Noonday Press, 1992); John Urry, *Consuming Places* (London: Routledge, 1995); Mark Gottdiener, *The Theming of America: Dreams, Visions, and Commercial Spaces* (New Haven, CT: Yale University Press, 1997); Charlene Spretnak, *The Resurgence of the Real: Body, Nature and Place in a Hypermodern World* (London: Routledge, 1997); and Philip Allmendinger, "Planning Practice and the Post-Modern Debate," *International Planning Studies* (vol. 3, no. 2, pp. 227–248, 1998).

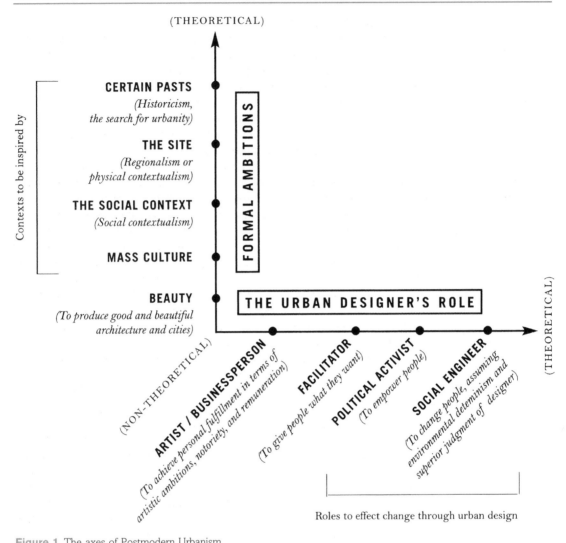

(THEORETICAL)

CERTAIN PASTS
*(Historicism,
the search for urbanity)*

THE SITE
*(Regionalism or
physical contextualism)*

THE SOCIAL CONTEXT
(Social contextualism)

MASS CULTURE

BEAUTY
*(To produce good and beautiful
architecture and cities)*

Contexts to be inspired by

FORMAL AMBITIONS

THE URBAN DESIGNER'S ROLE

(THEORETICAL)

(NON-THEORETICAL)

ARTIST / BUSINESSPERSON
*(To achieve personal fulfillment in terms of
artistic ambitions, notoriety, and remuneration)*

FACILITATOR
(To give people what they want)

POLITICAL ACTIVIST
(To empower people)

SOCIAL ENGINEER
*(To change people, assuming
environmental determinism and
superior judgment of designer)*

Roles to effect change through urban design

Figure 1 The axes of Postmodern Urbanism.

The reactions to modernist architecture and planning can be mapped along two axes, one indicating the formal ambitions of urban designers and the other the ways in which they perceive their role [Figure 1]. These axes meet at the point where urban designers aspire to realize their personal artistic and financial ambitions, with little or no theoretical justification entering the mix, and the axes diverge along the designers' respective theoretical paths. The formal ambition axis moves from producing good and beautiful built forms to drawing inspiration from mass culture, the social context, the site, and the past. The urban designer's role axis proceeds from the business-person and artist to the facilitator, political activist, and social engineer. Although the reactions

to modernist architecture and planning might be mapped along these axes, such an exercise would ultimately reveal little since theory is often a mask or justification for personal ambitions or vice versa.

Rather than chart the rhetoric of these various approaches, then, this chapter peers beyond it, by reviewing and assessing the major themes which fall along the axes of postmodern urbanism as inscribed within the larger postmodern reflex. These overlapping themes include contextualism, historicism, the search for urbanity, regionalism, anti-universalism, pluralism, collage, self-referentiality, reflexivity, preoccupation with image/decor/scenography, superficiality, depthlessness, ephemerality, fragmentation, populism, apoliticism, commercialism, loss of faith, and irony. The critique of postmodern urbanism advanced in this chapter is organized as follows: Form Follows Fiction; Form Follows Fear; Form Follows Finesse; Form Follows Finance; and The Result. The concluding section, On Balance, presents certain correctives of postmodern urbanism as well as promising initiatives that have emerged in the 1990s.

The challenge to the modern project and the decline of the public realm to which modern urbanism was accomplice called for new responses from urban designers. Whereas "modernism from the 1910s to the 1960s . . . responded to the challenge of establishing social order for a mass society; post-modernism since the 1960s . . . responded to the challenge of placelessness and a need for urban community" (Ley 1987, 40). In contrast to modern urbanism's insistence upon structural honesty and functionality, postmodern urbanism sought to satisfy needs that were not merely functional and to convey meanings other than the building tectonics. In architectural theory, Ada Louise Huxtable observed, there was "a search for meaning and symbolism, a way to establish architecture's ties with human experience, a way to find and express a value system, a concern for architecture in the context of society" (Huxtable 1981a, 73–74).

As modernism's minimalist tendencies grew ever more stifling, urban designers embraced maximalism and inclusivity, as expressed in the maxims "Less is a bore" (Venturi 1966) and "More is more" (Stern in Williams 1985). The parallel shift occurring in literature is evocatively portrayed by the protagonist in John Barth's *Tidewater Tales* (1986), a writer whose increasingly minimalist style ultimately blocks his ability to write or dream until circumstances (including the birth of his first child) re-ignite his creative juices, this time in a maximalist form. Likewise in urban design theory, universalism and purism were gradually supplanted by pluralism and contextualism while the role of the urban designer shifted from that of inspired genius, artist, or social engineer to that of a more humble, and at times servile, facilitator.

[. . .]

THE RESULT

A principal feature of postmodern urbanism is contextualism (historical, physical, social, and mass cultural), in contrast to modern urbanism's break from the past and the site. When contextualism is achieved in urban design, it is usually appreciated (successful) unless somehow inappropriate or regarded by the users as a patronizing gesture. In most cases, however, contextualism is not achieved, because of economic and political constraints, the invention of histories, shortcomings of urban designers (who may only be paying lip service to contextualism while pursuing more personal goals), and other reasons. In short, these goals usually prove elusive owing to urban designers' ironic failure to acknowledge the larger contexts in which they build. When contextualism is not achieved, the urban design initiative is usually not appreciated (unsuccessful), except in certain instances where people believe a place is historically, physically, or socially contextual (even if it is not) or don't care because the place succeeds for other reasons such as the standard of living it offers, its prestige, and/or its location.

The contextual attempts to gain inspiration from the site, the social context, and mass culture have more in common with attempts to gain inspiration from the past than may initially appear to be the case. Indeed, they converge where urban design draws from a fictionalized and media-massaged past or vernacular.[1] Like the historicist tendency, these others betray a sense of insecurity and/or confusion and suggest a desire for self-affirmation, self-expression, self-discovery, and "rootedness." And like historicism, these efforts also tend to be more rhetorical than real, largely because their premises contain denials and because the formulation and implementation of these agendas by elites subvert their initial claims. We might say that postmodern urban form follows fiction, finesse, fear, and finance as well as function. But then so did modern urban form.

Ultimately, despite its efforts to counter the negative aspects of modern urbanism, postmodern urbanism falls into many of the same traps. Despite its eagerness to counter the human insensitivity of modern urbanism, postmodern urbanism's preoccupation with surface treatments and irony makes it equally guilty of neglecting the human component. By denying transformations that have taken place, postmodern urbanism may even be accentuating the most criticized elements of modern urbanism such as the emphasis on formal considerations and elitism. Ingersoll has asserted:

To project a return to a "traditional" city and with it a future of "neovillagers" may be more of a fantasy than any science-fiction vision of a society dominated by robots. If the urban process is confined to aesthetic criteria alone, the social consequences, such as the elimination of emancipatory demands from the urban program, may be as unpleasant as those wrought by the functionalist fallacies of the postwar period . . . It is as if *urbs*, the bound city form of the past, could be considered without *civitas*, the social agreement to share that lost urban promised land.

(Ingersoll 1989a, 21)[2]

As Clarke has said, although its agenda suggests an antithesis, "postmodernism has a legacy from modernism it has yet to contradict" (Clarke 1989, 13). Although architects "may no longer be talking of the unadorned cube as the aesthetic model," he contends, their works are still divorced from the larger context, particularly social, in which they are situated (ibid.). Although this style may look different on the surface, it is just as fragmented as what it pretends to be criticizing, because flexible accumulation favors urban design interventions which distinguish themselves, thereby mitigating against contextualism.[3] The modernist refusal to acknowledge context, as epitomized in the reflecting glass wall (see Jameson 1984; Holston 1989; Harvey 1989, 88), might be interpreted as a refusal to acknowledge the emergent mass culture and culture of consumption.[4] But postmodern urbanism's continued denial of the conditions of a mass society, despite its efforts to acknowledge them through contextualism, merely exacerbates the problems of modern urbanism. This denial is epitomized by certain postmodernists' refusal to build any physical structure or place, only to design or theorize. Although justified as a form of resistance, this informed choice only perpetuates the conditions they oppose (Dutton 1986, 23).

ON BALANCE

While much ink has been spilled on pronouncing the banes of postmodern urbanism (along with postmodernism generally), there is also widespread sentiment that it offers a number of correctives to that which preceded it. Indeed, Relph has suggested that these reactions to modern urbanism have ushered in "a quiet revolution in how cities are made and maintained" with the result that "repressive architecture and planning by great corporate or government bureaucracies is being replaced by more sensitive and varied alternatives" (Relph 1987, 215; see Mangin 1985; Muschamp 1994b).

Although historicism can be "essentially elitist, esoteric, and distant" (Clarke and Dutton 1986, 2) and can devolve into kitsch, it can also provide a sense of security and "rootedness" when judiciously applied, as in the reconstruction of European central cities (Gleye 1983). The potentially creative component of borrowing from the past is suggested by folklorist Barbara Kirshenblatt-Gimblett, who maintains that "traditionalizing" or "restoring" (Kirshenblatt-Gimblett 1983, 208) is a universal behavior which entails a process of giving form and meaning by referring to something old while creating "new contexts, audiences, and meanings for the forms" (Kirshenblatt-Gimblett, 211).[5]

Other contextualisms have also succeeded to some extent in achieving an urbanism that is meaningful to more people (i.e. a more pluralistic urbanism). Efforts to design in a physically contextual manner have, for the most part, been an antidote to the modernist emphasis on the architectural object and disregard for the site. Its close cousin, regionalism, has also proven to be a welcome departure from the high modernist contempt for existing styles even though, like historicism, it may appear as a caricatured, mass-produced travesty of the regional context, and/or a neocolonialist undertaking (by developers, technocrats, and urban designers) to prevent the "natives" from becoming more cosmopolitan (like the earlier French colonial urban design).

Residential design in postmodern urbanism offers certain advantages over that of modernism. The Athens Charter maintained that instead of connected low-rise housing lining the streets, housing should be provided in high-rise buildings located in the center of large lots away from streets and from each other in order to maximize open green space and natural light in the homes. Secondly, it maintained that these buildings should be raised onto *pilotis* to open up views from the ground and endow large buildings with a sense of lightness. Finally, it recommended that roofs be flat to offer additional living space. Urban design theory since the 1960s spots, all three are simultaneously absent. On these "sites" (actually, what is the opposite of a site? They are like holes bored through the concept of city) public art emerges like the Loch Ness Monster, equal parts figurative and abstract, usually self-cleaning. **6.11** Specific cities still seriously debate the mistakes of architects – for instance, their proposals to create raised pedestrian networks with tentacles leading from one block to the next as a solution to congestion – but the Generic City simply enjoys the benefits of their inventions: *decks*, *bridges*, *tunnels*, *motorways* – a huge proliferation of the paraphernalia of connection – frequently draped with ferns and flowers as if to ward off original sin, creating a vegetal congestion more severe than a fifties science-fiction movie. **6.12** The roads are only for cars. People (pedestrians) are led on rides (as in an amusement park), on "promenades" that lift them off the ground, then subject them to a catalog of exaggerated conditions – wind, heat, steepness, cold, interior, exterior, smells, fumes – in a sequence that is a grotesque caricature of life in the historic city. **6.13** There *is* horizontality in the Generic City, but it is on the way out. It consists either of history that is not yet erased or of Tudor-like enclaves that multiply around the center as newly minted emblems of preservation. **6.14** Ironically, though itself new, the Generic City is encircled by a constellation of New Towns: New Towns are like year-rings. Somehow, New Towns age very quickly, the way a five-year-old child develops wrinkles and arthritis through the disease called progeria. **6.15** The Generic City presents the final death of planning. Why? Not because it is not planned – in fact, huge complementary universes of bureaucrats and developers funnel unimaginable flows of energy and money into its completion; for the same money, its plains can be fertilized by diamonds, its mud fields paved in gold bricks . . . But its most dangerous *and* most exhilarating discovery is that planning makes no difference whatsoever. Buildings may be placed well (a tower near a metro station) or badly (whole centers miles away from any road). They flourish/perish unpredictably. Networks become over-stretched, age, rot, become obsolescent; populations double, triple, quadruple, suddenly disappear. The surface of the city explodes, the economy accelerates, slows down, bursts, collapses. Like ancient mothers that still nourish titanic embryos, whole cities are built

on colonial infrastructures of which the oppressors took the blueprints back home. Nobody knows where, how, since when the sewers run, the exact location of the telephone lines, what the reason was for the position of the center, where monumental axes end. All it proves is that there are infinite hidden margins, colossal reservoirs of slack, a perpetual, organic process of adjustment, standards, behavior; expectations change with the biological intelligence of the most alert animal. In this apotheosis of multiple choice it will never be possible again to reconstruct cause and effect. They work – that is all. **6.16** The Generic City's aspiration toward tropicality automatically implies the rejection of any lingering reference to the city as fortress, as citadel; it is open and accommodating like a mangrove forest.

7. POLITICS

7.1 The Generic City has a (sometimes distant) relationship with a more or less authoritarian regime – local or national. Usually the cronies of the "leader" – whoever that was – decided to develop a piece of "downtown" or the periphery, or even to start a new city in the middle of nowhere, and so, triggered the boom that put the city on the map. **7.2** Very often, the regime has evolved to a surprising degree of invisibility, as if, through its very permissiveness, the Generic City resists the dictatorial.

8. SOCIOLOGY

8.1 It is very surprising that the triumph of the Generic City has not coincided with the triumph of sociology – a discipline whose "field" has been extended by the Generic City beyond its wildest imagination. The Generic City *is* sociology, happening. Each Generic City is a petri dish – or an infinitely patient blackboard on which almost any hypothesis can be "proven" and then erased, never again to reverberate in the minds of its authors or its audience. **8.2** Clearly, there is a proliferation of communities – a sociological zapping – that resists a single overriding interpretation. The Generic City is loosening every structure that made anything coalesce in the past. **8.3** While infinitely patient, the

Generic City is also persistently resistant to speculation: it proves that sociology may be the worst system to capture sociology in the making. It outwits each established critique. It contributes huge amounts of evidence for and – in even more impressive quantities – against each hypothesis. In *A* tower blocks lead to suicide, in *B* to happiness ever after. In *C* they are seen as a first stepping stone toward emancipation (presumably under some kind of invisible "duress," however), in *D* simply as passé. Constructed in unimaginable numbers in *K*, they are being exploded in *L*. Creativity is inexplicably high in *E*, nonexistent in *F*. *G* is a seamless ethnic mosaic, *H* perpetually at the mercy of separatism, if not on the verge of civil war. Model *Y* will never last because of its tampering with family structure, but *Z* flourishes – a word no academic would ever apply to any activity in the Generic City – because of it. Religion is eroded in *V*, surviving in *W*, transmuted in *X*. **8.4** Strangely, nobody has thought that cumulatively the endless contradictions of these interpretations prove the richness of the Generic City; that is the one hypothesis that has been eliminated in advance.

9. QUARTERS

9.1 There is always a quarter called Lipservice, where a minimum of the past is preserved: usually it has an old train/tramway or double-decker bus driving through it, ringing ominous bells – domesticated versions of the Flying Dutchman's phantom vessel. Its phone booths are either red and transplanted from London, or equipped with small Chinese roofs. Lipservice – also called Afterthought, Waterfront, Too Late, 42nd Street, simply the Village, or even Underground – is an elaborate mythic operation: it celebrates the past as only the recently conceived can. It is a machine. **9.2** The Generic City had a past, once. In its drive for prominence, large sections of it somehow disappeared, first unlamented – the past apparently was surprisingly unsanitary, even dangerous – then, without warning, relief turned into regret. Certain prophets – long white hair, gray socks, sandals – had always been warning that the past was necessary – a resource. Slowly, the destruction machine grinds to a halt; some random hovels on the laundered Euclidean plane are saved, restored to a

splendor they never had . . . **9.3** In spite of its absence, history is the major preoccupation, even industry, of the Generic City. On the liberated grounds, around the restored hovels, still more hotels are constructed to receive additional tourists in direct proportion to the erasure of the past. Its disappearance has no influence on their numbers, or maybe it is just a last-minute rush. Tourism is now independent of destination . . . **9.4** Instead of specific memories, the associations the Generic City mobilizes are general memories, memories of memories: if not all memories at the same time, then at least an abstract, token memory, a deja vu that never ends, generic memory. **9.5** In spite of its modest physical presence (Lipservice is never more than three stories high: homage to/revenge of Jane Jacobs?) it condenses the entire past in a single complex. History returns not as farce here, but as *service*: costumed merchants (funny hats, bare midriffs, veils) voluntarily enact the conditions (slavery, tyranny, disease, poverty, colony) – that their nation once went to war to abolish. Like a replicating virus, worldwide, the colonial seems the only inexhaustible source of the authentic. **9.6** 42nd Street: ostensibly the places where the past is preserved, they are actually the places where the past has changed the most, is the most distant – as if seen through the wrong end of a telescope – or even completely eliminated. **9.7** Only the memory of former excess is strong enough to charge the bland. As if they try to warm themselves at the heat of an extinguished volcano, the most popular sites (with tourists, and in the Generic City that includes everyone) are the ones once most intensely associated with sex and misconduct. Innocents invade the former haunts of pimps, prostitutes, hustlers, transvestites, and to a lesser degree, artists. Paradoxically, at the same moment that the information highway is about to deliver pornography by the truckload to their living rooms, it is as if the experience of walking on these warmed-over embers of transgression and sin makes them feel special, alive. In an age that does not generate new aura, the value of established aura skyrockets. Is walking on these ashes the nearest they will get to guilt? Existentialism diluted to the intensity of a Perrier? **9.8** Each Generic City has a waterfront, not necessarily with water – it can also be with desert, for instance – but at least an edge where it meets another condition, as if a position

PART FIVE

Typology and Morphology in Urban Design

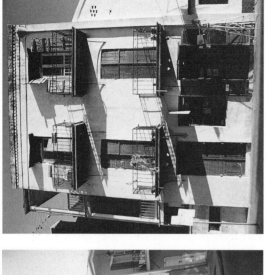

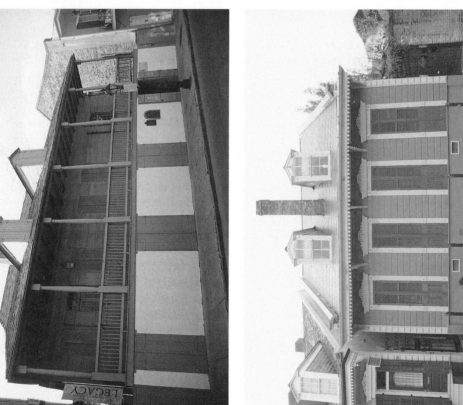

Plates 5-8 Residential Building Types in the French Quarter of New Orleans: the plantation-style French Colonial House, circa mid-eighteenth century; the vertically-oriented Townhouse, built throughout the nineteenth century; the very adaptable Creole Cottage, circa early nineteenth century; and the diminutive Single Shotgun, circa late nineteenth century. (Photos: M. Larice)

INTRODUCTION TO PART FIVE

The use of figure-ground drawings, square mile maps, street sections, aerial photos, and computer-generated drawings in the study and comparison of urban form patterns is so common nowadays that we often forget how new these tools actually are. The emergence of typological and morphological practice in the late 1950s provided urban designers with a research arm that was particularly suited to exploring issues of urban spatial form. While the two terms are different in many senses, they both approach the study of form through scientifically rational goggles. Typology refers to the study of categorized form types in architecture, and increasingly in urban design and landscape architecture as well. As opposed to building type, which refers to functionality, architectural typologies refer to the form characteristics of buildings. Morphology on the other hand is the study of larger urban structures, patterns and form issues. Some authors have taken to combining the terms into a new term, "typomorphology," which focuses on larger scale urban form patterns. Others refer to this topical material as "tissue studies." Opposed to the image studies, environmental behavior research, and phenomenological approaches of previous authors in this reader, morphologists are interested primarily in the tangible physical world of objects and less interested in subject experience or the social use of space. Many study urban form longitudinally, looking at change over time. Others look at predominant form types to help guide present-day design efforts.

The study of morphology and typology gained popularity in the late 1970s with rediscovered interest in the benefits of traditional urbanism. Author-theorists working in Europe, such as Aldo Rossi, Saverio Muratori, Gianfranco Caniggia, M.R.G. Conzen, T.R. Slater, Léon and Robert Krier, and Philippe Panerai, created and drove the exploration of these new research interests, many arguing for the reconstruction of the European City. Researchers at the University of Birmingham's Urban Morphology Research Group are a central locus for interest in this field of inquiry. In the United States, interest in morphological methods was championed by Colin Rowe and Fred Koetter in their book *Collage City* (1978). In this work the authors demonstrated the power of figure-ground drawings, or "black plans" as they are sometimes referred, as an analytical and critical tool in comparing traditional and modern urban spaces. Their goal was the undoing of a half-century of design practice and education. In pointing out the flaws in modern design practice, the book influenced a generation of urban designers and illustrated how morphological methods could be used as a starting point for design. Over the next twenty-five years, morphological practice would assume many forms, both as a primary research tool and as a device to inform design practice. Typological practice has been particularly influential in the development of prescriptive design guidelines and urban design codes in use among the New Urbanists. In the academies, morphological methods are woven into much urban design research nowadays.

This part of the reader begins with a collection of Léon Krier's musings on the failure of modern town planning. His "Critiques" are presented in the form of humorous cartoons pointing out the various flaws in zoning, industrialization, reconstruction and urban form. Of note is the relatively famous diagram illustrating how the public features and monuments of the city should be integrated with the private domestic blocks to form an integrated urban pattern. The text following these cartoons is a treatise on the benefits of traditional urban form founded in the use of small blocks and streets that shape space. The second reading by Anthony Vidler calls for a new typology of urban form based in the traditional city itself. Echoing the criticisms of others, Vidler also comments on the "fragmentation produced by the elemental, institutional, and mechanistic typologies of the recent past." While it appears he is a proponent of all things traditional, this is true only in the sense of urban structuring and not in the design of architecture itself, which should remain vital and engaged with the present. The final reading in Part Five discusses the emergence of typomorphology as a rigorous and necessary research arm of urban design knowledge and practice. Written by Anne Vernez Moudon, the reading explores the three traditions of urban typomorphology practiced in Italian, French, and Anglo-American contexts. Each of these "schools" contributes something of relevance to our understanding of the built environment, together suggesting that time, form, and scale are all at play.

Readers interested in the topics of typology and morphology should refer to the following works for supplemental reading: Colin Rowe and Fred Koetter, *Collage City* (Cambridge, MA: MIT Press, 1978); Aldo Rossi, *The Architecture of the City* (Cambridge, MA: MIT Press, English translation 1982, first Italian text 1966); Robert Krier, *Urban Space* (London: Academy Editions, 1979, first German text 1975); Michael Southworth and Eran Ben-Joseph, *Streets and the Shaping of Towns and Cities* (New York: McGraw-Hill, 1997); and Spiro Kostoff, *The City Shaped: Urban Patterns and Meanings Through History* (Boston, MA: Little, Brown and Bulfinch Press, 1991).

"Critiques" and "Urban Components"

from *Houses, Palaces, Cities* (1984)

Léon Krier

Editors' Introduction

In *Houses, Palaces, Cities*, editor Demetri Porphyrios collects the various critiques, cartoons, and ideas of the European neo-traditionalist Léon Krier. Krier is an architect, theoretician, and author from Luxembourg, who championed the rationality of the traditional pre-industrial city as a model for current development and as a means for saving urban life from the destructive onslaught of modernist planning and design. His ideas were influenced by a number of historically based ideas and urban theories, including those of Plato, the Renaissance humanists, Camillo Sitte, French rational architecture, and Ferdinand Tönnies' concepts of *Gemeinschaft* and *Gesellschaft*.

During the late 1960s and into the 1970s, Krier along with a group of designers and writers, such as Maurice Culot, Pierluigi Nicolin, Philippe Panerai, Jacques Lucan, Jean Dethier, Antoine Grumbach, and Robert Delevoy, became engaged in debates over the future of European urbanism, producing a series of statements on "anti-industrial resistance" and "the reconstruction of the European city." In 1978, at a Palermo conference called "The City in the City," the group produced a statement outlining their principles, which would later become the Movement for the Reconstruction of the European City. In the late 1970s and 1980s a series of international exhibitions and design competitions were held (Roma Interrotta in Rome, Les Halles and Parc de la Villette in Paris, International Bauausstellung in Berlin), where members of the Movement were able to showcase their ideas for wider dissemination. In the 1980s, Krier designed several projects and new towns in Europe, served as a design consultant for the town of Seaside, Florida (where he also built a neo-classically inspired white tower house for himself), and began advising Prince Charles in the 1990s on concepts of neo-traditional design and planning, including the design for the neo-traditional village of Poundbury in Dorset.

As its spokesperson, Krier outlined the Movement's key driving principles: historical centers should be preserved; urban space should organize city form; typological and morphological study should guide design process; streets, squares, and residential *quartiers* should be reconstructed; fragmented parts of the city and the functions of urban living should be reintegrated into a coherent wholeness; design should be based on aesthetic beauty and imitation of the best pre-industrial cities; and cities should be articulated into domestic and public spaces with an awareness of fabric, monumentality, and traditional building techniques. The Movement is a stark departure from the principles and processes of modern urbanism, in the rejection of functional zoning, megastructure building, industrial processes, and the destruction of urban redevelopment. For Krier, Porphyrios suggests, "the chief task of urbanism today is to challenge the peculiar falsehood of modern industrial consciousness and to defend practical reason against the domination of universal technique" (from "Cities of Stone," the introduction to *Houses, Palaces, Cities*). The Movement for the Reconstruction of the European City was not without its critics, including a new generation of modern architects and theorists such as Rem Koolhaas and other Post-Urban designers.

In this reading, Krier outlines his opposition to modernist design, planning and urban development in a series of cartoon-illustrated critiques, many of them humorous in delivery. Several of his "Critiques" are presented here, while others, such as "Critique of the Megastructural City" and "Critique of Modernisms," have not been included. In a section titled "Urban Components," he diagrams the desired form of the *Civitas* as the aggregation of the *Res Publica* (the city's public and civic monuments) and the *Res (Economica) Privata* (which provides the domestic fabric and formal structuring elements for the city). He champions the human-scaled city block as "the most important typological element in the composition of urban spaces, the key element of the urban pattern" – an element which has continued to grow due to changes in the mode of urban production, the increasing scale of design, and the concentration of political, economic and cultural power. His stated goal is that urban blocks should be as small as is "typologically viable" and form as many streets and squares as possible.

In addition to a long list of built projects, written works by Léon Krier include the following: "Manifesto: The Reconstruction of the European City or Anti-Industrial Resistance as a Global Project," edited with Maurice Culot, in *Counterprojects* (Brussels: Archives d'Architecture Moderne, 1980), *Quatremére de Quincy: De l'imitation*, edited with Demetri Porphyrios (Brussels: Archives d'Architecture Moderne, 1980), "Forward Comrades, We Must Go Back," *Oppositions 24* (Spring 1981), "The New Traditional Town: Two Plans by Léon Krier for Bremen and Berlin-Tegel," *Lotus* (vol. 36, 1982) and "The Reconstruction of the European City," *Architectural Design* (vol. 54, nos. 11/12, 1984).

A key text of the neo-rational architects is Robert Delevoy *et al.* (eds), *Rational: Architecture: Rationelle: The Reconstruction of the European City* (Brussels: Archives d'Architecture Moderne, 1978). For historical information on the rise of neo-rational design see: Geoffrey Broadbent, *Emerging Concepts in Urban Space Design* (New York: Van Nostrand Reinhold International, 1990); and Nan Ellin, "Urban Design Theory on the European Continent," in *Postmodern Urbanism* (New York: Princeton Architectural Press, 1996).

Other authors writing in a similar spirit on the use of typology and morphology for purposes of urban design and reconstruction, include: Robert Krier, "Typological and Morphological Elements of the Concept of Urban Space," in *Urban Design/Stradtraum* (London: Academy Editions, 1978), *Urban Space* (London: Academy Editions, 1980), and *Architectural Composition* (London: Academy Editions, 1988); Andres Duany, Elizabeth Plater-Zyberk, Alex Kreiger, and William Lennertz, *Towns and Town-Making Principles* (New York: Rizzoli and Harvard Graduate School of Design, 1991); John A. Dutton, *New American Urbanism: Re-forming the Suburban Metropolis* (Milan: Skira, 2001); and Andres Duany, Elizabeth Plater-Zyberk, and Robert Alminana, *New Civic Architecture: Elements of Town Planning* (New York: Rizzoli, 2003). Both Krier and the later New Urbanists were influenced by Werner Hegemann and Elbert Peets' book *The American Vitruvius: An Architect's Handbook of Civic Art* (New York: Architectural Book Publishing, 1922, reissued New York: Princeton Architectural Press, 1989).

CRITIQUES

TOWN AND COUNTRY

Against the global destruction of the city and countryside that we are witnessing, we propose a global philosophical, political and technical project of reconstruction. One cannot destroy the city without also destroying the countryside. City and countryside are antithetical notions. The reconstruction of the territory must be defined in a strict physical and legal separation of city and countryside. First of all we must drastically reduce the built perimeters of the city and precisely redefine rural land in order to establish clearly what is city and what is countryside. Any notion of functional zoning must be abolished. There can be no industrial zones, pedestrian zones, shopping or housing zones. There can only be urban quarters which integrate all the functions of urban life. The notions of metropolitan centre and periphery must be abolished. [Figures 1–4].

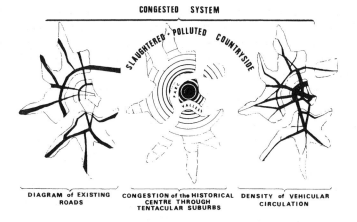

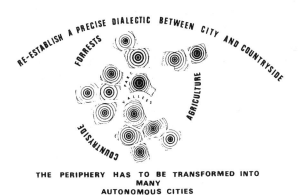

Figure 1 Town and Country

CRITIQUE OF ZONING

The industrial project reduces the essential functions of life to three distinct categories: production, consumption and reproduction (material and intellectual, education, propaganda, health and culture). Isolation, fragmentation, separation and finally the *territorial regrouping* of functions encourage this process of reduction and the internal rationalisation of functions. In urban practice, this *fragmentation* is realised through *functional zoning* (administrative, cultural, industrial, commercial, residential zones, etc). *The first imperative of zoning is to transform every part of the territory (city or*

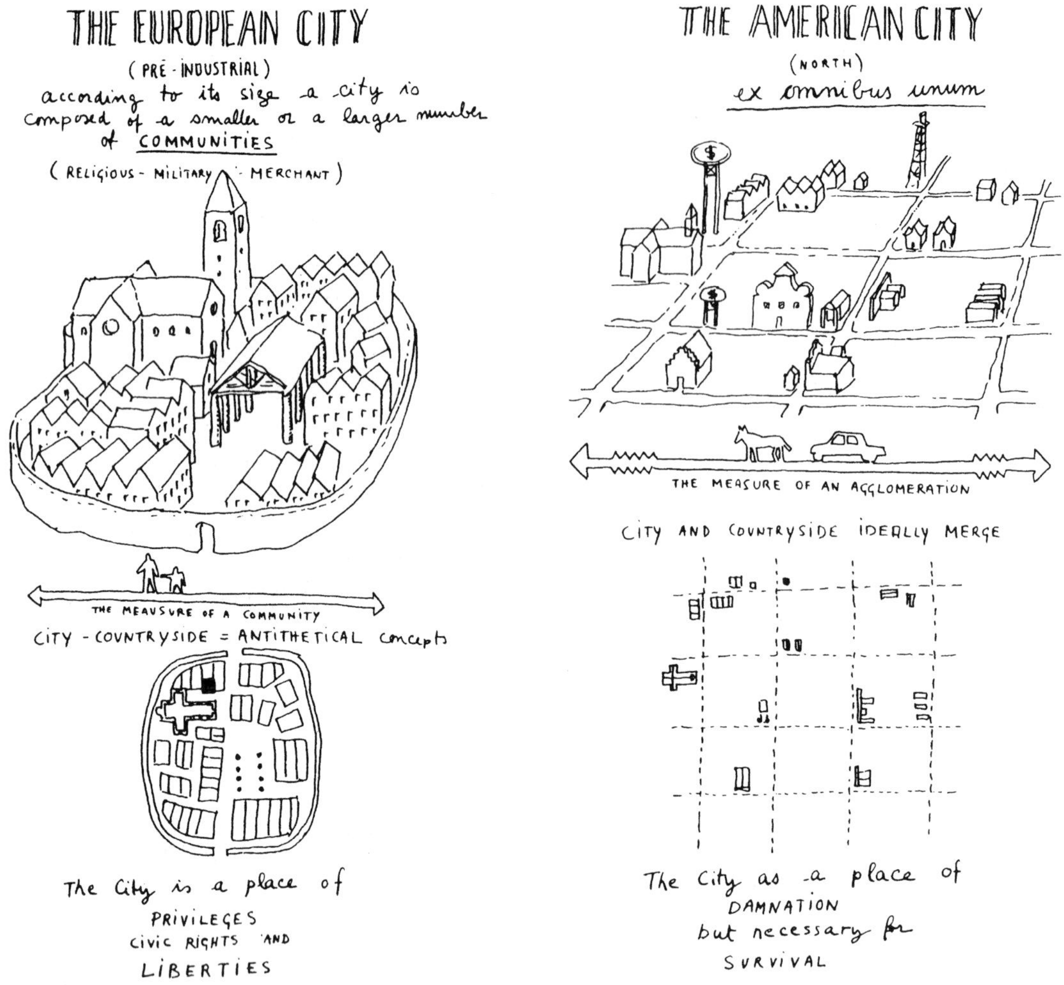

THE EUROPEAN CITY
(PRÉ - INDUSTRIAL)
according to its size a city is composed of a smaller or a larger number of COMMUNITIES
(RELIGIOUS - MILITARY - MERCHANT)

THE MEASURE OF A COMMUNITY
CITY - COUNTRYSIDE = ANTITHETICAL concept

The City is a place of
PRIVILEGES
CIVIC RIGHTS AND
LIBERTIES

THE AMERICAN CITY
(NORTH)
ex omnibus unum

THE MEASURE OF AN AGGLOMERATION
CITY AND COUNTRYSIDE IDEALLY MERGE

The City as a place of
DAMNATION
but necessary for
SURVIVAL

Figure 2 Town and Country

countryside) in such a way that every citizen can finally only accomplish: only a single task; in a defined place; in a determined manner; at the exclusion of all other tasks. *The second imperative of zoning* is the daily and effective mobilisation of society in its entirety (all social classes of all ages: infants, children, adults, the elderly, the rich, the poor). The production and use of roads and means of transport become the principal activities of an industrial economy. The railroads, and the roads of concrete, asphalt and earth are the arterial system, the common gathering place and the cement of an atomised society. The means of private and public mechanical transport – trains, planes, automobiles – are the principal instrument of this mobilisation; the necessary extension of the human body. The goal of the industrial plan is to guarantee, at least for a few generations, maximum consumption of units of time and energy. Mechanical transport becomes, in other words, the principal function in the industrial metabolism of man with nature. The authorities of bridges and roads and the industries of transport and energy are sacred cows, [Figures 5–7].

The politics of industrial infrastructure has been based on the territorial separation of functions.

City
&
LANDSCAPE

A CITY needs APPROX

so much Land

for its nutrition

The Columbus Factor

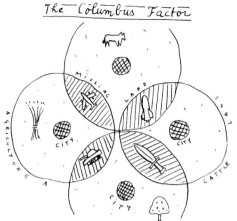

X Number of Cities need X times more _Land_
For whatever _Land_ they are missing they are going
to bash in their own heads and rather than reduce
their own numbers, they are going to invade,
conquer and subjugate far _Lands_, continents
and peoples

Figure 3 Town and Country

1850 ~ 1950
THE FORMATION of the INDUSTRIAL ANTI-CITY
(INDUSTRIAL CITY = CONTRADICTIO in TERMINI)

INDUSTRIAL
SUBURB.
BANLIEU·FAUBOURG
VORORT·
SATELLITES·TRABANTEN
BESIEGE
the CITY

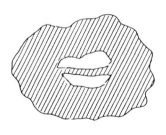

The CITY is
FINE
WITHOUT SUBURB
_____
SUB~URB
UNTHINKABLE

WITHOUT the CITY

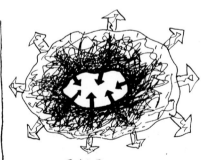

SUBURBS
FIRST
DESTROY the

LANDSCAPE & FORESTS
AND THEN
the
CITY

Figure 4 Town and Country

FIVE

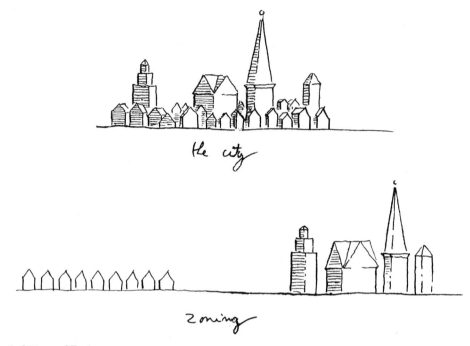

Figure 5 Critique of Zoning

All industrial states independently of their political ideology are promoting and imposing the functional *zoning* of cities and countryside with equal brutality and pseudo-scientific arguments in spite of all the resistance of urban or rural populations. Functional Zoning is not an innocent instrument; it has been the most effective means in destroying the infinitely complex social and physical fabric of pre-industrial urban communities, of urban democracy and culture. Functional Zoning of city and countryside has been an authoritarian project corresponding nowhere to a democratic demand. Zoning is the *abstraction* of city and countryside. *One cannot destroy the cities without also destroying the countryside.* Zoning is the *abstraction* of communities; it reduces the proudest communities to mere statistical entities, expressed in numbers and densities. Zoning, dictated by big industry and their financial and administrative empires, can be fought only by democratic pressure that demands the reconstruction of urban communities where *residence*, *work* and *leisure* are all within walking distance. Functional Zoning based on infinite territorial sprawl has resulted in maximum energy

consumption. The slavery of mobility to which every citizen has been condemned forces him to waste both time and energy in daily transport, while at the same time it has made him into a potential and involuntary agent of energy waste. Neither *private* nor *public* transport policies can effectively curtail the waste of material or social energy caused by functional zoning. An intelligent energy consumption policy is possible only by integrating the main urban functions into urban quarters (*districts*) of limited territorial size. Those energy-saving policies which do not recognise this condition are doomed to lead to totalitarian measures of control and social coercion [Figures 8 and 9].

CRITIQUE OF INDUSTRIALISATION

The belief in unlimited technical progress and development has brought the most '*developed*' countries to the brink of physical and cultural exhaustion. The fever of short-term profit has ravaged cities and countryside. Industrial production,

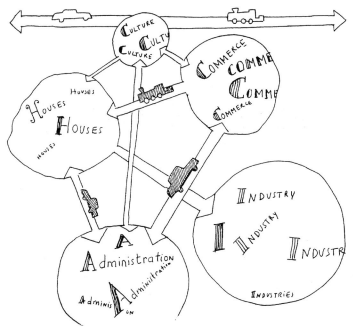

The INDUSTRIAL Anti-CITY is DECOMPOSED into ZONES

Figure 6 Critique of Zoning

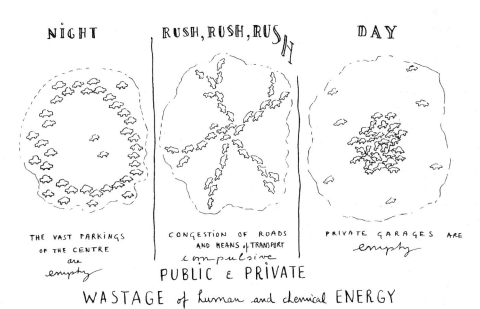

MOTORIZED ~ TRAFFIC

THE EFFECTS OF FUNCTIONAL ZONING

NIGHT RUSH, RUSH, RUSH DAY

THE VAST PARKINGS OF THE CENTRE are empty

CONGESTION OF ROADS AND MEANS of TRANSPORT compulsive

PRIVATE GARAGES ARE empty

PUBLIC & PRIVATE

WASTAGE of human and chemical ENERGY

Figure 7 Critique of Zoning

The CITY of the PEDESTRIAN

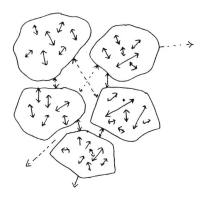

MINIMUM DISTANCES
FOR MAXIMUM ACHIEVEMENT & PLEASURE

Figure 8   Critique of Zoning

The (Anti-) CITY of the MOTORCAR

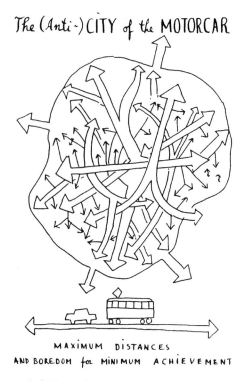

MAXIMUM DISTANCES
AND BOREDOM for MINIMUM ACHIEVEMENT

Figure 9   Critique of Zoning

that is, the extreme development of productive forces, has destroyed in less than two hundred years those cities and landscapes which had been the result of thousands of years of human labour, intelligence and culture. Industrialisation of building must be considered as a total failure. Its ulterior motive has never been the professed proletarisation of material comfort but instead the maximisation of short-term profits and the consolidation of economic and political monopolies. Industrialisation has not brought any significant technical improvement in building. It has not reduced the cost of construction. It has not shortened the time of production. It has not created more jobs. It has not helped to improve the working conditions of the workers. It has on the contrary destroyed a millenary and highly sophisticated craft. It has proven incapable of finding solutions for the typological, social and morphological complexity of the historical centres. And although building today is still organised largely according to forms of artisanal production, craftsmanship as an autonomous culture has been destroyed by the industrial and social division of labour. A culture of

building and architecture must be based on a highly sophisticated manual tradition of construction and not on the formulation of 'specialist professional bodies.' Industrialisation has in the end only facilitated centralisation of capital and of political power, whether private or public [Figures 10–12].

THE IDEA OF RECONSTRUCTION

The challenge to our generation is to refuse to build now. To protest against the transformation and destruction of the cities serves no purpose if we do not have a global alternative plan of reconstruction in our hands. A critique which has no project is but another face of a totally fragmented society, of which the city is only a model. A critique without a vision gazes as impotently at the future as the historian without a project gazes at the past. Professional criticism has killed critique in the same way as historiography has killed history. The project which our generation must elaborate has to fight the destruction of urban society on all levels, cultural, political, economic. Only with this project of

Figure 16 The zoning of modern cities has resulted in the random distribution of both public and private buildings. The artificiality and wastefulness of zoning has destroyed our cities.

the vast municipal perimeter blocks in Moscow could not just as well have been built as a multitude of small urban blocks with a familiar scale of streets and squares! [Figure 17]

What I will try to criticise is an *historical tendency* illustrated by the fact that *larger and larger building programmes* (resulting from the concentration of economic, political and cultural power) have resulted almost *'naturally' in larger and larger building blocks!* The Palace of Justice in Brussels has the size of a medieval parish; the length of the Karl Marx Hof equals the diameter of the centre of Vienna from wall to wall: a single building gesture resulting from a single programme, executed by one architect.

My aim is, however, not just to describe an irreversible historical fatality but to establish an hypothesis: the social and cultural complexity of a city has necessarily to do with its physical and structural complexity and density. The size of an ideal urban block cannot be established more precisely than the ideal height of the human body. One can, however, deduce through comparison and

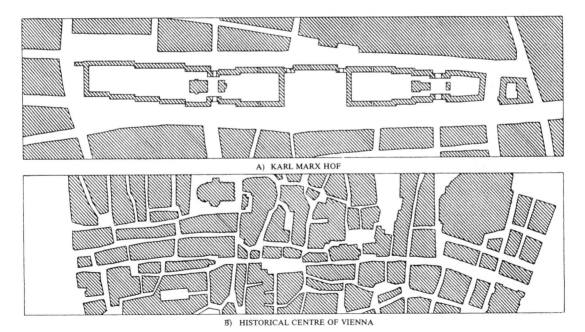

A) KARL MARX HOF

B) HISTORICAL CENTRE OF VIENNA

Figure 17 This comparison at the same scale shows how through zoning the city becomes decomposed, not only physically and functionally but above all *socially*. The social, cultural, economic complexity and density of pre-industrial Vienna could not be contrasted more violently with the social and cultural emptiness of the "höfe." A city is reduced to a mere 'artistic' gesture.

experience sizes of urban blocks which are more apt to form a complex urban pattern than others.

My main affirmations as regards urban design will be: *urban blocks should be as small in length and width as is typologically viable; they should form as many well defined streets and squares as possible in the form of a multi-directional horizontal pattern of urban spaces.*

Orientation

Stübben recommends north–south orientation for rectangular blocks in order to reduce north exposure to the smallest facade and to have east–west exposure for most facades.

The most inspired contribution in relation to orientation is Cerda's 45° rotation of the Ensanche grid in Barcelona in a northeast–southwest and southeast–northwest direction, thus avoiding any north facades. Furthermore, each facade is reached by the sun both in summer and winter.

The dialectic of building block and urban space

The building block, 'insula', 'pâté de maison' or 'ilôt', 'Haüser-Block', must be identified as the most important typological element in the composition of urban spaces, the key element of any urban pattern. It belongs to a European tradition of building cities in the form of streets and squares. As a typo-

logically fixed element it can generate urban space but it can also remain undefined and merely result from the order of an urban pattern (of streets and squares) [Figure 18].

The three diagrams (types of urban space) describe the three possible dialectical connections of building block and public space. These three polemical categories have all participated in the formation of the European City, either following each other chronologically or overlaying and transforming each other in the process. They hardly ever occur as exclusive systems but complement each other to form a highly differentiated urban environment.

The building block is either the instrument to form streets and squares or it results from a pattern of streets and squares.

But before coming to its specific urban characteristics, an *insula* has to be defined in a more general territorial and geographical sense. The block is primarily a plot of land defined all around by a multitude of planned and unplanned paths, roads and streets. This is as true for the very large geographical blocks (including agricultural land, forests, mountains) as it is true for urban blocks.

Though the rural block need not be of any specific size, I want to stress that urban blocks ought to have well defined qualities of *size, volume, orientation, typology, order and complexity* in order to become *urban*. Although the size and nature of urban blocks vary enormously, I want to define a very limited range of principles not only for analysis but as a basis of urban design philosophy [Figures 19, 20 and 21].

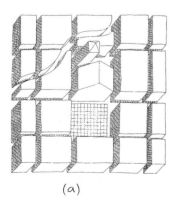
(a)

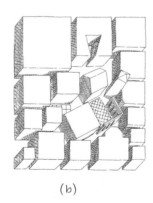
(b)

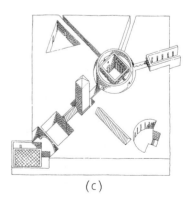
(c)

Figure 18 Three types of urban space.
(a) The urban blocks are the result of a pattern of streets and squares. The pattern is typologically classifiable.
(b) The pattern of streets and squares is the result of the position of the blocks. The blocks are typologically classifiable.
(c) The streets and squares are precise formal types. These public rooms are typologically classifiable.

Figure 19 Geographical and urban blocks.

The size of a building block

In the European city, the smallest and typologically most complex building blocks are to be found in the urban centres. They tend to grow larger and typologically simpler towards the periphery before finally dissolving into single free-standing objects. This tendency is more obvious where the sections of the centre and periphery correspond to different times of construction (pre-industrial and industrial). One can conclude that: *small blocks are the result of the maximum exploitation of urban ground caused by great density of activities, high cost of urban ground, etc; and that a great number of streets on a relatively small area correspond to the maximum length of commercial facade.*

If the main cause for small urban blocks and for a dense urban pattern is primarily economic, it is this very same reason which has created the intimate character of a highly urban environment. Such an environment is the basis of urban culture, of intense *social*, *cultural* and *economic* exchange. If this hypothesis is true, the opposite is also true, i.e. suburban or peripheral areas or city extensions are generally characterised by vast urban blocks (Berlin, Barcelona, etc). In pre-industrial cities the outer ring of urban blocks often included agricultural land, fields, large gardens. In the case of city extensions due to low cost of land, the blocks of the periphery often included large gardens, municipal parks, etc.

The high density large block

As the vast blocks of the periphery became more and more part of urban centres, the gardens were built up with residential premises or artisans' workshops. It is this internal densification and exploitation of the urban blocks which lead to their final destruction and the savage criticisms by Le Corbusier and Gropius.

Building block, form of property, form of street

The traditional *insula*, formed by an addition of urban houses, is characterised at ground level by a great number of entrances. The street is used not only as a space of distribution and orientation but as a space of economic and social exchange. There is a strict relationship between building type, form of property and the form of the public space, the street.

The twentieth-century perimeter block is still able to form streets and squares but it tends more and more to become an autonomous organism with its own system of distribution, corridors, 'rues

Figure 20 The illustration clearly shows the hypothesis that in urban centers the blocks are smallest and that they grow larger towards the periphery where they often contain large gardens and fields.

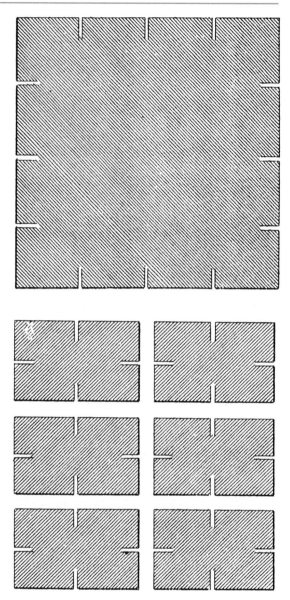

Figure 21 Small urban blocks increase public frontage and accessibility.

intérieures', access balconies, all competing with the streets.

The number of entrances on the street is not dependent any more on the number of residential units contained in the block. The relationship between building type and street becomes dictated purely by external legislations about fire, etc. The street is reduced purely to the function of access. The tarmac becomes more important than the public space [Figure 22].

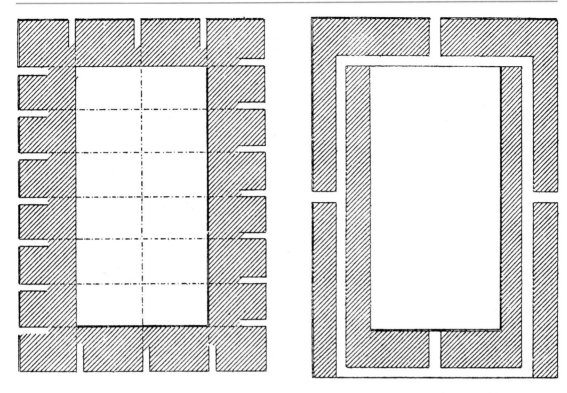

Figure 22 In modern blocks an internal corridor system competes with the main function of the street, a meeting place of public and private realms.

The limits of the perimeter block

Perimeter blocks tend by their very nature to be very large, including gardens and even parks. However useful or beautiful they might be as isolated examples, as places of quiet green (Bahnhofstrasse, Zurich), if they are understood as a repetitive system, the street pattern they form becomes a spatial megastructure which is socially disruptive. The tendency to design these huge blocks into single architectural objects with one door is the very cause of their institutional barrack-like character (Karl Marx Hof, etc.). Their courts usually degenerate as they require too much servicing. The large perimeter block was only the last step in the dissolution of the urban fabric.

The street and the scale of the building block

We have shown that towards the periphery of the city the blocks not only tend to grow larger but they are also generally separated by wider and longer streets; this phenomenon seems to control both the development of planned city extensions and of incremental urban growth.

It is, however, not true that in the traditional city centres wide streets are necessarily lined by large urban blocks. If a street is to be important and lively within a multidirectional urban pattern, it has to be drained by as many streets as possible. In the history of modern urban planning there seems, however, to have existed an almost 'natural' tendency to front large urban spaces with large urban blocks.

High urban density and the modern critique of the building block

High density and increased exploitation of urban ground have been wrongly identified as being responsible for the inhuman condition of the nineteenth-century city. Instead the badly lit light-wells, the polluted streets and the endless corridorial spaces inside the vast blocks were in fact the result of a *wrong typological choice: the large urban block.*

One could easily demonstrate that even higher densities can be reached with smaller blocks without the disadvantage of light-wells and badly lit courts. Certain central areas of Manhattan or the 'Spanish Quarter' in Naples are good examples. The savage attacks of Le Corbusier and Gropius in the 1920s against the nineteenth-century block, an attack which psychologically prepared for the destruction of the traditional European city, used a global critique whereas only a technical criticism would have been necessary [Figure 23].

The nineteenth-century institutional building block as functional and social labyrinth

For quite different reasons from those mentioned in connection with residential blocks, the nineteenth-century institutional buildings formed blocks of unprecedented size. The Palace of Justice in Brussels, the Hospitals, the British Museum etc., often reach the size of a whole parish or an entire urban district. These institutional monoliths formed veritable labyrinthine islands within the fine structure of streets and squares and they contributed to the explosion of the social and physical fabric of the traditional city.

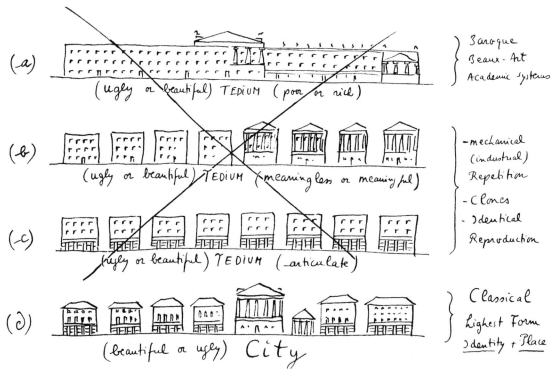

Figure 23 (a) As architectures became reduced to stylistic systems in the nineteenth century, it was capable of articulating virtually any large mass of building that resulted from functional zoning and concentration. Neither housing nor administration offices were socially complex and rich enough to suggest high artistic results over a long period of time. Architecture, as purely theatrical decor, could then be abandoned all too easily.

(b) A first articulation of an amorphous functional mass occurs by reducing it to smaller blocks.

(c) Functional mix is the basis for the articulation of an urban fabric into semi-public and private functions.

(d) This pattern of solids (blocks) and voids (streets) becomes articulated into public buildings (monuments) and urban fabric.

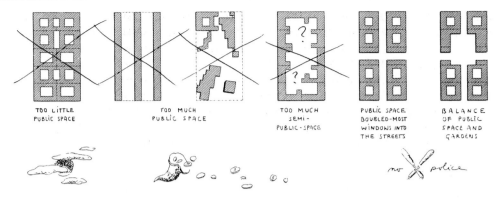

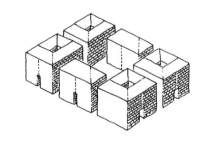

Figure 24 The Berlin Block was not wrong but its measures were wrong. The apartment on the street was not wrong but that on the court was wrong. The length of the street is not the problem but the length of the block is the problem. Car traffic should remain in the existing street system. The oversized blocks should be broken down into small blocks by means of pedestrian streets and squares.

Like castles, abbeys or palaces in the pre-industrial cities, they formed secluded organisms, using their own privatised system of distribution, corridors, cloisters, balconies, courts, etc. These semi-public 'rues intérieures' became in Kafka the symbolic spaces of institutional repression, the building masses themselves became the symbols of usurped political and cultural power.

These buildings contain an alternative distributive system competing with the traditional street. The number of doors opening to the street is minimal, reducing therefore the street (the public space) to a mere access route. We find here a first and definite break in the dialectic between building type and type of public space [Figures 24 and 25].

In religious abbeys or royal palaces this seclusion from the city was quite conscious. Religious or aristocratic life isolated itself from the trading and manufacturing city. Instead, the design of institutional monsters was not at all a typological necessity. It had rather to do with the architectural representation of the new bourgeois power which, in the construction of extravagantly vast and overwhelming structures, symbolised its own aims.

We have shown that both the size of the nineteenth-century residential block and the twentieth-century perimeter blocks were not dictated by a typological necessity. The rooms of the Palace of Justice in Rome and of the Karl Marx Hof in Vienna could have been distributed without problems into a multitude of smaller urban blocks

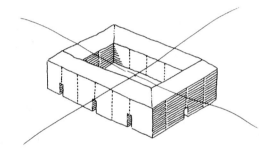

Figure 25 Insula Tegeliensis.

organised by public streets and squares and these could have found a close dialectical relationship with the existing city.

It is the centralisation of functions (of political and cultural power) and land which have resulted in a specific typological choice based on large building programmes. Historically, larger and larger building programmes have resulted in larger and larger buildings and this tendency reaches its apogee today in synthetic megastructures (the whole city has become one big programme, i.e. one building). The Berlin University, the comprehensive schools, Milton Keynes Town Centre etc., are characteristic examples of this tendency. Today we pretend that the specific typological choice of megastructural conglomerates against a multitude of building types is a stylistic one. In fact, a millennial culture of urban building types and spaces with precise measures is swept away and sacrificed to an obsession with *building systems* apparently designed to solve all the problems of the city by means of an industrialised kit of parts. The unavoidable result is the destruction of time and place, of Architecture and the City.

"The Third Typology"
from *Oppositions 7* (1976)

Anthony Vidler

Editors' Introduction

Prior to industrialization, modernism, and the rise of self-conscious design, cities and their buildings were derived largely from intuitive experiences, traditions, local culture, and available resources. In general, new forms arose slowly in an evolutionary and incremental manner as needs dictated. In contrast, designers of the Modern Movement transcended cultural and local resource constraints to provide solutions to social problems through innovative, expressionistic and often revolutionary approaches to urban form. Modernist designers and planners provided a radical break with the traditional urban forms of the pre-industrial world, focusing instead on the creation of rationalized spaces and buildings. Arising in Europe in the 1960s, approximately at the same time as the US freeway revolts and Jane Jacobs' call for planning and design reform, a number of designers began questioning the modernist break with traditional urbanism, suggesting a need to reintroduce precedent in urban form-making. Rather than the modernist need for self-expression, creative destruction, and the "new, new thing," these empirically-based designers and researchers began questioning the results of modernist design practice. Emanating particularly from Europe, architects such as Aldo Rossi, Carlo Aymonino, Robert and Leon Krier noted the strange contrasts and disunities of modernist design in contexts of older traditional cities, suggesting instead a return to more eternal urban forms. These "neo-empiricists" argued that historic form types had proven themselves over time and could be used to create designs that were socially satisfying, more legible, and blended more harmoniously with traditional urban settings.

Theories of typology can be traced back to concepts of Platonic ideal form and to the Enlightenment practice of botanical categorization and encylopedic method. In the late eighteenth century, Quatremère de Quincy provided an early definition of "type," which he suggested was more a rule for creating form, rather than a visual model of what was to be created. Developing from this early encyclopedic understanding, the practice of typology requires classification of physical form elements into groups based on varied factors, such as geometry, use, period of time, symbolism, construction type, style, etc. This type of form analysis has become useful in understanding prevalent form patterns within specified geographic domains, as well as providing a starting point for design interventions. Designers often use typology as a kit of tools to be referenced when they come upon a project site that emulates conditions similar to where the original referent was found – then adapting the lesson of the original type to the new site condition. Typological method in design has been particularly useful for the Italian School of urban morphology, described by Anne Vernez Moudon in the following reading.

This short reading by Anthony Vidler, from *Oppositions* 7, discusses the rise of urban form typologies and introduces a new typology based on the existential form of the city. Vidler describes two earlier typologies that were architecturally-based and provided legitimization for design practices of the day. A first typology is found in the Enlightenment rationalization of neo-classical architecture through nature, described by Marc Antoine Laugier in his notion of the primitive hut. A second typology rationalizes modernist architectural practice through processes of mechanization and mass-production, as can be found in Le Corbusier's notion of the "machine for living." Both of these typologies utilize external rationales to validate the architectural design advocacy of their time. In describing a new third typology, Vidler eschews external validations of typology, opting instead for one that is internally self-referential – the city itself. This new typology arises for many of the same reasons described above: the desire for continuity in urban form, the importance of function in form-making, and as a critique of an unsatisfactory modernism. It is predicated by the large-scale, structural elements of the city (the street, avenue, square, and arcade – and perhaps infrastructure and background buildings as well), rather than merely architectural objects found in the city. In describing this new typology, the author cautions against its use in nostalgic or historical replication, suggesting that types drawn from the city should be interpreted within our current time and local contexts. In many ways, Vidler's third typology can be read as advocacy for empirically-based research in urban form and a contemporary urban design practice associated with the importance of public space and everyday urbanism.

Anthony Vidler is Dean of the Cooper Union School of Architecture in New York City. He is a historian who has largely focused on French architectural history since the Enlightenment, and is an avid critic of contemporary and modern architecture and urbanism. His publications include: "The Idea of Type: The Transformation of the Academic Ideal 1750–1830," in *Oppositions 8* (Spring 1977), *The Writing on the Walls: Architectural Theory in the Late Enlightenment* (New York: Princeton Architectural Press, 1987), *Claude-Nicolas Ledoux: Architecture and Social Reform at the End of the Ancien Régime* (Cambridge, MA: MIT Press, 1990), *The Architectural Uncanny: Essays in the Modern Unhomely* (Cambridge, MA: MIT Press, 1992), and *Warped Space: Art, Architecture and Anxiety in Modern Culture* (Cambridge, MA: MIT Press, 2000).

Primary works on typology include: Giulio Carlo Argan, "On the Typology of Architecture," *Architectural Design* (no. 33, December 1963); Alan Colquhoun, "Typology and Design Method," *Arena: Journal of the Architectural Association* (vol. 83, no. 913, 1967); Aldo Rossi, *The Architecture of the City* (Cambridge, MA: MIT Press, 1982, original in Italian 1966); Rafael Moneo, "On Typology," *Oppositions 13* (Summer 1978); Leon Krier, *Rational Architecture* (Brussels: Archives d'Architecture Moderne, 1978); Rob Krier, *Urban Space* (New York: Rizzoli, 1979); Piergiorgio Gerosa, "Architectonic Elements of the Urban Typology," *Lotus International* (no. 24, 1979); Karen A. Franck and Lynda H. Schneekloth (eds), *Ordering Space: Types in Architecture and Design* (New York: Van Nostrand Reinhold, 1994); Micha Bandini, "Typology as a Form of Convention," *Architectural Association Files* (no. 6, 1984). Texts which describe typology and its use in practice include: Geoffrey Broadbent, *Emerging Concepts in Urban Space Design* (London: Van Nostrand Reinhold International, 1990); and Douglas Kelbaugh, *Repairing the American Metropolis: Common Place Revisited* (Seattle, WA: Washington University Press, 2002).

From the middle of the eighteenth century, two distinct typologies have informed the production of architecture.

The first, developed out of the rationalist philosophy of the Enlightenment, and initially formulated by the Abbé Laugier, proposed that a natural basis for design was to be found in the model of the primitive hut. The second, growing out of the need to confront the question of mass production at the end of the nineteenth century, and most clearly stated by Le Corbusier, proposed that the model of architectural design should be founded in the production process itself. Both typologies were firm in their belief that rational science, and later technological production, embodied the most progressive "forms" of the age, and that the mission of architecture was to conform to, and perhaps even master these forms as the agent of progress.

With the current questioning of the premises of the Modern Movement, there has been a renewed interest in the forms and fabric of pre-industrial cities, which again raises the issue of typology in architecture. From Aldo Rossi's transformations of the formal structure and typical institutions of the eighteenth-century city, to the sketches of the brothers [Leon and Rob] Krier that recall the primitive types of the Enlightenment *philosophes*, rapidly multiplying examples suggest the emergence of a new, third typology.

We might characterize the fundamental attribute of this third typology as an espousal, not of an abstract nature, nor of a technological utopia, but rather of the traditional city as the locus of its concern. The city, that is, provides the material for classification, and the forms of its artifacts provide the basis for re-composition. This third typology, like the first two, is clearly based on reason and classification as its guiding principles and thus differs markedly from those latter-day romanticisms of "townscape" and "strip-city" that have been proposed as replacements for Modern Movement urbanism since the fifties.

Nevertheless, a closer scrutiny reveals that the idea of type held by the eighteenth-century rationalists was of a very different order from that of the early modernists and that the third typology now emerging is radically different from both.

The celebrated "primitive hut" of Laugier, paradigm of the first typology, was founded on a belief in the rational order of nature; the origin of each architectural element was natural; the chain that linked the column to the hut to the city was parallel to the chain that linked the natural world; and the primary geometries favored for the combination of type-elements were seen as expressive of the underlying form of nature beneath its surface appearance.

While the early Modern Movement also made an appeal to nature, it did so more as an analogy than as an ontological premise. It referred especially to the newly developing nature of the machine. This second typology of architecture was now equivalent to the typology of mass production objects (subject themselves to a quasi-Darwinian law of the selection of the fittest). The link established between the column, the house-type, and the city was seen as analogous to the pyramid of production from the smallest tool to the most complex machine, and the primary geometrical forms of the new architecture were seen as the most appropriate for machine tooling.

In these two typologies, architecture, made by man, was being compared and legitimized by another "nature" outside itself. In the third typology, as exemplified in the work of the new Rationalists, however, there is no such attempt at validation. The columns, houses, and urban spaces, while linked in an unbreakable chain of continuity, refer only to their own nature as architectural elements, and their geometries are neither scientific nor technical but essentially architectural. It is clear that the nature referred to in these recent designs is no more nor less than the nature of the city itself, emptied of specific social content from any particular time and allowed to speak simply of its own formal condition.

This concept of the city as the site of a new typology is evidently born of a desire to stress the continuity of form and history against the fragmentation produced by the elemental, institutional, and mechanistic typologies of the recent past. The city is considered as a whole, its past and present revealed in its physical structure. It is in itself and of itself a new typology. This typology is not built up out of separate elements, nor assembled out of objects classified according to use, social ideology, or technical characteristics: it stands complete and ready to be de-composed into fragments. These fragments do not re-invent institutional type-forms nor repeat past typological forms: they are selected and reassembled according to criteria derived from three levels of meaning – the first, inherited from meanings ascribed by the past existence of the forms; the second, derived from choice of the specific fragment and its boundaries, which often cross between previous types; the third, proposed by a re-composition of these fragments in a new context.

Such an "ontology of the city" is indeed radical. It denies all the social utopian and progressively positivist definitions of architecture for the last two hundred years. No longer is architecture a realm that has to relate to a hypothesized "society" in order to be conceived and understood; no longer does "architecture write history" in the sense of particularizing a specific social condition in a specific time or place. The need to speak of function, of social mores – of anything, that is, beyond the nature of architectural form itself – is removed. At this point, as Victor Hugo realized so presciently in the 1830s, communication through the printed word, and lately through the mass media has released architecture from the role of "social book" into its specialized domain.

This does not of course mean that architecture in this sense no longer performs any function, no longer satisfies any need beyond the whim of an "art for art's sake" designer, but simply that the principal conditions for the invention of object and environments do not necessarily have to include a unitary statement of fit between form and use. Here it is that the adoption of the city as the site for the identification of the architectural typology becomes crucial. In the accumulated experience of the city, its public spaces and institutional forms, a typology can be understood that defies a one-to-one reading of function, but which, at the same time, ensures a relation at another level to a continuing tradition of city life. The distinguishing characteristic of the new ontology beyond the specifically formal aspect is that the city, as opposed to the single column, the hut-house, or the useful machine, is and always has been political in its essence. The fragmentation and re-composition of its spatial and institutional forms thereby can never be separated from the political implications.

When a series of typical forms are selected from the past of a city, they do not come, however dismembered, deprived of their original political and social meaning. The original sense of the form, the layers of accrued implication deposited by time and human experience cannot be lightly brushed away; and certainly it is not the intention of the Rationalists to disinfect their types in this way. Rather, the carried meanings of these types may be used to provide a key to their newly invested meanings. The technique, or rather the fundamental compositional method suggested by the Rationalists is the transformation of selected types – partial or whole – into entirely new entities that draw their communicative power and potential critical force from the understanding of this transformation. The City Hall project for Trieste by Aldo Rossi, for example, has been rightly understood to refer, among other evocations in its complex form, to the image of a late eighteenth-century prison. In the period of the first formalization of this type, as [Giovanni Battista] Piranesi demonstrated, it was possible to see in *prison* a powerfully comprehensive image of the dilemma of society itself, poised between a disintegrating religious faith and a materialist reason. Now, Rossi, in ascribing to the city hall (itself a recognizable type in the nineteenth century) the affect of prison, attains a new level of signification, which evidently is a reference to the ambiguous condition of civic government. In the formulation, the two types are not merged: indeed, city hall has been replaced by open arcade standing in contradiction on prison. The dialectic

is clear as a fable: the society that understands the reference to prison will still have need of the reminder, while at the very point that the image finally loses all meaning, the society will either have become entirely prison, or, perhaps, its opposite. The metaphoric opposition deployed in this example can be traced in many of Rossi's schemes and in the work of the Rationalists as a whole, not only in institutional form but also in the spaces of the city.

This new typology is explicitly critical of the Modern Movement; it utilizes the clarity of the eighteenth-century city to rebuke the fragmentation, de-centralization, and formal disintegration introduced into contemporary urban life by the zoning techniques and technological advances of the twenties. While the Modern Movement found its hell in the closed, cramped, and insalubrious quarters of the old industrial cities, and its Eden in the uninterrupted sea of sunlit space filled with greenery – a city become a garden – the new typology as a critique of modern urbanism raises the continuous fabric, the clear distinction between public and private marked by the walls of street and square, to the level of principle. Its nightmare is the isolated building set in an undifferentiated park. The heroes of this new typology are therefore to be found not among the nostalgic, anti-city utopians of the nineteenth century nor among the critics of industrial and technical progress of the twentieth, but rather among those who, as the professional servants of urban life, direct their design skills to solving the questions of avenue, arcade, street and square, park and house, institution and equipment in a continuous typology of elements that together coheres with past fabric and present intervention to make one comprehensible experience of the city.

For this typology, there is no clear set of rules for the transformations and their objects, nor any polemically defined set of historical precedents. Nor should there be; the continued vitality of this architectural practice rests in its essential engagement with the precise demands of the present and not in any holistic mythicization of the past. It refuses any "nostalgia" in its evocations of history, except to give its restorations sharper focus; it refuses all unitary descriptions of the social meaning of form, recognizing the specious quality of any single ascription of social order to an architectural order; it finally refuses all eclecticism, resolutely filtering its "quotations" through the lens of a modernist aesthetic. In this sense, it is an entirely modern movement, and one that places its faith in the essentially public nature of all architecture, as against the increasingly private visions of romantic individualists in the last decade. In it, the city and typology are reasserted as the only possible bases for the restoration of a critical role to an architecture otherwise assassinated by the apparently endless cycle of production and consumption.

"Getting to Know the Built Landscape: Typomorphology"

from *Ordering Space: Types in Architecture and Design* (1994)

Anne Vernez Moudon

Editors' Introduction

While many urban designers and planners are currently working with the concept of type as a means of understanding urban places and developing design models, typomorphological theory is less well developed and used in North America than in Europe. The following review of the three European schools of thought on typomorphology by Anne Vernez Moudon, excerpted from Karen A. Franck and Lynda H. Schneekloth (eds), *Ordering Space: Types in Architecture and Design*, analyzes their contributions to knowledge about built urban form. The Italian School, which began in the 1940s in response to the failures of modern architecture to fit in with the built fabric of existing Italian cities, is primarily concerned with looking for traditional typomorphological patterns that can inform current design approaches so that the new will harmonize with the old. The Versailles School uses typomorphological methods to develop socio-critical critiques of cities and how they evolved over time. The English or Conzenean School – referring to its central figure, geographer M.R.G. Conzen – has no prescriptive motivation. Its focus is strictly on research, analyzing and describing urban form and explaining how it came to be.

Anne Vernez Moudon teaches urban design and planning at the University of Washington, Seattle. Her book *Built for Change: Neighborhood Architecture in San Francisco* (Cambridge, MA: MIT Press, 1986), which presents a thorough analysis of the evolution and development of physical form of San Francisco's Alamo Square neighborhood from the mid-nineteenth century through the 1980s, provides an excellent example of how the Conzenean method of typomorphological analysis can be used for urban design research. Moudon has also written a number of papers on typomorphological theory besides the one reprinted here, including "Urban Morphology as an Emerging Interdisciplinary Field," *Urban Morphology* (vol. 1, pp. 3–10, 1997).

For additional writings on typomorphology that speak to the use of type in a range of disciplines, in both research and theory, look to the many other excellent essays in Franck and Schneekloth's *Ordering Space: Types in Architecture and Design*. Another edited volume that contains essays focused on the typomorphology of suburban places is Kiril Stanilov and Brenda Case Scheer (eds), Suburban *Form: An International Perspective* (New York: Routledge, 2004).

An early book on typomorphology that comes from the architectural field is Aldo Rossi's highly theoretical book *The Architecture of the City*, a translation by Diane Ghirardo and Joan Ockman of *L'Architettura della Citta* (Cambridge, MA: MIT Press, 1982), which analyzes the history and meaning of cities using typomorphological methods. More recently, architect Douglas S. Kelbaugh, who is both a New Urbanist and a critical regionalist and currently serves as Dean of the Taubman School of Architecture and Urban Planning at the University of Michigan, has written an evocative book that challenges the current state of architecture and urban design education and practice, *Repairing the American Metropolis: Common Place Revisited* (Seattle,

WA: University of Washington Press, 2002). The book contains an excellent chapter called "Typology: An Architecture of Limits" that speaks of the use of typology in architecture and urban design practice, arguing that it provides an enduring, universal code of urban design that can help link the uniqueness of local place to the larger world of human culture. John Ellis' article "Explaining Residential Density Places," *Places* (vol. 15. no. 2, pp. 34–43, 2004), shows how the concept of type is useful for understanding potential residential densities associated with different building forms, and hence a method for developing urban design framework plans. Another article that theorizes the use of typomorphology in the landscape architecture field is Katherine Crew and Anne Forsyth, "LandSCAPES: A Typology of Approaches to Landscape Architecture," *Landscape Journal* (vol. 22, no. 1, pp. 37–53, 2003). The Urban Morphology Research Group has compiled a wealth of information on the Conzenean typomorphological method and its website provides up-to-date information on research linking urban morphology and planning practice (www.gees.bham.ac.uk/research/umrg/). The group also publishes the journal *Urban Morphology*, which highlights international research and debates on the subject. Further resources on typomorphology can be found in the extensive bibliography included at the end of this reading.

The concept of type is in good currency in the fields of planning and design in North America: streets, buildings, open spaces, neighborhoods, etc., are commonly organized in classes.[1] Yet the theories framing the nature, purpose, and applications of type in these fields remain vague and flawed with ambiguity. The definition and use of type to characterize urban form, its building, and open spaces are particularly weak; most rely on functional or aesthetic criteria (Moudon 1987). In a strident critique of the use of type in North American architecture, Bandini called typological work a collection of "easily appropriated icons" – a potpourri of images of buildings randomly selected by architects who find them inspiring (Bandini 1984, 81). This apparent shallowness contrasts with the numerous and complex definitions of urban form and building type that have been debated and refined in Europe for several centuries (Goode 1992; Tice 1993). Clearly, serious gaps in interpretation have occurred as the concept is transported from one continent to the next, translated from one language and culture to others, and transformed from discipline to discipline. These gaps characterize a state of affairs that this chapter begins to unveil. The focus is on typomorphology, an area of study by European architects and geographers which now spans the past four decades.

Typomorphological studies reveal the physical and spatial structure of cities. They are typological and morphological because they describe urban form (morphology) based on detailed classifications of buildings and open spaces by type (typology). Typomorphology is the study of urban form derived from studies of typical spaces and structures. Typomorphology is an unusual approach to urban form. First, it considers all scales of the built landscape, from the small room or garden to the large urbanized area. Second, it characterizes urban form as a dynamic and continuously changing entity immersed in a dialectic relationship with its producers and inhabitants. Hence, it stipulates that city form can only be understood as it is produced over time. Typomorphology accounts for what Italian urbanist Saverio Muratori called an "operational history of urban form," because it is a record of actions taken by planners, designers, and builders, both lay and professional, as they mold city form (Muratori 1959; Muratori *et al.* 1963). Typomorphology offers a working definition of space and building types, and serves as a rich launching ground for studying the nature of building design, its relationship to the city, and to the society in which it takes place.

A typomorphological approach to defining type differs from other approaches in three ways. First, type in typomorphology combines the volumetric characteristics of built structures with their related open spaces to define a built landscape type.[2] This approach is in opposition to the monumental, siteless typology of Durand, for instance. The element that links built spaces to open spaces is the lot or parcel, the basic cell of the urban fabric. Second, the inclusion of land and its subdivisions as a constituent element of type makes land the link between the building scale and the city scale. Third, the built landscape type is a morphogenetic, not a morphological, unit because it is defined by time – the time of its conception, production, use, or mutation.

This chapter reviews the work of three schools of thought on typomorphology which I have identified and researched following my own work *Built for Change* (Moudon 1986).[3] Centered in Italy, in France, and in England, these three schools have generated lively debates among students of the built landscape with architects, planners, sociologists, geographers, and others participating. For the most part, these disciplines and professions in North America have ignored or misinterpreted the deliberations on typomorphology in Europe and England.

The typomorphological schools of thought make different contributions to knowledge of the built landscape. They address different disciplinary and cultural issues and use different methods of inquiry. Until recently, the schools have had little contact with each other (Choay and Merlin 1986; Whitehand and Larkham 1992). Together, however, these schools elaborate the exciting beginnings of a scholarly approach to the built landscape which complements established design research. They outline a way of learning how cities are produced and built that can support the further development of design and planning theory.

MURATORI AND CANIGGIA IN ITALY

In Italy, typomorphological studies began in the 1940s at the instigation of Saverio Muratori (1910–1973), an architect who was profoundly disturbed by the devastating effects of modern architecture on existing habitats and cities. Muratori and his principal follower, Gianfranco Caniggia (1933–1987), analyzed the city building process in traditional Italian towns, making this analysis the foundation for a theory of design. Their analyses rest on extensive classifications of buildings and related open spaces extending from their original state to their various mutations over time. Muratori's and Caniggia's work had a major impact on design theory and practice in Italy and, indirectly, on the use of building types in architectural design in North America.

Muratori

Saverio Muratori saw that the roots of architecture lie not in the fantastic projections of the modernists, but within the more continuous tradition of city building which prevailed from antiquity until the 1930s. Teaching at the University of Venice in the 1950s, and then at the University of Rome after 1964, Muratori made the morphological study of existing cities a first, mandatory step in his architectural design studios. As a philosopher, researcher, and practitioner, he is recognized as the early pioneer of the typomorphological trend in Italian architecture, and the spiritual father of such well-known architects as Aldo Rossi and Carlo Aymonino. Muratori's course syllabus soon became seminal for Italian architects who, to this date, see urban morphological analysis as a necessary preparatory step for design (Muratori [1959] 1985). He also published two extensive "operational histories," one of the city of Venice and the other of Rome (Muratori 1959; Muratori *et al.* 1963).

For Muratori, the structure of cities could only be understood historically, with building typology as the basis of urban analysis. Urban form and structure, he stipulated, are an aggregate of many ideas, choices, and actions which are manifested in given buildings and their surrounding spaces (gardens, streets, etc.). These buildings and spaces, called *edilizia* in Italian and loosely translated as the built landscape, can be classified by type, which summarizes the essence of their Character. These different types become a *tipologia edilizia*, or a typology of buildings and related open spaces, which defines the essence of the building fabric.

Muratori's early emphasis on the typological process as the tool to understand city building explains why, in recent years, ideas and debates about building typologies have been developed more fully in Italy than anywhere else (Gerosa 1986). Unfortunately, however, much of the interesting polemic following Muratori's legacy has been lost to non-Italian audiences. Specifically, the elaborate work of Gianfranco Caniggia, one of Muratori's early assistants and the principal heir to the Muratorian tradition, remains little known outside of Italy. And even there, it has been kept out of the limelight for reasons that will be discussed in the next section.[4]

Caniggia

Gianfranco Caniggia first published an operational history of the city of Como in 1963, *Lettura di una città: Como*, with an introduction by Muratori. The research for the book had been conducted in Muratori's Centro Studi di Storia Urbanistica (Caniggia [1963] 1984b). Caniggia subsequently carried out numerous empirical studies of cities in Italy, Sicily, North Africa, and northern Europe, often in collaboration with other planners and architects and as preambles to preservation efforts.[5] However, Caniggia's is the work of an architect, not a historian. His own publications seek not to document the historical process of a city's development, but to isolate the fundamental principles of city making (Caniggia 1984a, [1976] 1985; Caniggia and Maffei 1979). They are meant to teach these principles to guide the identification of the elements and rules that mark genesis and then the transformation of city fabric.

Caniggia explains the human environment as made of "built objects," all related one to the other. He identifies built objects at four different scales: the building (*edificio*), the group of buildings (*tessuto* or building fabric), the city (*città*), and the region (*territorio*).[6] Each object is described as a complex entity made of elements, structures, systems, and organisms. Thus the built environment is an organism made of components that are themselves organisms. Caniggia stresses the modularity of the environment (how objects fit one into the other) and its scalar dimension (how objects – organisms at one scale fit into objects at other scales) as two important principles of the structure of the environment. Objects relate one to the other, and must be understood in relation to other objects at different scales. All built objects that are affected by planning and design activity must be studied from the scale of the single building to the scale of the territory within which buildings are set [Figure 1].

Caniggia stands out in the group of typomorphologists introduced here because he clearly states that the physical city is not an object but a process: cities are built incrementally with many small elements being juxtaposed. An understanding of the formation and transformation of cities is guided by the analysis of the mutation of the type through both time and space. For him as an architect, the analysis of urban form proceeds from the small to the large elements of the environment (Caniggia and Maffei 1979, 57–74, 122–65).

Caniggia, like Muratori, does not use the word *morphology*, because, in his theoretical construct of the human environment, urban form per se is *not* an object of study. Instead, he calls himself a "*tipologo*," because he believes that the establishment of *procedural typologies* (*tipologia processuale*) is the basis for understanding the making and hence the design of the city and its architecture. He defines *type* as the *conceptual* existence of an object in the form of the "experience of this object," apart from its physical existence or its phenomenological being ("experience" meaning cultural experience, and *not* the individual experience of an existential nature which is a more commonly used definition in Anglo-Saxon cultures).

Procedural typologies can be defined at all scales of the human environment: for buildings and their ancillary spaces (*edilizia*), the urban fabric, the city, and the territory. Caniggia focused on the scale of the *edilizia*.[7] There, a base type is identified in terms of its volumetric characteristics, its position relative to the street, and its solar orientation. The base type is then reviewed over time for possible mutations or adaptations. The type is therefore defined in formal terms, in terms of its relation to scales above and below, and in terms of its evolution over time. Most types of buildings in Italy have roots in the Etruscan or Roman cities, and their mutations are reconstructed through medieval times. Caniggia identified the elementary Roman *domus* as the base type which evolved into a courtyard house, then into a row house, and finally into a linear house.

Focusing on the processes by which cities are made, Caniggia portrays an extremely dynamic picture of the built world, whose production is the result of a dialectic, or an active relationship, between human action and "environmental reaction." According to him, this human action is directed either by a "spontaneous conscience" (*coscienza spotanea*), which is an immediate understanding of what is necessary to make a building, or by a "critical conscience" (*coscienza critica*), which is a self-conscious thought process guiding the building activity which may not refer to cultural heritage. The spontaneous conscience yields *basic structures* (read: vernacular, common houses), while the critical conscience leads to *specialized structures*

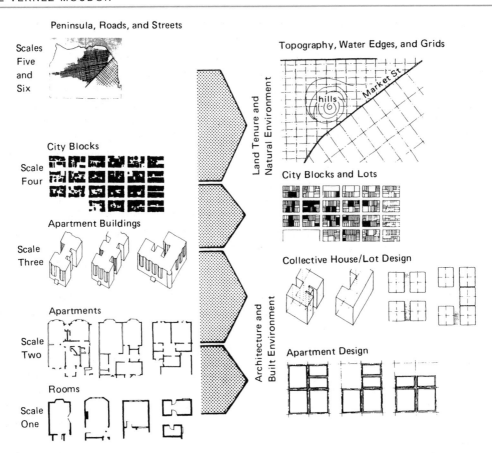

Figure 1 Modularity in the built landscape (source: A.V. Moudon 1986, *Built for Change: Neighborhood Architecture in San Francisco*, MIT Press, p. 124). This diagram shows how a typical turn-of-the-century apartment building in San Francisco fits into its host fabric. Reading from the bottom up: Rooms are grouped to form apartments, which are then grouped to form the apartment building; the land subdivision pattern organizes the position of buildings within the block; blocks fit into the city according to the layout of the streets; and the network of streets fits into the landscape.

(read: monuments) (Caniggia and Maffei 1979, 39–57).

Debates surrounding the Muratorian School

The relative obscurity of the Muratorian School beyond Italian borders contrasts with the immense influence it has had on an entire generation of architects who became internationally known. It was Muratori who led Rossi, Aymonino, Scolari, Gregotti, and others to the historical city as a source of knowledge and inspiration. Muratori's condemnation of the modernist city was an early subject of research by architects Aymonino (Aymonino *et al.* 1966; Aymonino 1976) and Rossi (1981, [1966] 1982). They established that the modernist and the traditional city differed in at least two areas: in the ways individual buildings related to the city as a whole, and in the ways individual buildings were designed. (Interestingly, however, none of the Italian typomorphologists analyzed the modernist city systematically.)

Rossi continued Muratori's argument against buildings designed to respond directly and solely to programmatic needs, advocating instead a formal composition of space based on materials and on generic functions and related spatial needs. Rossi's principal concern was to demonstrate the power of what he called the autonomy of architecture. Elaborating on Muratori's case against functionalism,[8] he claimed that built forms are themselves embodiments of people and their societies, and therefore can be understood, and ultimately shaped, outside of the realm of the social sciences.[9]

Aymonino shed light on what he termed the "reversed" relationship between building and city which modernism introduced. Explaining how the existence of the city was based on a dialectical relationship between building typology and urban morphology, he noted how the compact building types of the medieval city are the "servants" of urban form – pieces of space defining a collective fabric. As the modern city develops, however, new building types emerge that are largely independent of urban form (e.g., theaters, libraries). In the modern city, he claimed, the relationship between typology and morphology has been reversed, with building types defining individual environments that do not serve a collective urban form, such as malls and office parks (Aymonino 1976).

Aymonino's and Rossi's work clearly empathized with Muratori's and Caniggia's thinking. However, these famous students did part from their master in their interpretation of the crisis of modernism. Aymonino and his colleagues accepted the reversed relationship between building and city as part of an irreversible change in the socioeconomic forces that shaped the city. Muratori and Caniggia, on the other hand, saw it as an aberration, a temporary crisis in the way cities are produced. This difference in interpretation led to a parallel, yet irreconcilable, difference in the way urban analysis related to the development of a design theory. If, according to Aymonino and his colleagues, the relationship between building and city has been broken in the contemporary city, then the analysis of the traditional city can no longer inform the design of new buildings. But if, according to Muratori and Caniggia, the traditional relationship between building and city must be restored in the contemporary city, then the design of new buildings must rely on the analysis of the traditional city. This disagreement generated an intense debate on the nature of building typology and its value to architectural design and theory. The basic question became: can there be and should there be any continuity between existing and new building types?

Building typology and design theory

Historian Guilio Carlo Argan (1965) structured the debate by highlighting what he identified as the two "moments" in the design process: (1) the *typological moment*, when the rules of design and building used in the past (and thus yielding types which have been called *a posteriori*) are identified and understood, and (2) the *moment of invention*, when the artist answers the historical and cultural questions through a critical approach (yielding so-called *a priori* types). Muratori and Caniggia scorned a priori building types as arbitrary inventions by architects; they believed that the architect's creative work must be harnessed by common building traditions. But Aymonino, Rossi, and others thought that designers, in creating anew, were free to interpret the historical city as they wished. Justifying the architect's freedom from past conventions, Aymonino wrote:

> [U]rban analysis does not provide a structure for architectural intervention. In fact, it is wrong to assume a direct relationship of cause and effect between the two: this leads to the academic embalming of architecture, shown clearly in the projects of Muratori's and his School.
> (Aymonino 1976, 176)[10]

In contrast, Muratori and later Caniggia defined architectural design intervention as conditioned by what they call preexisting structures. These include the existing built environment as well as the building traditions and living practices which shaped it. Caniggia specifically stated that the architect is a *technician* organizing the human environment (*tecnico della structurazione del spazio antropico*). As a technician, the architect must fit his work into the growth and transformation processes that take place in any city, and witness the dialectic between buildings and their fabric. He believed that architects and planners need to overcome the crisis of modern architecture through a critical examination of the process of formation and transformation of the human environment. This critical examination cannot be based superficially on style and experience, but must rely on knowledge of the historical processes shaping urban form.

These distinct positions lead to two radically different approaches to design theory: one that rests entirely on the history of city building and its analysis, and the other that is defined solely by the architect, and which may or may not borrow from this history.

So far, in Italy and in other parts of Europe, the strict disciplinarian doctrine which Muratori and Caniggia advocated and practiced has been less popular in design circles than the liberal stand of Aymonino and his colleagues. The commercial success of the designs of Rossi and Gregotti have no doubt precipitated this trend. Today, Muratorian urban analyses are performed by designers primarily as a predesign exercise for sensing the logic and tradition of the site. But only in cases of preservation projects do urban analyses have an actual impact on the designs proposed.

Interpretation of the Italian work in North America

The intricacies and subtleties of the Italian discourse never reached North America. Early reviews of the work sidestepped the heart of the debate. Historian Anthony Vidler and architect Rafael Moneo focused on the use of building typology in architecture.[11] They did not dwell on the relationship between building types and urban form. Nor did they discuss the tension between analysis and design and the two moments of the design process described by Argan (Vidler 1976; Moneo 1978). Vidler pointed to three stages in the definition of typology which culminated with Aldo Rossi's writings.[12] He saw Rossi's primary contribution as having designed building types that were no longer based on concepts of functional organization (which the French School calls the abstract plan types of the modernist approach), but on actual constructions found within the traditional city fabric (which the French call consecrated types).

Concentrating on the downfall of modernism and interested in the consequences of neorationalist proposals for architectural design, Vidler was particularly curious about replacing the functionally-based building types of the moderns with form-specific types of traditional buildings. Moneo was less impressed with what he called functionally indifferent building types, and complained that

the Italian work emphasized the attributes of urban form and "reduced" typological studies to the field of urban analysis (Moneo 1978, 35–36). Thus by limiting their inquiries to the architectural scale, these writers missed an opportunity to introduce the breadth of typomorphological studies to the Anglo-Saxon world and to begin exploring the relationships between buildings and cities in this context.

The subsequent notoriety of Rossi's *The Architecture of the City* (published in English in 1982, 16 years after its publication in Italian, and six years after Vidler's discussion of this work) also contributed in oversimplifying the typomorphological debate. In spite of its provocative views, Rossi's book remains a personal statement about understanding the city through its architecture. *The Architecture of the City* principally influenced architects in English-speaking countries and generated only curiosity about the relationship between buildings and cities; it did not demonstrate convincingly the value of systematic urban analysis for urban design. And by the 1980s, Rossi's projects and drawings had become more prominent in architectural circles than the theoretical underpinnings first described in the book (Moudon 1987).

The plan and implementation strategy of the City of Bologna's restoration work did capture the attention of the few North American architects and planners with community development interests (Cervellati *et al.* 1977; Comune di Bologna 1979). The project was the labor of Italian architects who collaborated with Caniggia and hence operated within the theoretical tenets of a typomorphological approach and beyond the particular case study. However, the impact of this work remained small, limited as it was by the perceived uniqueness of the city, and its particular social and historical heritage.

The legacy

The most important contribution of the Muratorian School lies in its attempt to build a theory of design based on traditional processes of city building. It reads city form as a historical settlement process, a territorial conquest to control space with materials and building techniques. The research identifies basic organisms (elements and processes) that underlie the formation and transformation of the built landscape. It recognizes that sociopolitical forces shape the design and production of cities and act as a framework within which architects and planners must work. The approach is based on the notion of a dynamic relationship between human action and environmental reaction which matches in an interesting way the one used in studies of person – environment relations in English-speaking countries.

Muratori's and Caniggia's primary publications serve as textbooks for architecture students to read and analyze the city building process before they begin the design process. Caniggia's texts are synthetic and abstract, centered on the typological process as a tool to record the mutation of a base type of *edilizia*, the smallest element of the built landscape, over time. The typological process therefore becomes a link between analysis and design: as types of buildings and territories are shown to have permeated centuries of urbanization, they are proven to be generic and therefore must be continued in contemporary design.

While Muratori is increasingly recognized as the father of typomorphology, his work as well as Caniggia's remains little known outside of Italy. In Italy itself, the work has been trivialized in many ways by architects who have treated the traditional building of the city as an anachronism. A few young historians of the city are emerging, however, whose research is based on Caniggia's teachings. Gian Luigi Maffei and Paolo Maretto have published challenging histories of the building of Florence, Venice, and Genoa which add a new, scholarly dimension to Caniggia's work (Maffei 1990a, 1990b; Maretto 1986). These exemplary books illustrate the power of applying the typomorphological approach to the history of cities.

CONZEN AND THE URBAN MORPHOLOGY RESEARCH GROUP IN ENGLAND

M.R.G. Conzen's work is available in English, and hence accessible to readers of this volume. However, because its significance has yet to be fully appreciated in either geographical or design and planning circles, the work needs to be an integral part of this chapter.[13] Conzen's contribution is especially important in the context of typomorphology

because it excludes the prescriptive dimension of planning and design which underlies the Italian and French work. The focus is strictly on research intended to describe, analyze, and explain how urban form is made.[14] As a geographer, the freedom Conzen gained from not having to concern himself directly with the future city and its design has allowed him to concentrate fully on studying the actual city, the processes for building it, and on developing methods for analyzing it. As a result, his approach offers the most thorough, detailed, and systematic typomorphological method of the three schools.

Conzenean philosophy and method

M.R.G. Conzen first studied cultural geography at the Geographical Institute of the University of Berlin, where urban morphology became a subject of study in the late nineteenth century (Whitehand 1981). He later trained as a town planner in England, where he practiced as such until he accepted an academic position in geography at the University of Newcastle upon Tyne.

Conzen's townscape is a palimpsest of society and culture on which features of particular periods stand out while others are obliterated over time. His empirical research has focused primarily on the reading of the town plan. However, he describes his complete method as three pronged, to include the *town plan* (primarily a two-dimensional cartographic representation of a town's physical layout), the *building fabric* (made of buildings and related open spaces), and the pattern of *land and building utilization* (detailed land use) (Conzen 1968, 113–16). All three analytical components are interrelated genetically and functionally. The corresponding documents needed to explain urban form include: the town plan, the distribution plan of urban building types, and the distribution plan of urban land uses. Conzen's work itself has concentrated almost exclusively on the study of the town plan. In spite of representing a town in only two dimensions, the town plan embodies, for all intents and purposes, all the essential characteristics of urban form.

In an approach he calls *town-plan analysis*, Conzen identifies three fundamental elements of the town plan: the streets, the plots, and the buildings, which all fit one into the other as a precise puzzle. Caniggia, and later the Versailles School, also use the town plan and its elements in their research, yet Conzen's clear identification of the plan and of its basic elements as analytical tools sets an important point of departure for typomorphological analysis.

According to Conzen, the town plan is to be analyzed over time in an evolutionary fashion. The fundamental unit of analysis is the individual plot. It is the basic element of the pattern of land subdivision and acts as an organizational grid for the urban form. Conzen further introduces the concept of *compositeness* of the town plan to describe the variations in the forms, uses, and configurations found in different parts of the city. The composite town plan is made of different units called *plan units*, which are best noted in the variations typically found in street, lot, building size, and shape. Thus the different plan units are due to differences in the socioeconomic roots of the settings as well as to the different periods of building. Plan units contribute to the stratification of the townscape, *stratification* meaning literally storage into layers, the formation and deposit into strata.

The definition of the plan unit as a unique combination of types of street patterns, buildings, and lot configurations is also an important contribution. In Conzenean terms, the plan unit itself identifies a type of what Caniggia calls the urban fabric (Caniggia has not, however, spelled out clearly the characteristics of its components). Conzen and Caniggia's research thus become complementary, with Caniggia providing an approach to the definition of building types and Conzen to the types of urban fabrics.

Conzen's own studies focus primarily on medieval towns, and they reach a climax in the analysis of the town of Alnwick, Northumberland (Conzen 1960), which covers the origin of the city and its growth and transformation until the twentieth century. The study illustrates Conzen's methodological contributions. Regional soil structure, ancient road network, the old town's site, topography, and surrounding field structure all explain the town's layout. Urbs, suburbs, and original plot structure – still readily visible in today's fabric – are reconstructed as well. At the center of the analysis is the formation of the *burgage*, the basic plot of land that is narrow and deep. A detailed study of a burgage along one of Alnwick's streets illustrates many of the transformations that are apparent in subsequent studies of other medieval towns.

Conzen also introduces the concepts of *market colonization*, or the gradual development of the original open-air marketplace at the center of town, and *fringe belt*, a zone of atypical buildings and land uses on either side of a town's walls. These concepts encapsulate phenomena that can be found in other cities in other times. Fringe belts are common occurrences around areas of intense development such as contemporary downtowns, and market colonization is visible today in many commercial malls. These phenomena occur at the scale of the plan unit because they engender special types of urban fabrics.

The Urban Morphology Research Group

Following Conzen's research contributions, several historical geographers in the 1980s formed the Urban Morphology Research Group at the University of Birmingham. The Group's mission is to conduct research in urban morphology and to integrate it with more traditional concerns in the field of geography. It has also worked to facilitate access to M.R.G. Conzen's writings and graphic studies that have not been widely distributed.

Individuals in the Urban Morphology Research Group have different specialties. T.R. Slater's focus is closest to Conzen's in its emphasis on the town-plan analysis of medieval towns (Slater 1987). J.W.R. Whitehand is concerned about the effects of the building and development industries on urban form (Whitehand 1987, 1992). His prolific writings on the fringe belt and building cycle concepts rely on the identification of transformation of building types – the mutations of existing types or the emergence of completely new types.[15] He and P.J. Larkham are now turning to the study of suburban areas, thus testing Conzenean methods on more recent urban forms [Figures 2 and 3]. P.J. Larkham has applied the method to preservation projects. He and others have assembled a glossary of terms used in Conzenean analysis which illustrate the group's commitment to morphological study (Jones and Larkham 1991).

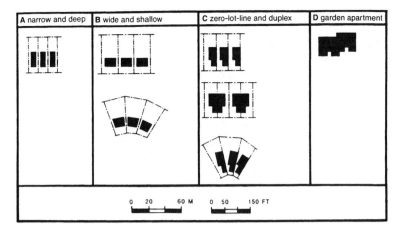

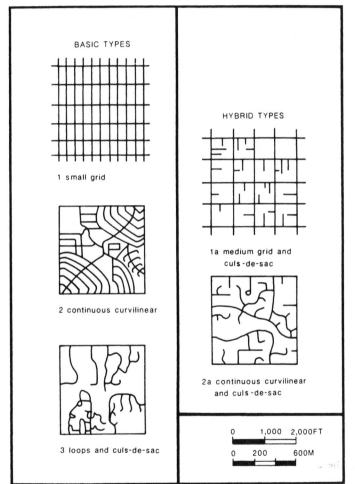

Figure 2 Elements of U.S. suburban residential forms: houses, lots, and streets (source: A.V. Moudon 1992b), "The Evolution of Twentieth-Century Residential Forms: An American Case Study." In *Urban Landscapes: An International Perspective*, eds. J.W.R. Whitehand and P.J. Larkham, Routledge, pp. 173–6). These illustrations show two levels of resolution in the built landscape. The simple lines and shapes outlining houses, lots, and streets illustrate a low level of specificity in describing the types.
(a) Houses and lots (at top).
(b) Street pattern (at bottom).

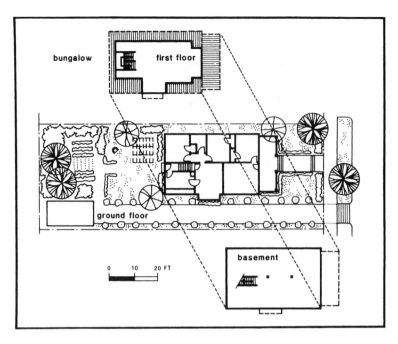

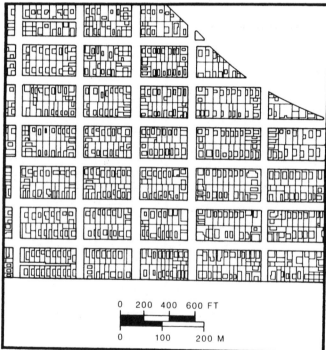

Figure 3 Elements of U.S. suburban residential forms: plan units (source: A.V. Moudon 1992b, "The Evolution of Twentieth-Century Residential Forms: An American Case Study." In *Urban Landscapes: An International Perspective*, eds. J.W.R. Whitehand and P.J. Larkham, Routledge, pp. 182, 185).

(a) Plan unit and house plan typical of suburban residential development until the 1930s (above). It integrates Street Type One (small grid) and House Type A (narrow and deep) shown in Figure 2.

(b) Plan unit and house plan typical of development between the 1930s and the 1960s (next page). It combines Street Type Two (continuous curvilinear) and House Type B (wide and shallow). A higher level of specificity in defining types is used than in Figure 2; double lines describe the width of streets, and details of the material quality of buildings and related open spaces area included.

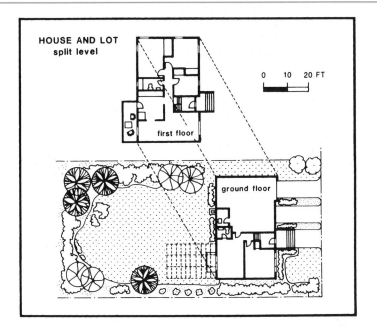

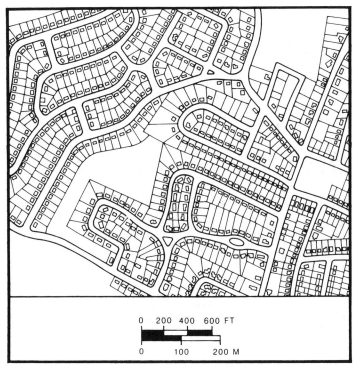

Figure 3 (*cont'd*)

International and interdisciplinary outreach

To broaden the scope of the Conzenean approach, and, in so doing, to affirm the importance of studying urban morphology, the Birmingham Group is seeking to expand the number of towns studied, to extend research to more recent cities, and to pursue cross-cultural comparisons (Whitehand 1988). This outreach program, if continued, would assemble material on the variety of extant building, space, and urban fabric types and would be the first international and longitudinal data base on the city building process. It would be rich ground for research and would further strengthen the links between morphological research and planning and conservation practices (Slater 1984).

T.R. Slater (1990) has edited a book, *The Built Form of Western Cities*, which includes analyses of industrial towns, and makes several references to research in Italy and the United States.[16] A chapter by M.P. Conzen reports on comparative studies of nineteenth-century American towns, using some of the concepts developed by his father. Discussing the nature of the morphology of these towns, M.P. Conzen reiterates the importance of the cadastre and the building fabric in understanding the town plan. He notes how little detailed empirical work has been done on town morphology in the United States: the few studies of extant building types (notable exceptions including Kniffen's work [Upton and Vlach 1986]) have generally been eclipsed by the more popular, but amorphological work on the spatial structure of urban *land uses* (see also Conzen 1980).

J.W.R. Whitehand and P.J. Larkham (1992) recently edited a second international volume, *Urban Landscapes: An International Perspective*. A chapter by D. Holdsworth reconstructs the development of office buildings in downtown Manhattan using computer simulation techniques. My own chapter offers a typology of U.S. suburban residential form, identifying basic house and street types as well as suburban plan units.

Finally, a doctoral thesis sponsored by the Group compares Conzen's method with the work of Caniggia at the scales of the building and the urban fabric (Kropf 1993). Beyond the obvious importance of making parts of Caniggia's contribution accessible to English-speaking readers, the thesis makes methodological headway in the definition of type. Kropf clarifies the distinction between levels of *resolution* (the different scales which are clearly recognizable in the built landscape) and levels of *specificity* (the different levels of detail at which type can be defined). For instance, elements such as streets, buildings, and open spaces are at one scale or level of resolution and plan units or urban fabrics are at another. Types of streets can be established at different levels of specificity. For instance, street width and block size may be the characteristics used to differentiate one type of street from another, or those characteristics plus the number of vehicular lanes, arborization, drainage, etc. could be used to identify the types. Kropf introduces the notion of *outline* as a tool for defining type in the built landscape. Building types are commonly identified by their graphic outline, as are most other elements such as rooms, streets, yards, lots, and so forth. Outlining appears to be a standard means of describing various types of spatial elements in the built landscape.

The legacy

Conzen's approach has been called morphogenetic rather than morphological because it stresses not only the elemental structure of the city but its temporal dimension and its evolution. Morphogenesis and the morphogenetic approach are more accurate terms for describing the methods used than typomorphology. They are accepted in geography (Vance 1977, 1990).

Conzen's methodological contribution lies in the strength of the town-plan analysis, the definition of its elements and plan units. It confirms and clarifies the work of French and Italian typomorphologists. Their methods and findings being similar, they begin to define a systematic way to describe the built landscape. Recent efforts to expand the scope of cities studied and to spur comparative work all begin to consolidate a bona fide field of morphogenetic analysis of the built landscape which promises to provide practical applications in city planning and design. So far, however, assessments of Conzen's work by the few urban designers and planners who know it remain mixed. They lament the work's thoroughness, and question its direct usefulness to design beyond the management of the historic urban landscape (Samuels 1988, 1990; Bandini 1988, 1992).

The Birmingham Group clearly is looking for applications of the morphogenetic approach which transcend historic landscapes and address general issues of what they term "townscape management," an activity akin to, yet different from, urban planning. With its emphasis on managing the existing city according to its historic evolution, townscape management is the city planning equivalent of adaptive reuse of buildings. The Conzenean approach begins to provide an analytical basis for facilities management planning, which is itself a growing subfield of city planning.

THE VERSAILLES SCHOOL IN FRANCE

The Versailles School of Architecture emerged from the widespread institutional reform that took place after the students' and workers' riots in 1968.[17] The school followed the Muratorian philosophy which had preceded it, believing that modernism had created an unmendable break from the past and that the roots of architecture had to be rediscovered in past traditions. However, the French work emerged in a special intellectual climate. Whereas debates in Italy and in England involved, respectively, architects and geographers, in France, sociologists, historians, geographers, and planners all worked together with architects to achieve an improved understanding of the city. The resulting approach to typomorphology not only is oriented to issues of design and geography but also can incorporate literary and social science perspectives. In this sense, the Versailles School stands between the Italian and the British schools, and addresses issues of both design and the city-building process.

Intellectual climate contributing to the formation of the school

The work of the Versailles School is part of France's long history of applying typological study to architectural design. Quatremère de Quincy, Abbé Laugier, and Durand were the first to experiment with architectural types. French hegemony in the field of urban geography and the legacy of a Lavedan and a Poëte left important marks in the design community as well.[18] The Cartesian thinking necessary for good classification still remains ingrained in the culture. But the relationship between building types and urban form was not established in France until the early 1970s.

French intellectuals of the 1960s became highly critical of the institutions and professions responsible for the reconstruction of the war-damaged country. A policy of massive housing production based on selected aspects of modern design theories devastated the French urban landscape, perhaps more so than anywhere else in Europe. Twenty years after the end of World War II, thousands of HLM (*habitations à loyer modéré*) grouped in so-called satellite towns on the periphery of cities, and of Paris in particular. Sociologist-philosopher Henri Lefebvre was strongest in condemning the focus on housing production, with all its paraphernalia of efficiency and pseudoscience, as destructive of French social practices (Lefebvre 1968, 1970).[19] Lefebvre was first to claim that appropriation, or the domination of material space including the city itself, was the ultimate goal of social life. He argued that contemporary construction and house production methods crushed people's natural instincts for appropriation and weakened the relationship between people and their environments.[20]

Lefebvre influenced many students, particularly architects and urbanists who turned to the traditional city for theoretical inspiration. Among them were Jean Castex (an architect), Philippe Panerai (an architect-urbanist), and Jean-Charles Depaule (a sociologist) who constituted the original core of the Versailles School of Architecture.[21] Lefebvre's teachings fostered interdisciplinary work and a *rapprochement* with the social sciences, and encouraged the search for a socially responsive and responsible architecture.

Work in urban history also influenced urban morphology at the time of the 1968 reforms. Historian André Chastel and his team headed by Françoise Boudon were the first to focus on how ordinary buildings are built and rebuilt over long periods of time (Boudon *et al.* 1977). Subsequent research in the provinces as well as in Paris continues this tradition (see, for instance, *Typologie opérationnelle de l'habitat ancien* 1979; Fortier 1986).

LADRHAUS: a dual purpose

The Versailles team's work now spans two decades of uninterrupted research and includes four books, as well as studies of many cities, and critical essays on urban design and practice. The original group of researchers expanded and formed LADRHAUS (Laboratoire de recherche: Histoire architecturale et urbaine-Sociétés or Research Laboratory: Architectural and Urban History – Societies).[22] The

French work is broader than the Muratorian and the Conzenean schools' in terms of both the subjects studied and the methods used. Of the four books produced by the group, one is a critical analysis of the roots and effects of the modern movement in the recent history of city building (Castex *et al.* 1977). This critique relies on the comparative study of carefully selected projects tracing the evolution of urban form from traditional, pre-nineteenth century street-and-block architecture to the straight, linelike architecture of the modern movement.[23] Two other books focus on individual cases studies: the City of Versailles (Castex *et al.* 1980) and the Bastides new towns (Divorne *et al.* 1985). These studies are explicit applications of typomorphological analysis. One book is a compendium of philosophical and methodological issues related to typomorphology (Panerai *et al.* 1980).

This published work is historical and descriptive, and thus in the same vein as Conzen's. Case studies rest on the explicit documentation of the evolution of typical buildings and their corresponding fabrics, as well as on analyses of their social history. The work is different from Muratori's and Caniggia's who, in their more direct search for a prescriptive design theory to set future design activity in the proper direction, could forgo explicitness in their descriptive work. Hence, in comparison with the French work, the early studies of Venice, Rome, and Como read as designers' reconstitutions of the city building process. Drawings are personalized, chronologies missing, and explanatory texts remain vague in their historical reference and laced with abstract theoretical design discussions. Indeed, in most of their publications, Muratori and Caniggia used their case study research to identify the basic principles and rules which, in their minds, were most useful as natural guides for the design of the future city.[24]

The French research, however, is also motivated by the need to identify the ingredients of good city design. Like Muratori and Caniggia, Castex and Panerai teach and periodically practice architecture and urban design. Hence the research addresses issues of urban design, particularly in the face of modernity and the urban crisis. This preoccupation is apparent in the identification of *architectural models*, defined as basic concepts governing the organization of urban space, which starts in the first book, *Formes urbaines: De l'îlot à la barre* (*Urban Forms: From the Block to the Slab*) (Castex *et al.* 1977)

and continues in the case studies of the Bastides and Versailles.

The dual purpose of descriptive research and identification of design models permeates all of the French work and adds complexity to the field of typomorphology. It calls for the development of an applied discipline to study the city as a physical entity – or, what is often called the city as "architecture."[25] And it demands that lessons be drawn from this discipline to serve the practice of urban design – to assess the effectiveness and the impact of different design approaches and theories on the city and urban life.

The attempt to treat typomorphology as a new and separate discipline, an eminently modern stand, contrasts with the more reflective and personal writings of Muratori's and Caniggia's. It also differs from Conzen's work; as a social scientist he could relate directly to the existing fields of geomorphology and cultural and urban geography. The Versailles School had to justify a new discipline in the light of other, established disciplines. And it had to prove its relevance to the practice of design to satisfy the design and planning professions. So, on one hand, the book *Eléments d'analyse urbaine* (*Elements of Urban Analysis*) (Panerai *et al.* 1980) stipulates that the knowledge derived from urban analyses enhances the ability to describe and discuss the city as a sociophysical phenomenon, and thus sets the design of the city within the broad, multidisciplinary intellectual framework of the humanities and the social sciences. And on the other hand, the case study research is carefully targeted to critique design theory in the context of how cities have been built.

Outlining a discipline for understanding the city and its design

The Versailles team is aware that theirs is the first attempt in France to document how "architecture" fares as a discipline in the analysis of the "urban crisis" already well documented in philosophy, sociology, psychology, and economics. Although their quest parallels Rossi's argument for the autonomy of architecture, their stand is less polemical than exploratory, relying on the close ties with Lefebvre, who early on advocated the need to know material space as well as the people inhabiting it.

The multidisciplinary background of the Versailles team members allows them to recognize that the city can be read in many ways, including the architectural way, even if it has yet to gain approval as a legitimate route to understanding the city.[26] Confronting these questions, *Eléments* (Panerai *et al.* 1980) is the principal work that engages directly with issues of historiography as well as methods in the social sciences.

In this work, the relationship between built space and social space is described as a dialectic between urban form and social action (discussed by Caniggia as well, but in less detail). Identifying built space as conceptually separate from social space, the authors explain how physical space is assumed, invested in, qualified, named, and eventually "practiced" by people in everyday life (much like a musical instrument, it would seem). They argue that while physical space has its own logic and organization, which are uncovered by morphological analysis, it becomes real and takes on meaning through social action. There are also discussions of the nearly identified phenomenon of the consumption of space which has inspired a number of research projects since then (Croizé, *et al.* 1991).

The study of material space, as it evolves and changes over time, introduces a social dimension to an otherwise static object. Social forces are embedded in the changes recorded in built space. For example, houses that are typical of a given period are inevitably transformed over time to respond to social change. Transformed house types in turn reflect the social forces at work. Thus the historical dimension of typomorphological studies insures the definition of a built landscape that integrates material space with the social forces that produce it.

Eléments critiques traditional macroscopic approaches in geography because they divide the city artificially into suburbs, center city, and fringes, and focus only on major, dominant land uses, disregarding the smaller scale at which the built landscape is actually produced and experienced.

An entire chapter in *Eléments* acknowledges the positive early influence of both Sitte's and Lynch's approaches. Called *picturesque analyses* (evidently because they are based on perception and firsthand experience), these approaches are thought to complement the understanding of urban form. With the exception of M.P. Conzen who, in a 1978 article, discusses the parallel existence of the objective (material) and subjective (perceived, experienced) structure of form without favoring one over the other, no other typomorphologist surveyed has discussed explicitly the possible interrelationships between these two analytical approaches. [27]

Method of typological analysis

In *Eléments*, the methodological components of typological analysis are framed within the historical evolution of the method and includes recent Italian work. The reader is exposed to the different ways of establishing types, whereas both Conzen and Caniggia promote only their own. A type is defined as an "abstract object built through analysis" that reproduces the properties that are deemed essential by the analyst of a family of *real* objects. Second, building classification systems can be used to two different ends: to seek *exemplary* specimens or to define *groups* or *families* of similar specimens. The identification of groups of similar specimens yields elements that are common to all (e.g., a California bungalow), while exemplars represent outstanding specimens within the groups (e.g., a house by architects Greene and Greene).

Completing the historical argument started by Vidler, the Versailles School notes that modern classification techniques date from the Enlightenment when the natural sciences embarked on systematic observations of the plant and animal worlds.[28] The first industrial revolution then brings the Encyclopedists and, among them, Quatremère de Quincy, who first made the important distinction between the type as a *model* to be replicated, and the type as a *rule* to be followed. The differences between a posteriori and a priori types are stressed. While early classifications of buildings and parts of buildings are descriptive, resting primarily on formal and stylistic criteria, by the end of the eighteenth century the French *polytechnicien* J.N.L. Durand proposed building typologies that are both *descriptive* of the characteristics of extant buildings and spaces and *analytical* or critical of these characteristics.[29] In what constitutes a further breakthrough, Durand's typologies become *generative*: guiding the reinterpretation of building types described and applying the concepts to other sites and contexts.[30] For the first time in history also, buildings are conceived as separate from their site and context. Durand emerges as an eminently modern thinker (a point made less clearly by Vidler in his "Third Typology" [Vidler 1976]).

The Versailles School identifies two categories of building types in use today. There are *consecrated* types of buildings that can be found repeatedly in various periods of history, such as Roman villas and cathedrals. They correspond to Vidler's first and third typologies (the archetypes and the traditional urban types). These types are a mix of basic functional programs and specific spatial configurations. And there are *typical plans*, Vidler's second typology (the prototypes). Trademarks of modernism, typical plans are standards or norms meant to guide replication, related not to tradition but to future production. Consecrated types include not only vernacular settings (called *architecture banale*), but also high-style architecture (called *architecture savante*). Consecrated types thus can be monumental, but they differ from typical plans in that they always relate to the fabric of the city. Furthermore, they are form specific and often functionally indifferent (as per Moneo 1978).[31]

As illustrated by the work of Durand, the move from consecrated types to typical plans or standards occurred gradually, over a long period of profound changes in the practice of architecture and building. It included an enlargement in typological scale which has been particularly significant since the nineteenth century, and is evident in the emergence of mass-produced terraced buildings in England and large public buildings in France following the French Revolution.

The process of defining types is addressed, albeit succinctly. It includes four steps. The first step is the choice of the scale at which the analysis will be conducted. The level likely to be the most appropriate for architectural design is the building or the parcel. Another level includes the group of buildings and related parcels, as, for instance, the city block or group of blocks (this is similar to Caniggia's *tessuto* and elaborated by Conzen's concept of *plan units*). The choice of level or scale of the typological analysis will necessarily limit the scope of the study.

A second step is the classification of building types, which involves the selection of criteria on which the typological process rests – for example,

volume, function, architectural style, etc. The classification process that follows is the result of trial and error usually based on comparisons and analyses of analogues. A third step elaborates on the tools available for refining the classification process: exemplars, rules, and variations are introduced as concepts that support the analogous and comparative classification process. And a final step relates one type to the other, thus generating a *typology*.

A critical history: the other side of design theory

Review of the hand-picked case studies suggests a new approach to design theory. The Italians debated the relationship between typomorphological analysis and design theory, whereas the French critique the history of design theory. Whether they followed the Muratorian tradition or not, Italian architects generally shared a dialectical view that opposed the traditional to the modern city. When they asked what the contemporary city can and should be, and what architects and urban designers should do, the answer was either continuity or discontinuity between past and present. Although sparked by the same angst that the Italians experienced about the mission and role of the architect, French researchers differ from the Italians in that they identify many different kinds of traditional cities – as for instance, the Bastides as planted towns and Versailles as a new town with both monumental and traditional characteristics. As a result, they do not consider modernism only in opposition to the traditional city. Modernism is not a temporary state of crisis, but a set of new design principles that have gradually infiltrated the city-building process over a relatively long period of time. For the Versailles School, the present is not a complete break from the past, and the past offers several different models for the future. In this sense, the French work does not associate issues of continuity or discontinuity in the built landscape with past and future. Since both states have existed in the past, both are likely to be possible in the future.

The differences between the Italian and the French contributions can now be illustrated simply by building on Argan's argument. The Italians only distinguished between a posteriori and a priori types, the former representing the traditional way of making the city and the latter being primarily the concoction of elite designers to shape the future. The French argue that there exist types which today are a posteriori but originated as a priori types. They reflect explicit, elite theories, as for instance the residential tower. These types thus represent discontinuities which occurred in the past. They must not only be included in urban analysis, but they must be evaluated for their relative effectiveness. This pluralistic view complicates the study of history: it demands that city building be studied along with the history of design theory. And it demands that the history of design theory be not only operational, as Muratori claimed, but also *critical.*

While the history of design theory is a well-developed subject in Italy, the focus on the history of urban design theory is particularly strong in France thanks to the work of Françoise Choay (1965, 1980). However, whereas Choay's interest lies in the history of consciously articulated ideas and concepts about the city, the Versailles School focuses on the history of applications of design theory. Thus the critical history of design theory has itself two dimensions: the history of *design theory as ideas* (e.g., one can refer here to the Athens Chartes or to the Cité Radieuse as the ideals and principles of modernist design) and the history of *design theory as practiced* (e.g., the case of the Unité d'habitation, or any of the new towns built according to the modernist principles).[32]

The Versailles School studies theories that are culturally defined and theories that are elite driven (for instance, the theories behind the design of the Bastides and the popular neighborhoods of Versailles, versus those used in the monumental Versailles of the king and his court). These two different origins of theories generate different architectures. One is ordinary, the aforementioned *architecture banale* or the architecture of everyday life, and the other is scholarly, *architecture savante*, or high-style architecture. They deplore the fact that elite architectures all tend to sever their relationship to the city and to become monumental, a phenomenon that they recognize in the study of the City of Versailles as well as in the study of the emergence of the modern movement (Castex *et al.* 1977).

The particular cases studied show that good models used to design the city oscillate back and forth between the need to control and provide order in city design and the need to create environments that respond to the needs and actions of their immediate inhabitants. This puts in question the value of a global composition of the city (an underlying concern and general direction in the evolution of urban design theory), proposing instead an emerging definition of city form through the incremental acts of many people.[33] The search for formal models which allow this incremental, participatory process of designing the city points to simple urban blocks subdivided into several parcels. Street hierarchies work well as long as the superblocks created are further subdivided into autonomous blocks with clear, legible public access and their own sets of independent parcels.

The legacy

The Versailles School favors a separate discipline for studying the built landscape that serves to evaluate design theory. The novel aspect of this stand forces the School to discuss methods and philosophy in a multidisciplinary context, which neither the Conzenean nor the Muratorian schools had to do. The development of an applied discipline paves the way for a systematic approach to design evaluation. Also, the simultaneous investigation of traditional and elite city-building processes invites a critical review of design theory in light of its actual achievements. The work recognizes the need for mixing tradition and innovation in the way cities are designed, and for keeping monumentality under control.

The Versailles work has taken solid roots in both design practice and research in francophone countries. Typological and morphological investigations are fully integrated into the growing discourse on the built landscape and its design. However, for all its outreach into the disciplines involved in the urban crisis, and in spite of its multidisciplinary origins, the Versailles work has made slow progress in the field of urban planning – a field that is separate from architecture in post-World War II France. When the Institut d'urbanisme of the University of Paris undertook a major research project on urban morphology and its applications to planning in 1985, it did not include any of the Versailles faculty, even though architects from other countries were invited to contribute to the project (Choay and

Merlin 1986). Since then, however, both Panerai and Castex have been teaching urban history and morphology at the Institut, and the final edition of the Institut's research on morphology contains further references to the Conzenean School (Merlin *et al.* 1988).

CONCLUSION

The three schools of typomorphology offer an intellectually challenging framework for thinking about the built landscape within the historical context of the city. Italy's provides a theoretical foundation for planning and design within age-old traditions of city building. England's offers a scholarly approach to researching how the built landscape is produced. And France's outlines a new discipline that combines the study of the built landscape with a critical assessment of design theory. Together these schools suggest an order for a formidable agenda of research, planning, and design that takes into account the relationships between space, time, habitat, and culture. In this order, type provides the essential conceptual framework for understanding the built landscape and intervening in it.

All three schools claim that the built landscape must be understood in three fundamental dimensions: time, form, and scale. The built landscape is in a constant state of evolution and change, subject to sociocultural forces constructing, using, and transforming space. So all typological work must be linked to a measure of time. Built and open spaces together constitute form. They are persistent; they dominate the definition of the built landscape as use and function come and go according to changing social practices and related needs. Since elements of form are highly sensitive to sociocultural forces operating over time, they are morphogenetic rather morphological. And several scales permeate the structure of the built landscape from the inhabited room to the city as a whole, and the block and district in between.

Together these three dimensions of time, form, and scale weave an intricate web of relationships between fields and disciplines which all too often remain separate. A focus on the formal dimension of the built landscape facilitates linkages between analysis and design, linkages that are tenuous when urban analyses address primarily economic or social dimensions. Yet, by the same token, the time dimension insures that form remains linked to sociocultural and historical forces. The marriage of space and time is the marriage of architecture and history, and architecture and the social sciences as advocated by Porter and Tigerman (1992). And the scalar dimension of the built landscape demands the integration of architectural and city planning approaches (Goode 1992).

Debates about typomorphology in the three schools illuminate the use of type in design theory. The schools differentiate between descriptive, analytical, explicative critical, and generative types. They are therefore able to separate conceptually the description, analysis, and critique of the historical and the existing city from the projection of the future city. They can learn to know the built landscape, to explain it, and to theorize about its production without worrying about its future design. The three schools provide the tools to monitor the emergence of new types and to relate them to theory, whether it is tradition-bound and culturally defined or consciously articulated. And they can evaluate the actual effects of past design theories on the existing built landscape.

The intellectual framework sketched out by the three schools is propitious for research and teaching about the built landscape (Moudon [1995]). Further, this material offers a basis for what the Birmingham Group defines as townscape management. Managing the built landscape is an ongoing process that includes planning, designing, and construction as continuous tasks performed by many different actors. A typomorphological approach yields a data base on the built landscape that can be used by various public entities charged with maintaining, upgrading, and modifying it. Public regulatory and capital improvement agencies responsible for urban planning and design, public works, transportation, parks and open space, housing, and community development need to work together to build on the wealth in urban infrastructure and amenities already in place. A shared data base can inform and guide future intervention.

This intellectual framework should also prove useful to such practitioners as Daniel Solomon (1992), Andres Duany and Elizabeth Plater-Zyberk, Peter Calthorpe (Katz 1994), Stanford Eckstut, John Kriken, and others who, in an intuitive way, have come to believe that solutions to good community design lie within the broad context of making the city. Their town plans, street and land subdivision layouts, and building codes as architectural strategies to balance community and individual needs belong together under the theoretical umbrella of typomorphology. The three schools of typomorphology offer such practitioners a rich data base on forms and form making processes. And more importantly, morphogenetic research grounds this design work in the history of city building. Types no longer need to be arbitrarily borrowed icons. They are structuring concepts which have been tested in the reality of city building. They are place-bound and time-bound, responding and adapting to new social, economic, and technological circumstances.

NOTES

1 See for instance Downing [1850] 1969; Pevsner 1976; Myers and Baird 1978; Rowe and Koetter 1978; Groth 1981, 1988; Upton 1981; Hull 1982, 1983; Boyer 1985; and Schön 1988, to cite a few cases in a broad range of applications.

2 I use *built landscape* as an umbrella term that includes urban form, city form, built environment, etc. Built landscape is attractive because it marries concepts of built and open spaces (which "built environment" does not), and because it connotes concrete material space (while "urban form" is more abstract). Italian and French architects often refer to "architecture" with a small "a" to depict the same phenomena.

3 This chapter is adapted from a manuscript in progress, tentatively entitled 'City Building.' The research was initiated in 1987 under an Individual Fellowship from the National Endowment for the Arts (Moudon 1987).

4 Such influential Italian historians as Leonardo Benevolo and Manfredo Tafuri only paid lip service to Muratori's work and ignored Caniggia's until after the mid-1980s (Tafuri 1989).

5 Published volumes of this work are available for the town of Venzone (Sartogo n.d.), the cities of Naples (Ciccone 1984), Florence (Maffei 1981; Malfroy and Caniggia 1986), and Venice (Maretto 1986). Caniggia was also an active practitioner; he had an office in Rome in partnership with Francesca Sartogo (Caniggia 1984c).

6 Caniggia also studied the development of pre-Etruscan settlements in various regions of Italy. His theories explaining the pattern of these settlements go beyond the concerns of this chapter, but they do establish further links between urban and regional form.

7 Canniggia's work on the types of elements that make up buildings and on the spatial organization of roads and settlements is not included in this discussion.

8 Functionalism, the prevailing approach to architectural design in the postwar period, stipulates that architecture is best understood and practiced in a multidisciplinary context: the psychological, social, and economic components of buildings have to be considered as external forces, to be handled by appropriate professionals.

9 Rossi is not interested in the systematic study of the city's origins and evolution or in its operational history. Theoretical or methodological aspects of typology or morphology by and large are absent from *The Architecture of the City* (Lawrence 1985). Rossi wants to break away from the Muratorian tradition; the "master" is not mentioned in the book.

10 Argan's own position is ambiguous. He says that a building typology is not a mere classification but the definition of an aesthetic purpose. The classification of buildings has three dimensions: the shape of the building, its major building elements, and its decorative elements. He argues that in studying typology, the designer considers history as a source of information for the new project to be "naturally connected to the past." Yet in this process, the designer has freed himself from the conditioning influences of the past as a model, accepting it instead as a completed process: precedents need to be understood, not copied mindlessly (Argan 1965).

11 Although Vidler's and Moneo's writings were most influential, other writings in the architectural literature do refer to the Italian typological work. See entire issues of the *Journal of Architectural Education* in 1982 and *Casabella* in 1985; Colquhoun 1969; Ungers 1979; Anderson 1982; Castex and Panerai 1982; Porphyrios 1984; Brown 1986; and Broadbent 1990.

12 Vidler traces the first typology back to the Enlightenment, when architectural typologies

exemplified by the work of Abbe Laugier classified the different elements of buildings as geometric forms related to natural elements (the column as a tree, for instance). These types were *archetypes* or ideal types to be emulated. Later on, Durand expanded the notion of type to describe special public programs, their different plan configurations and facade compositions, from which designers could choose. The second typology belonged to the modernists who advocated building types fit for mass production. Theirs were *prototypes* or first expressions of a type. A third typology identified by Vidler and Moneo (although Moneo did not use the term) was developed in the 1960s by the Tendenza, the then little-known neorationalist group championed by Aldo Rossi. The Tendenza identified building types based on urban vernacular traditions.

13 Fortunately, Conzenean ideas have recently been enjoying a revival in England. Yet the work remains largely unknown in France, Italy, and the United States. Geographer James Vance at the University of California, Berkeley, is one of the few proponents of Conzen's method, and the historian Spiro Kostof, on the same campus as Vance, referred to Conzen in his publications (Kostof 1991, 1992).

14 In principle, geographers are charged with studying elements of the landscape and generating knowledge that designers and planners can then use. However, this particular focus generally has been neglected by the discipline, leaving a gap that only a few social scientists and designers have been attempting to fill. Why geographers have left this gap and why designers have not moved into this field more forcefully is worth another paper. Aspects of this subject are addressed by Whitehand (1981, 1987).

15 The connections that Whitehand establishes among traditional measures of urban development, economics, and resulting city form are important for explaining the city-building process. Certainly, the descriptive powers of morphological studies can only be complimented and reinforced by economic arguments. Whitehand's pioneering work begins the difficult task of relating real estate and community development practices to city planning and design theory.

16 In 1987 Slater also began editing a newsletter which now reaches an impressive number of individuals and groups in Ireland, Germany, Spain, Switzerland, Poland, Austria, the United States, and elsewhere.

17 After 1968, reforms changed the education of architects and urbanists, and supported the development of an infrastructure to support extensive design and historical research. The old Ecole des Beaux-Arts was replaced by some eight *unités pédagogiques* (UP), still scattered around the periphery of Paris. Each *unité* represents an autonomous school of architecture, housing not only the staff to teach studios and other architectural subjects, but person-environment studies, urban design, and urban studies. The Versailles School of Architecture is known as UP 3, or third Unité pédagogique.

18 Recently, the works of geographers Roncayolo and Rouleau have reinforced the focus on urban historical architecture (Rouleau 1983, 1985).

19 Lefebvre taught at the Institut d'urbanisme of the University of Paris, where he influenced a number of designers and planners with a kind of urban sociology that included fundamental aspects of anthropology. Another influential person at the Institut d'urbanisme is philosopher Françoise Choay, whose seminal publications have focused on the roots of urban design theory but not on the city.

20 Lefebvre was also the director of the Institute of Urban Sociology, which conducted an influential study published in 1966 as *L'Habitat pavillonnaire* (*The single family detached dwelling*) (Raymond *et al.* 1979). Object of planners' and architects' derision, yet object of desire for 82 percent of French men and women at the time, the pavillonnaire symbolized the conflicts between people's choices and the values of professional urbanists.

21 In UP 8, founding member Henri Raymond was one of the researchers and authors of *L'Habitat pavillonnaire*. Another founding member, Bernard Huet, had spent a year at the Polytechnic in Milan, and was aware of Italian work in typomorphology. He became editor of the *Architecture d'Aujourd'hui* in 1974, at which time his student, Christian Devillers, published "Residential Typology and Urban Morphology." A few years later Huet himself published a

small manifesto in favor of historically grounded architecture (Huet 1978).

22 LADRHAUS keeps in close contact with groups having similar interests in Spain and Latin America. Many of the team's case projects have used environments that are familiar to the researchers: Versailles, various Parisian neighborhoods, and the Parisian fringe. Field trips with students led to special investigations, with several small projects carried out in Italy where the team also retains close intellectual ties. Over the past decade, Panerai and Depaule have been immersed in research on Cairo, Egypt, and other towns in North Africa.

23 The case studies include Haussmann's Paris, London's garden cities, Amsterdam's extensions, Ernst May's Frankfurt, and Le Corbusier's Cité Radieuse.

24 The strength of recent research by Maretto and Maffei is changing the nature of Italian work in typomorphology, bringing its scholarship up to par with the Conzenean and the Versailles schools (Maffei 1990a, 1990b; Maretto 1986).

25 This generalized use of the term "architecture" is now common in many parts of Europe to describe material space. Other terms such as built environment or landscape are often avoided because they are deemed to emphasize social rather than material space or to connote a narrow focus on the aesthetics of space.

26 They realize that their work will be questioned because it does not correspond to traditional disciplines and to commonly accepted categories of inquiry: "too historical for the theoretician, not mathematical enough in the methodologist's eyes, too empirical for the historian's taste" (Castex et al. 1977, 7). They add:

> Hence the apparent ambiguous nature of our work: a morphological study, yet one based on examples that are approached historically; an architectural study, yet one carried out at the scale of the urban fabric; a spatial study, yet one based on social concerns.
> (Castex et al. 1977, 9)

27 Conversely, proponents of the study of urban structures *as they are perceived* usually ignore or even dismiss the utility of "objective" urban form as a mere "geometric" exercise, which gives the false impression that urban form can be discussed "objectively," that is to say in its "true" dimension (Lynch 1981; Goodey 1985; Moudon 1992a).

28 The work of Carl von Unnee Linnaeus stands out as illustrative of this period. Applications of classification techniques in architecture are illustrated in the work of the Abbe Laugier, who did borrow from the natural sciences.

29 Durand's 1801 *Recueil et parallele des edifices de tout genre anciens et modernes* is a catalogue of buildings that represent the "basis of architectural culture" at the time (Panerai *et al.* 1980, 76).

30 In Durand's second volume, *Précis des leçons d'architecture données à l'Ecole polytechnique* (1802).

31 Saverio Muratori is noted as a pioneer in the quest to abandon typical house plans in favor of consecrated types. Further, Muratori's novel approach to typology, which anchors the common building to its site and groups parcels to define the elementary organization of the building fabric, is recognized as the first approach to establish a dialectic between building types and urban form.

32 Historian Corboz's terminology helps explain further the scope of the Versailles School (Corboz 1992). He differentiates between the *city of the geographer* (the Conzenean and the Versailles School's interest) and the *city of ideas* (Choay's primary interest). The city of the geographer is both built and used; its design is often governed by two, sometimes conflicting, processes. One is a set of culturally-bound traditions and the second is theories consciously elaborated by one of several elites (architects, bankers, etc.). The Versailles School studies both types of design processes and considers them both part of the history of design theory.

33 This concurs with Rossi, who denies the value of urban design in the term's narrow sense of designing the city because the city should evolve, rather than be designed (Rossi 1982, 116).

REFERENCES

Anderson, Stanford. 1982. Types and conventions in time: toward a history for the duration and change of artifacts. *Perspecta* 18: 109–18.

Argan, G.C. 1965. Sul concetto di tipologia architettonica. *Progetto e Destino:* 75–81.

Aymonino, C., M. Brusatin, G. Fabbri, M. Lena, P. Loverro, S. Lucianetti, and A. Rossi. 1966. *La Città di Padova, saggio di analisi urbana.* Rome: Officina edizoni.

Aymonino, Carlo. 1976. *Il significato della città.* Bari: Laterza.

Bandini, Micha. 1984. Typology as a form of convention. *AA Files* 6 (May): 73–82.

—— 1988. La contribution britannique à la morphologie urbaine. In *Morphologie urbaine et parcellaire*, eds. P. Merlin, E. d' Alfonso, and F. Choay, 81–86. Paris: Presses universitaires de Vincennes.

—— 1992. Some architectural approaches to urban form. In *Urban Landscapes: An International Perspective*, eds. J.W.R. Whitehand and P.J. Larkham, 133–69. London: Routledge.

Boudon, Françoise, Andrè Chastel, Hélène Couzy, and Françoise Hamon. 1977. *Système de l' architecture urbaine: le quartier des Halles à Paris.* Paris: Editions du Centre National de la Recherche Scientifique.

Boyer, Christine C. 1985. *Manhattan Manners: Architecture and Style 1850–1900.* New York: Rizzoli.

Broadbent, Geoffrey. 1990. *Emerging Concepts in Urban Space Design.* New York: Van Nostrand Reinhold.

Brown, Denise Scott. 1986. Invention and tradition in the making of American place. *The Harvard Architecture Review:* 162–71.

Caniggia, Gianfranco. 1984a. *Composizione architettonica e tipologia edilizia, 2. Il projetto nell' edilizia di base.* Venice: Marsilio Editori.

—— [1963] 1984b. *Lettura di una città: Como.* Como: Edizione New Press.

—— 1984c. *Moderno con moderno.* Venice: Marsilio Editori.

—— [1976] 1985. *Strutture dello spazio antropico.* Florence: Ed. Alinea.

Caniggia, Gianfranco and Gian Luigi Maffei. 1979. *Composizione architettonica e tipologia edilizia, 1. Lettura dell'edilizia di base.* Venice: Marsilio Editori.

Casabella. 1985. Issue on typology. 49 (January/February).

Castex, Jean, Jean-Charles Depaule, and Philippe Panerai. 1977. *Formes urbaines: De l'îlot à la barre.* Paris: Dunod.

Castex, Jean, Patrick Céleste, and Philippe Panerai. 1980. *Lecture d'une ville: Versailles.* Paris: Editions du Moniteur.

Castex, Jean and Philippe Panerai. 1982. Prospects for typomorphology. *Lotus International* 36: 94–99.

Cervellati, Pier Luigi, Roberto Scannavini, and Carlo de Angelis. 1977. *La nuova cultura della città. La salvaguardia dei centri storici, la riappropriazione sociale degli organismi urbani e l'analisi dello sviluppo territoriale nell' esperienza di Bologna.* Milan: Mondadori.

Choay, Françoise, ed. 1965. *L'Urbanisme: Utopie et réalitiés, une anthologie.* Paris: Editions de Seuil.

—— 1980. *La règle et le modèle: Sur la théorie de l'architecture et de l'urbanisme.* Paris: Editions du Seuil.

Choay, Françoise and Pierre Merlin. 1986. *A propos de la morphologie urbaine.* Tome 1, rapport de synthèse. Laboratoire "Théorie des mutations urbaines en pays développés," Université de Paris VIII, Institut durbanisme de l' Académie de Paris, E.N.P.C, mars.

Ciccone, F. ed. 1984. *Recupero e riqualificazione urbana nel programma staordinario per Napoli.* Milano: Dott. Antonino Giuffrè Editore, August.

Colquhoun, Alan. 1969. Typology and design method. In *Meaning in Architecture*, eds. C. Jencks and G. Baird. New York: George Braziller.

Comune di Bologna. 1979. *Risanamento Conservativo del centro storico di Bologna.* Bologna: Graficoop.

Conzen, M.P. 1978. Analytical approaches to the urban landscape. In *Dimensions of Human Geography*, ed. K.W. Butzer, 128–65. Chicago, Ill.: University of Chicago Department of Geography, Research Paper 186.

—— 1980. The morphology of nineteenth-century cities in the United States. In *Urbanization in the Americas: The Background in Comparative Perspective*, eds. W. Borah, J. Hardoy, and G. Stelter, 119–41. Ottawa: National Museum of Man.

—— 1990. Town-plan analysis in an American setting: cadastral processes in Boston and Omaha, 1630–1930. In *The Built Form of Western Cities*, ed. T.R. Slater, 142–70. Leicester, U.K.: Leicester University Press.

Conzen, M.R.G. 1960. Alnwick, Northumberland: a study in town-plan analysis. Publication No. 27. London: Institute of British Geographers.

—— 1968. The use of town-plans in the study of history. In *The Study of Urban History*, ed. H.J. Dyos, 114–30. New York: St. Martin's Press.

Corboz, André. 1992. L'urbanisme du XXe siècle, esquisse d'un profil. Genève: Fédération des Architectes Suisses.

Croizé Jean-Claude, Jean-Pierre Frey, and Pierre Pinon, eds. 1991. *Recherche sur la typologie et les types architecturaux.* Paris: L'Harmattan.

Devillers, Christian. 1974. Typologie de l'habitat et morphologie urbaine. *Architecture d' Aujourd'hui* 174 (July).

Divorne, Françoise, Bernard Gendre, Bruno Lavergne, and Philippe Panerai. 1985. *Les Bastides d' Aquitaine, du Bas-Languedoc et du Béarn, essai sur la régularité.* Brussels: Editions des Archives d'Architecture Moderne.

Downing, Andrew Jackson [1850] 1969. *The Architecture of Country Houses.* New York: Dover Publications.

Durand, J.N.L. 1801. *Recueil et parallèle des edifices de tout gente anciens et modernes, remarquables par leur beauté, par leur grandeur ou par leur singularité, et dessinés sur une même échelle.* Paris: An IX.

—— 1802. *Précis des leçons d' architecture données à l'Ecole polytechnique.* Paris: An X.

Dyos, H.J., ed. 1968. *The Study of Urban History.* New York: St. Martin's Press.

Fortier, Bruno. 1986. La rue Réaumur. *Les Annales de la recherche urbaine* 32: 23–28.

Gerosa, Pier Giorgio. 1986. Sur quelques aspects novateurs dans la théorie urbaine de Saverio Muratori. *Collection: Urbanisme et Sciences Sociales* 6. Strasbourg, France: Université des sciences humaines, Ecole d' architecture de Strasbourg.

Goode, Terence. 1992. Typological theory in the U.S.: the consumption of architectural authenticity. *Journal of Architectural Education* 46 (1): 2–13.

Goodey, Brian. 1985. The current condition of urban design: the significance of Lynch's final question: "How to make public the analysis of local place quality?" Unpublished paper based on a presentation to the seminar *L'analisi de 'luogo',* 23 March Istituto di Architettura e Urbanistica, Università degli Studi di Napoli, Italy.

Groth, Paul. 1981. Streetgrids as frameworks for urban variety. *The Harvard Architectural Review* 2 (Spring): 68–75.

—— 1988. Generic building and cultural landscapes as sources of urban history. *Journal of Architectural Education* 41 (3): 41–44.

Holdsworth, Deryck W. 1992. Morphological change in Lower Manhattan, New York, 1893–1920. In *Urban Landscapes: An International Perspective,* eds. J.W.R. Whitehand and P.J. Larkham, 114–32. London: Routledge.

Huet, Bernard. 1978. Small manifesto. *Rational Architecture Rationelle.* Brussels: Archives de l'Architecture Moderne 54.

Hull, Steven. 1982. *Alphabet City.* New York: Pamphlet Architecture.

—— 1983. *Rural & Urban Houses in North America.* New York: Pamphlet Architecture.

Jones, A.N. and P.J. Larkham. 1991. Glossary of urban form. Historical Geography Monograph No. 26. London: Institute of British Geographers.

Journal of Architectural Education. 1982. Issue on typology in design education. 35 (2).

Katz, Peter, ed. 1994. *A New Urbanism.* New York: McGraw-Hill.

Kostof, Spiro. 1991. *The City Shaped: Urban Patterns and Meanings Through History.* Boston: Little, Brown.

—— 1992. *The City Assembled: The Elements of Urban Form Through History.* Boston: Little, Brown.

Kropf, Karl S. 1993. The Definition of Built Form in Urban Morphology. Ph.D. dissertation, vols. 1, 2. Department of Geography, University of Birmingham, England.

Laugier, P. 1754. *Essai sur l'architecture.* Paris: Duchesne.

Lavedan, Henri. 1926. *Qu'est-ce que l'urbanisme?* Paris: Que sais-je.

Lawrence, Roderick J. 1985. Architecture of the city reinterpreted: a critical review. *Design Studies* 6 (3): 141–49.

Lefebvre, Henri. 1968. *Droit à la ville.* Paris: Editions Anthropos.

—— 1970. *Du rural à l'urbain.* Paris: Editions Anthropos.

Lynch, Kevin. 1981. *Theory of Good City Form.* Cambridge, Mass.: MIT Press.

Maffei, Gian Luigi. 1981. *La Progettazione edilizia a Firenze 1910–1930.* Venice: Marsilio Editori.

—— 1990a. La casa fiorentina nella storia della città. Venice: Marsolio Editori.

—— ed. 1990b. La casa rurale in Lunigiana. Venice: Marsolio Editori.

Malfroy, Sylvain and Gianfranco Caniggia. 1986. *L'Approche morphologique de la ville et du territoire.* Zurich: Eidgenössische Technische Hochschule, Lehrstuhl für Städtebaugeschichte, October.

Maretto, Paolo. 1986. *La casa veneziana nella storia della città, dalle origini all'ottocento.* Venice: Marsilio Editori.

F
I
V
E

Merlin, Pierre, E. d'Alfonso and F. Choay, eds. 1988. *Morphologie urbaine et parcellaire*. Paris: Presses universitaires de Vincennes.

Moneo, Raphael. 1978. On typology. *Oppositions 13* (Summer): 23–45.

Moudon, Anne Vernez. 1986. *Built for Change: Neighborhood Architecture in San Francisco*. Cambridge, Mass.: MIT Press.

——— 1987. The research component of typomorphological studies. AIA/ACSA Research Conference, Boston, November.

——— 1992a. A catholic approach to organizing what urban designers should know. *Journal of Planning Literature* 6 (May): 332–49.

——— 1992b. The evolution of twentieth-century residential forms: an American case study. In *Urban Landscapes: An International Perspective*, eds. J.W.R. Whitehand and P.J. Larkham, 170–206. London: Routledge.

——— [1995]. Teaching urban form. *Journal of Planning Education and Research* 14 (2): 123–33.

——— Manuscript in progress. City Building.

Muratori, Saverio. 1959. Studi per una operante storia urbana di Venezia. Rome: Instituto Poligraphico dello Stato.

——— [1959] 1985. Una lezione di seminario per la preparazione alla missione di architectti e per la formazione di docenti in una scuola di architettura. Transcribed by Guido Marinucci. Edizione dei Corsi di Composizione Architettonica di R.e S. Bollati, Facoltà di Architettura di Reggio Calabria.

Muratori, Saverio, Renato Bollati, Sergio Bollati, and Guido Marinucci. 1963. *Studi per una operante storia urbana di Roma*. Rome: Consiglio nazionale delle riceche.

Myers, Barton and George Baird. 1978. Vacant lottery. *Design Quarterly* 108.

Panerai, Philippe, Jean-Charles Depaule, Marcelle Demorgon, and Michel Veyrenche. 1980. *Eléments d'analyse urbaine*. Brussels: Editions Archives d'Architecture Moderne.

Pevsner, Nikolas, Sir. 1976. *A History of Building Types*. London: Thames and Hudson.

Poetë, Marcel. [1929] 1967. *Introduction à l'urbanisme*. Paris: Editions Anthropos.

Porphyrios, Demetri, ed. 1984. Leon Krier – houses, palaces, cities. *Architectural Design* 54 (7/8).

Porter, William L. and Stanley Tigerman. 1992. Administrators' essays on architectural history. *Journal of Architectural Education* 46 (1): 46–50.

Quatremère de Quincy, A.C. 1977. Type. With an introduction by Anthony Vidler. *Oppositions* 8 (Spring): 147–50.

Raymond H., N. Haumont, M.G. Raymond, and A. Haumont. 1979. *L'Habitat pavillonaire*. 3rd ed. Paris: Centre de Recherche d'Urbanisme.

Rossi, Aldo. 1964. Aspetti della tipologia residenziale a Berlino. *Casabella-continuità* 288 (June): 10–20.

——— 1981. *A Scientific Autobiography*. Cambridge, Mass.: MIT Press.

——— [1966] 1982. *The Architecture of the City*. Cambridge, Mass.: MIT Press.

Rouleau, Bernard. 1983. *Le Traçé des rues de Paris, formation, typologie, fonctions*. Paris: Editions du Centre Nationale de la Recherche Scientifique.

——— 1985. *Villages et faubourgs de l' ancien Paris: Histoire d'un espace urbain*. Paris: Editions du Seuil.

Rowe, Collin, and Fred Koetter. 1978. *Collage City*. Cambridge, Mass.: MIT Press.

Samuels, Ivor. 1988. La morphologie urbaine: de la recherche à la profession. In *Morphologie urbaine et parcelaire*, eds. P. Merlin, E. d' Alfonso, and F. Choay, 87–93. Paris: Presses universitaires de Vincennes.

——— 1990. Architectural practice and urban morphology. *In the Built Form of Western Cities*, ed. T.R. Slater, 415–35. Leicester, U.K.: Leicester University Press.

Sartogo, Francesca. n.d. *Venzone come e perché*. Roma: Alba Centro Stampa.

Schön, Donald A. 1988. Toward a marriage of artistry and applied science in the architectural design studio. *Journal of Architectural Education* 41 (4): 4–10.

Slater, T.R. 1984. Preservation, conservation and planning in historic towns. *The Geoaraphical Journal* 150 (3): 322–34

——— 1987. Ideal and reality in English Episcopal medieval town planning. *Transactions of Institute of British Geographers* 2: 191–203.

——— ed. 1990. *The Built Form of Western Cities*. Leicester, U.K.: Leicester University Press.

Solomon, Daniel. 1992. *ReBuilding*. New York: Rizzoli.

Tafuri, Manfredo. 1989. *History of Italian Architecture 1944–1985*. Cambridge, Mass.: MIT Press.

Tice, James. 1993. Themes and variations: a typological approach to housing design, teaching, and research. *Journal of Architectural Education* 46 (3): 162–75.

Suburbanization is the spread of suburban development patterns across a region or a nation – that is, the proliferation of sprawl forms of urbanization across a region or a nation.

The terms *sprawl* and *suburbanization* will be used interchangeably throughout this book.

WHAT MAKES SPRAWL?

The aforementioned definitions of sprawl and suburbanization still do not tell the whole story. While we now have an idea of what sprawl looks like and what its principal traits are, we still don't know why it is the way it is or exactly what goes into its construction. While sprawl development owes its existence to many factors, it is important to understand four essential ingredients of suburbanization:

- Land ownership and use
- Transportation patterns
- Telecommunications technology
- Regulations and standards.

The sections that follow deal with each factor in turn.

Land ownership and use

Most of the land in the United States – about 70 percent – is privately held.[19] Under American law, each parcel of land comes with a bundle of rights related to ownership, including water and air rights and the rights to sell the land, pass it along to heirs, use it, or develop it. These privately held entitlements give land value and marketability. As long as land remains privately owned – and its rights remain unencumbered – it is susceptible to being subdivided and built upon.

Land itself – along with the rights attendant to its ownership – can be bought and sold like any other commodity. Without a highly developed system of private land ownership and a viable market for land, sprawl as we know it would be virtually impossible. The concept of private land ownership is the foundation upon which the private home is built. It wasn't always that way, however. Native Americans viewed the land as something held in common. It was the early settlers from Europe who brought the concept of individual land ownership to the United States.

With the arrival of the Europeans, land ownership quickly achieved great importance in the New World. Since then, it has remained a basic tenet of the American ethos that being a landowner is the key to being a successful, fully vested member of society. In fact, during the decades following the ratification of the U.S. Constitution, many states limited voting rights to landowners. Thus, you not only had to be a white male over age twenty-one, but also had to own land to be counted as a real citizen. As Kenneth T. Jackson tells us in *Crabgrass Frontier:*

> The idea that land ownership was a mark of status, as well as a kind of sublime insurance against ill fortune, was brought to the New World as part of the baggage of the European settlers. They established a society on the basis of the private ownership of property, and every attempt to organize settlements along other lines ultimately failed. The principle of fee-simple tenure enabled families to buy, sell, rent, and bequeath land with great ease and a minimum of interference by Government. It became . . . "the freest land system anywhere in the world."[20]

Today, the American Dream is still to own one's own home on one's own piece of land. More than two-thirds of Americans own their own homes, and many have most of their money tied up in that very investment – which is also their shelter.[21] Purchasing a home is often the biggest investment Americans will make in their lifetimes.

Real estate markets

A colossal industry has been built around real estate, not only around simply buying and selling the land or its rights (or both) but also around deliberately increasing the value of the land by building on it. This is why most land gets developed: to increase its value and create wealth. Along with the basic bundle of entitlements, increasing land value is every landowner's right in America, just as making a profit is every individual's right whether that person owns land or not. The private ownership of land and the huge, almost liquid, market for it are vital to the very survival of suburban sprawl.

In 1997, the U.S. real estate industry produced revenues totaling over $240 billion.[22] That same year, related industry revenues for private construction totaled nearly $400 billion.[23] The size of the real estate and construction industries gives them significant influence in what gets built, where, and in what quantity. The real estate development industry delivers its products in response to demand – demand for houses, demand for offices, demand for shops, demand for hotel rooms, and so forth. Wheat fields in Kansas are wheat fields and not housing partly because of real estate markets. There is a smaller market for housing in the middle of rural Kansas than exists in a big metropolitan area, but there is a market for land on which to grow wheat. It is market demand that initially establishes what, where, and how much of everything gets built. The industry simply delivers the product.

But market demand is also determined to some degree by the product industry delivers. There was little demand for personal computers before the first one was invented and brought to market. Similarly, there was no recognizable market for indoor suburban shopping malls before the first one was built in Southdale, Minnesota, in 1956. When a successful formula like the indoor suburban shopping mall comes along, it can rapidly develop into a new market, exploding across the country-side. Financing feeds a growth industry, be it business or real estate. If a formula is successful and therefore profitable, it becomes easy to finance. The tendency is then to repeat the same formula many times, which partly explains the repetitiveness of suburban development. Single-family homes, shopping centers, and office parks in their current forms have been very successful models.

The cost of land

But sprawl would not have its current attributes if land were scarce and expensive. The existence of a large market for land development helps to explain U.S. patterns of urbanization overall, not just suburban sprawl. The unique pattern of sprawl can be partly attributed to the abundance and relatively low cost of land, which is necessary to allow dispersed, low-rise development to occur. Tall vertical cities like New York result, among other things, from the high cost of the land under the buildings. To justify the higher cost of the land, a developer has to build to a much higher density in Manhattan than in a typical suburban environment.

Why does land cost more in some center cities than it does in the suburbs? The higher cost can be traced to two factors: clustering and access. It is widely accepted that the monetary advantages of clustering (also known as the economies of agglomeration) are among the primary forces driving urbanization in general.[24] Businesses benefit economically by being able to shop for goods and services in a cluster. Furthermore, employees benefit from being able to shop for jobs in that same cluster and employers benefit from the large labor pool that results. The gathering of the labor pool in turn causes housing, stores, and other uses to be drawn into the resulting conurbation.

The second factor is access. Many cities originated by gathering around some major means of access to other, more distant markets in order to reduce transportation cost. Businesses originally needed to be as close as possible not only to one another, but also to a central import-export node, such as a harbor, river port, or rail station. This need reinforced clustering, which in turn drove up land cost and density. A typical result was the late-nineteenth or early-twentieth century U.S. manufacturing city with a high-density central business district gathered near port facilities and rail termini.

The equation changes when access becomes much more widespread, as happened with the introduction of cars, trucks, and pervasive high-speed roadway networks. Widespread access means that cheap land far from any city center becomes a usable commodity for businesses and homes alike. Without a compelling need to cluster, homes and businesses will naturally begin to spread out. As Terry Moore and Paul Thorsnes have written in *The Transportation/Land Use Connection*:

> Business firms . . . respond to land prices by spacing themselves as widely as possible. Spacing reduces competition for land which reduces its price. No other reasons (such as proximity to a port) exist to cause competition for land, and businesses reduce cost by occupying lower-priced land.[25]

What is true for businesses is also true for home owners. Reduced land cost means that single-family homes on relatively generous individual plots of land within commuting distance of work suddenly become an affordable commodity for many Americans. This demonstrates the very close relationship between land and transportation in defining modern patterns of human settlement. To have any worth, land must be served by some means of transportation, whether a transit stop, a highway interchange, or even just a lane or an alleyway connecting to a larger roadway system. When land is both accessible and inexpensive, building at much lower densities can be profitable. This combination is fundamental to sprawl development.

Transportation patterns

Land and market forces alone could not establish the low-density membrane that characterizes sprawl. History and economics tell us that without a transportation system capable of serving this pattern, sprawl simply would not exist. Without automobiles and paved roadways, we would inhabit an entirely different world.

Mode choice

Two major transportation factors determine development patterns: mode (or modal) choice and the physical layout or pattern of the transportation system itself (sometimes known as the transportation network). Mode choice refers to the availability of different kinds of transportation. Transportation modes consist of everything from walking to automobiles, railroads, boats, and air travel. When you travel from your home to work, what choices do you have for making the trip? Can you, for example, choose between walking, biking, driving, and riding public transportation? Sometimes a trip may involve multiple modes – for instance, driving to the train from home or taking a bus from the train to work. The transfer between each mode is called a mode change.

In the suburbs and beyond, mode choices typically are few. In many instances, the car is the only choice. When trip origins and destinations are highly dispersed over a wide area (the result of a continuum of low-density development), the private automobile is often the only adequate mode of transportation. When alternative choices are available,

a discouraging number of different mode changes may be required. A traveler may have to change from bus to bus to rail and back to bus again. All things being equal, the more mode changes that are required, the greater the disincentive will be to choosing an alternative to the automobile.

Travel time and cost also affect mode decisions. Travel time may be influenced by congestion on the roadways or the number of transit stops. Cost may be affected by the cost of passage as well as by the availability of reasonably priced parking. The automobile can sometimes seem to be the least expensive mode due to the tendency to ignore both the cost of the car and the overall cost of the auto/roadway transportation system – even though that cost is actually quite substantial in both dollars and externalities.

Modal choices can vary significantly, depending on the kind of trip taken. Trips generally can be categorized as either local or long distance. Local trips (also referred to as daily trips) are less than 100 miles one way. These trips basically fall into two categories: work trips or commuting (travel to and from work), and nonwork trips (errands, shopping, school, and so forth).

Local work trips

Commuting trips in contemporary suburbs are almost invariably beyond walking distance and mostly have been since the days of railroads and streetcars. In those cases where a major urban core or other high-density employment center is involved, bus or rail transit may be available for trip making, but a car may be needed to get to the train or bus. As suburban patterns develop farther and farther from major urban centers, the car becomes the only real mode choice for most commuter trips. More than 70 percent of all commuting trips in the nation have nothing to do with downtown; rather, they are to and from suburban and exurban destinations.[26] For this type of commuting trip, the automobile often is the only option.

Local nonwork trips

As with work trips, the car is usually the only choice for suburban nonwork trips because of the low density and horizontal separation of uses. This means that many basic errands are generally too far to walk, and the "trip-ends" – or origins and destinations – are too dispersed for any form of mass transit to make sense. These disparate origins and destinations do not usually lend themselves to any sort of fixed-route transit system. Bicycles might work, but because of lack of suitable roads, distance, weather, or other reasons, biking is often ruled out.

Furthermore, local trips may include a number of stops on each trip with varying numbers of people and bundles to be picked up or dropped off. This succession of stops is called trip chaining. A typical example might be a short journey where a parent takes a child to a music lesson and then drives on to drop off the dry cleaning, make a stop at a hardware outlet, and then do the grocery shopping. Interestingly, local nonwork trips are by far the largest segment of all travel. According to the U.S. Department of Transportation, nonwork trips make up more than 75 percent of total person miles traveled in the United States.[27]

Long-distance travel

In sheer numbers, the amount of local travel in the United States is overwhelming when compared to long-distance travel. More than 99 percent of all person trips made in the United States on all modes are local, accounting for more than 75 percent of all person miles traveled in the nation[28] Thus, it is tempting to dismiss the contribution of long-distance travel in defining the pattern of sprawl as relatively trivial. This might be justified were it not for the issue of the central import-export node described earlier in this chapter.

In the early part of the twentieth century, long-distance passengers and freight accessed a typical city by means of one or more centrally located import-export nodes, such as a rail station or a port. This phenomenon reinforced the dense clustering of early-twentieth-century manufacturing cities. Roadways and airports have significantly changed that pattern. Clustering is still an urbanizing force, but highway interchanges and multiple regional airports have replaced rail terminals and harbors as primary import-export nodes, radically altering the geographic scale and pattern of clustering.

Today, about 83 percent of the value of all freight in the United States is shipped by truck and plane, while 97 percent of all passengers travel by air and by road when taking a long trip.[29] Ports and rail termini are still used for heavy bulk cargo, but almost all people and most valuable goods travel in planes, cars, and trucks. As suburbs have spread and air and roadway travel has increased, the center-city train depot has become increasingly less relevant. Long-distance trips are now more likely to be made from one suburban area to another, with the car being the only practical way to get to and from the airport at either end of the trip.

In the end, almost all contemporary transportation choices use the car somehow in the process. The car is often the choice for local trips, commuting, and long-distance travel. Because automobile travel accounts for 92 percent of the total person miles traveled in the United States, the roadway system is by far the nation's foremost transportation network.[30] This means that auto-dependent development is basically self-perpetuating. Any new land development that hopes to succeed has to hook into the transportation pattern that connects everything else, which means extending the pattern of automobile dominance and limited mode choice.

The transportation network

The nation's roadway network is one of the most powerful forces determining the shape of metropolitan regions across the United States. Railroad, water, and air transportation have never been able to match the access granted by roadways. Combine this omnipresent network with automobiles and trucks, and once-simple roadways are converted to a high-speed transportation system that often outmatches railroads in travel time and accessibility. Mode choice could never be so dramatically skewed toward the automobile if it were not for the universal presence of roadways. It is this vast network that has made decentralization possible on a truly gigantic scale. It also, to a very large extent, defines the look and feel of our suburban world. As James S. Russell recently observed in *Harvard Design Magazine:* "What unites suburbia is not shared public space, or a coherent architectural vision, but a vast civil-engineered network of roads."[31]

As noisy, congested, and chaotic as the nation's roadway system can appear, it actually possesses a very intricate hierarchical structure. Under ideal circumstances, all of its component roadways are designed to function together as a unified structure of greater and lesser arteries and veins – like the human circulatory system, only made to move vehicles instead of blood cells. As extensive as it is, the entire network ultimately comprises just a few distinct roadway types. Together, this system of expressways, arterial roads, collector roads, local streets, and cul-de-sacs makes up almost the entire public environment of our suburbs.

But roads alone don't describe all of the system. All of the automobiles need someplace to park, and these parking areas define suburban sprawl as much as, if not more than, the roadway system. Garages, carports, and driveways adorn every contemporary residential subdivision, and shopping centers, malls, and office parks offer great fields of surface parking. Sometimes, these areas are landscaped but often only minimally to avoid blocking clear views of commercial signs. These areas could be designed differently (the cars could be – and sometimes are – placed underground or in structures), but surface parking is the most economical way to build parking as long as land is inexpensive enough. And, as we have seen, it is the roadway and car transportation system that has helped to make the cost of accessible land inexpensive enough to decentralize clustering patterns while at the same time leaving enough room to park most of the cars at grade.

Telecommunications technology

Electronic telecommunications are rapidly transforming the world around us. As William J. Mitchell recently wrote in his foreword to Thomas Horan's *Digital Places*:

> Digital telecommunications networks [will] transform urban form and function as radically as ... mechanized transportation networks, telegraph and telephone networks and electrical grids [have] done in the past. These networks ... loosen many of the spatial and temporal linkages that have traditionally bound human activities together in dense clusters, and they ... allow new patterns to emerge.[32]

While current thinking on how telecommunications may affect future urban patterns, we cannot know for certain what kind of world will finally emerge from this new revolution. However, we can clearly see what electricity, telephones, and computers have already done.

Electricity

The electrical grid is as pervasive as the roadway system (in fact, it often follows it). This shared ubiquity has freed businesses and homes to locate just about anywhere and be assured of a power source to run the machinery necessary for modern living and commerce. Before the twentieth century, no such widespread infrastructure existed. The machinery that electricity runs includes much of the infrastructure of modern telecommunications, such as computers, remote telephones, modems, fax machines, printers, copiers, and the like. Without available electrical power everywhere, most of the machines of modern commerce would not exist, nor would the systems that run our homes.

Corporations could not do business in a low-density suburban environment without at least electricity and telephone. Initially, it was the telephone that greatly reduced the need for businesses to share a common location – such as a major city – in which communications are facilitated by proximity.

Telecommunications and computers

Although telephones have been in existence since the end of the nineteenth century, the truly exponential advances in telecommunications and information systems have occurred only in the past fifty years. This technology revolution has made suburban sprawl possible on a scale that could never have been envisioned in the early twentieth century. The development of mainframe computers connected by telephone lines in the 1950s and 1960s meant that information could be readily and simultaneously shared by a network of remote facilities. Until then, major corporations had struggled to keep all of their operations under one roof to realize what economists refer to as "economies of scale of production."

Businesses realize economies of scale when the average cost of a unit of output (anything from a camshaft to a bank statement) falls as the total volume or scale of output increases. To realize these economies, businesses typically massed people and machines together under one roof. This phenomenon, together with clustering, has historically been one of the primary forces shaping urbanization by centralizing urban development.[33] For example, many service industry businesses once realized economies of scale by having everyone in one building in a big city. This meant paying a single rent check while simplifying management, information sharing, and communications. At the same time, the businesses were clustered near their customers and vendors, giving them ready access to both.

Together with roadways, cars, and airports, advances in telecommunications and computer technology have substantially changed how these forces work, allowing many companies to abandon older models and to decentralize, relocating major portions of their businesses to suburban locations or even to other parts of the country or overseas – wherever land or labor or both cost less. Many major corporations also realized that they no longer needed to have even their headquarters downtown. Now, the head office could move closer to the suburban homes of the CEO and other corporate officers, while links with customers and vendors could be handled electronically. Combined with roadways, automobiles, and airports, the growth of electronic telecommunications helps explain why more than 80 percent of the employment growth in the United States between 1980 and 1990 was in suburban and exurban locations – not downtown.[34]

Wiring the home

Suburban housing owes much to telephone, radio, and television. The telephone allowed instant communication between worker and household, even if the worker was miles away, making it easier to manage business and domestic affairs in two locations. The spread of residential subdivisions far from any theater district or concert hall also has been helped considerably by radio and television. Countless channels of programming have brought entertainment right into the home. You no longer need to get in your car to go to the movies. By 2000, average daily household television viewing was approaching eight hours per day, the number of television sets was climbing toward an average of 2.5 per household, and more than 80 percent of American households owned a VCR. Between 1995 and 1998, the number of households connected to the Internet increased from less than 10 percent to more than 50 percent. Greater than 80 percent of the nation spends each evening watching television.[35] Thus, even suburban movie houses have had to transform themselves into huge multiplex entertainment centers to survive. The advent of home entertainment centers and digital TV and recording media may make further inroads into the cinema business.

The Internet and beyond

There is little question that modern information systems have vastly expanded the freedom of location in our society. Businesses and residences can now situate themselves practically anywhere. Employees don't even have to show up at the office to go to work anymore. The number of telecommuters (those who spend at least part of their days working from home via computer and telephone) quadrupled from 4 million to almost 16 million between 1990 and 1998.[36] By 2000, the number had jumped to nearly 24 million.[37] The fiscal 2001 appropriations bill for the U.S. Department of Transportation requires that every federal agency give at least 25 percent of its eligible workforce the option of working outside the office by fall 2001.[38]

The future of the retailing industry also may be changing as on-line retailing becomes more popular. Books, music, computers, and an ever expanding list of other items can now be ordered electronically on-line and sent via air express right to your door. The volume of "e-retailing," as it is called, increased at an annual rate of 67 percent from 1999 through the end of 2000.[39] Since then, there has been an industry shakeout, but people are still buying on-line. It is impossible to predict where all of this will ultimately lead, but these communications developments undeniably have given us more freedom than ever to choose where we live and work. It seems almost as certain that they have made decentralization increasingly easier.

Regulations and standards

Another key factor that helps determine the final pattern of suburbanization is the battery of regulations, codes, and standards that govern development in communities across the United States. The result of a century-long interdisciplinary effort, this vast compendium of rules forms the "genetic code" of sprawl. Various parts of the compendium can be found in the subdivision regulations and zoning codes of most U.S. municipalities. Other segments can be found in the roadway manuals and standards issued by state and federal governments. Still other sections appear in the myriad privately published standard planning and design reference works for engineers, surveyors, planners, architects, and landscape architects.

These works set forth guidelines for minimum roadway widths, street patterns, parking layouts, lot grading, and many other items, right down to steps, curbing, and residential swimming pools. Although the patterns established by the genetic code can be very hard to make out from the monotony and chaos we see on the ground, they are very much a part of the suburban world around us. It is this genetic code that forms the pattern that we can see from the air, and it is this same compendium of rules, regulations, and standards that makes sprawl development in Georgia look just like sprawl development in California or New Jersey.

Zoning and building codes

As we have seen, the fact that land is private is fundamental to its development potential. How land gets developed – where and for what use – is largely determined by real estate markets. Even density and form are determined to some extent by market forces. But once the market for development has been established (or is on the way to becoming so), publicly regulated land use controls, or zoning, also can become a critical determinant in how land can be developed.

In most American communities where there is a market for new development, zoning controls land

– and to some extent its value – by regulating both land use and density. Use districts are established together with height and bulk regulations, the number of units or square feet of building allowed per acre, and setbacks that buildings are required to observe from the street and from one another. Zoning can govern what landowners can build on their own land as well as clarify expectations about what can be built next door. For example, a home owner who decides to put an addition on a house may find that zoning controls the size and location of the new construction. The addition may be limited in height, square footage, how far it may extend toward the boundaries of the lot, and what its use may be. After market forces have been established, zoning can ultimately be a major factor in determining what any given development will consist of and what it will look like. Some early subdivisions were built in rural areas that had no zoning. As communities grew, zoning was often put in place with the endorsement of home owners for their own protection.

Without some sort of formal control over how land in a particular district is used, each landholder in the district is continually at risk from neighbors. If you invest in building a house, you don't know for sure that a tannery or a pulp mill won't get built next door someday. This is one reason for controls: to provide reasonable expectations for the continued value of a given piece of land and thereby create a relatively stable marketplace. Regulations also exist to protect the public. Building a tannery or a pulp mill in the middle of a residential neighborhood can endanger public health and welfare. Crowding wooden residential structures too close together without adequate ventilation or emergency access can be both a health hazard and a fire hazard. Thus, zoning bylaws, subdivision regulations, and related codes continue to be considered necessary and effective for protecting public health and welfare.

In suburban areas, separation of land uses can be far more extreme than in older, urban-core areas. In older cities, compatible land uses are often mixed together. But, as Reid Ewing notes in his definition of sprawl shown earlier in this chapter, classic suburban zoning partitions all land uses into distinctly separate districts, which often are defined by roadways. Large arterial roadways and highway corridors, for example, are often zoned commercial or light industrial. Typically, industrial uses will be buffered from any residential uses by roads, landscaping, and open spaces or by intervening commercial uses (or both). Commercial uses, in turn, are themselves buffered from adjoining residential uses, often by roadways and landscaped areas. While some contemporary planners may rue the extreme to which the separation of uses has been taken, it should be remembered that many people still prefer quiet residential streets with nothing but houses on them. To these people, dictating otherwise would disrupt the character of the neighborhood and threaten property values.

But the codes do not stop at simply separating uses. Frequently, they also distinguish between different varieties of the same kind of use. For example, housing is sometimes separated into multifamily and single-family detached housing or even into different kinds of single-family housing, usually based on lot size (for example, half an acre, one acre, or two acres per unit). It has been argued that such finer gradations may discriminate by segregating people by economic class, with the plots in large-lot zoning areas available only to wealthy people.

In many ways, horizontal zoning seems very rational, both from the private as well as the public perspective. Because it protects public health and welfare as well as property values, a lot of people support it. Yet, as we have seen, it is this very horizontal separation of uses – ruling out other possible outcomes – that helps to define sprawl. As Andres Duany, one of the founders of New Urbanism, writes in *Suburban Nation*:

> The problem is that one cannot easily build Charleston any more, because it is against the law. Similarly, Boston's Beacon Hill, Nantucket, Santa Fe, Carmel – all of these well-known places, many of which have become tourist destinations, exist in direct violation of current zoning ordinances. Even the classic American Main Street, with its mixed-use buildings right up against the sidewalk, is now illegal in most municipalities.[40]

Many contemporary land use codes have ruled out older, walkable cities and downtowns in favor of horizontally separated zoning districts – in other words, sprawl zoning. This is why you can't walk to a corner store; it's usually too far away. It is true that not everyone wants to live over a store, but Duany's argument is that current codes don't allow any choice in the matter.

On the other hand, lest zoning be blamed for too much, it is useful to observe that places like Nantucket and Beacon Hill have zoning bylaws and other regulations that actually work to protect their historic character and require that new development fit the existing pattern. Also note that a sprawling city like Houston, Texas, has no zoning. In theory, Houston could have built itself like Manhattan or Colonial Williamsburg, or even a medieval Italian hill town, but it did not turn out that way. Even without zoning, Houston exhibits many of the sprawl characteristics of other metropolitan areas. It is spread out and generally low density, ranking fourth in degree of sprawl (ahead of Los Angeles and Miami) out of twenty-eight metropolitan areas ranked by the Surface Transportation Policy Project.[41]

The requirements of finance

To some degree, Houston makes up for its lack of zoning with other types of regulations, but market and financing factors have also helped fill the gap. As noted earlier, markets help define real estate product and most housing, offices, and shopping malls require financing to get built. In the case of housing, much of the financing takes the form of residential mortgages. Many home mortgages are guaranteed by government agencies who, over the years, have developed their own preferences and standards for what should be built. Those standards are reflected in many contemporary zoning codes and contribute to the genetic code of suburban development written into the manuals of many design and development professionals and sometimes even into covenants contained in the deeds of various projects.

The banks and insurance companies that finance many suburban commercial and residential projects have similar standards. Their requirements can dictate the size of the project, the uses that may be included, the number of parking spaces needed, and even the materials to be used in construction. To finance, build, and sell their real estate products, developers in places like Houston have

Environment, 1998 (Boston: EPA, 1998), chapter on sprawl, as posted on there Web site: www.epa.gov/region01/ra/soe98.html.

9 Sierra Club Web site: www.sierraclub.org/sprawl.

10 Natural Resource Defense Council Web site: www.nrdc.org/cities/default.asp.

11 Reid Ewing, "Is Los Angeles-Style Sprawl Desirable?" *Journal of the American Planning Association*, vol. 63, no. 1 (Winter 1997): 2–4.

12 Ewing provides a table listing seventeen urban planners, theorists, and authors, starting with William Whyte and including Anthony Downs, Constance Beaumont, Richard Moe, and others. Ewing, "Is Los Angeles-Style Sprawl Desirable?," 3.

13 Ibid., 4.

14 For example, Anthony Downs as cited in Kenneth A. Small, "Urban Sprawl: A Non-Diagnosis of Real Problems," *Metropolitan Development Patterns 2000 Annual Roundtable* (Cambridge, MA: Lincoln Institute of Land Policy, 2000), 27.

15 For example, see the previous note and also Susan M. Wachter, "Cities and Regions: Findings from the 1990 State of the Cities Report", in *Metropolitan Development Patterns 2000 Annual Roundtable* (Cambridge, MA: Lincoln Institute of Land Policy, 2000), 22.

16 See Irving M. Copi, *Introduction to Logic* (New York: Macmillan, 1968), 89–114.

17 *Merriam-Webster Online Dictionary*, www.m-w.com.

18 See Kenneth T. Jackson, "Suburbanization," in *The Reader's Companion to American History*, ed. Eric Froner and John A. Garraty (Boston: Houghton Mifflin, 1991), 1040–1043; see also Kenneth T. Jackson, *Crabgrass Frontier: The Suburbanization of the United States* (New York: Oxford University Press, 1985).

19 U.S. Department of Agriculture (USDA), Natural Resource Conservation Service (NRCS), *Land Ownership, 1992*, www.nhq.nrcs.usda.gov/cgibin/kmusser/mapgif.pl?mapid=2788. The figure excludes Alaska. Including Alaska and excluding nonland acreage, this figure falls to about 60 percent. See National Wilderness Institute, *State by State Government Land Ownership* (1995), www.nwi.org.

20 Jackson, *Crabgrass Frontier*, 53.

21 U.S. Census Bureau, *Housing Vacancy Survey: First Quarter 2001*, Table 5. *Homeownership Rates for the United States: 1965 to 2001*, www.census.gov/hhes/www/housing/hvs/q101tab5.html.

22 U.S. Census Bureau, *1997 Economic Census* (Washington, D.C.: U.S. Census Bureau, 2000), Table 1.

23 U.S. Department of Commerce, *U.S. Industry and Trade Outlook 2000* (Washington, D.C.: U.S. Department of Commerce, 2000), 6-2.

24 Terry Moore and Paul Thorsnes, *The Transportation / Land Use Connection: A Framework for Practical Policy* (Chicago: American Planning Association, 1994), 9.

25 Ibid., 20.

26 Alan E. Pisarski, *Commuting in America II: The Second National Report on Commuting Patterns and Trends* (Lansdowne, Va.: Eno Transportation Foundation, 1996), 72.

27 U.S. Department of Transportation (DOT), Bureau of Transportation Statistics (BTS), *1995 Nationwide Personal Transportation Survey* (Washington, D.C.: DOT, 1995), 11. See also Jane Holtz Kay, *Asphalt Nation: How the Automobile Took Over America and How We Can Take It Back* (New York: Crown, 1997), 21.

28 U.S. Department of Transportation (DOT), Bureau of Transportation Statistics *Transportation Statistics Annual Report, 1999* (Washington, D.C.: DOT, 1999), 36.

29 Ibid., 37, 46.

30 Ibid., 37.

31 James S. Russell, "Privatized Lives," *Harvard Design Magazine*, no. 12 (Fall 2000): 24.

32 William J. Mitchell, "The Electronic Agora," foreword in Thomas A. Horan, *Digital Places: Building Our City of Bits* (Washington, D.C.: Urban Land Institute, 2000), xi.

33 Moore and Thorsnes, *The Transportation / Land Use Connection*, 10 and following.

34 Pisarki, *Commuting in America II*, 25.

35 Robert Putnam, *Bowling Alone: The Collapse and Revival of American Community* (New York: Simon & Schuster, 2000), 222–23, 228.

36 Horan, *Digital Places*, 33.

37 Jonathan Glazer, "Telecommuting's Big Experiment," *New York Times*, May 9, 2001.

38 Glazer, "Telecommuting's Big Experiment."

39 U.S. Commerce Department, "Retail, E-Commerce Sales in First Quarter 2001," *U.S.*

Commerce Department News, May 16, 2001, www.census.gov/mrts/www/current.html.

40 Andres Duany, Elizabeth Plater-Zyberk, and Jeff Speck, *Suburban Nation: The Rise of Sprawl and the Decline of the American Dream* (New York: North Point Press, 2000), xi.

41 See Barbara McCann *et al.*/Surface Transportation Policy Project/Centre for Neighborhood Technology, *Driven to Spend: The Impact of Sprawl on Household Transportation Expenses* (Washington, D.C.: 2001), chap. 3, fig. H, www.transact.org/reports/driven/.default.html. See also Haya El Nasser, "A Comprehensive Look at Sprawl in America," *USA Today*, February 22, 2001. *USA Today* used a different ranking methodology in which the Houston-Galveston-Brazoria region ranks number 234 in a range of scores from 26 to 536.

42 *Webster's New World Dictionary, College Edition* (New York: The World Publishing Co., 1966), 1455.

43 Sometime between 1960 and 1970, depending on source and definitions used, see U.S. Census Bureau/Campbell Gibson and Emily Lennon, *Historical Census Statistics on the Foreign Born Population of the United States, 1850–1990*, (Washington, D.C.: U.S. Census Bureau, 1999). See also U.S. Census Bureau, *Selected Historical Census Data: Urban and Rural Definitions and Data*, www.census.gov/population/www.censusdata/ur-def.html. See also Pisarski, *Commuting in America II*, 18–19.

44 The New York CMSA includes parts of New York, New Jersey, and Pennsylvania. CMSA data from U.S. Census Bureau, Census 2000 PHC-T-3 Tanking Tables for Metropolitan Areas, Table 1, www.census.gov/population/cen2000/phct3/tab01.pdf. Additional New York City data from U.S. Census Bureau, *Table 22: Population of the 100 Largest Urban Places: 1990*, www.census.gov/population/documentation/twps0027/tab22.txt. Released June 1998. New York City data for 2000 from Susan Sachs, "New York City Tops 8 Million for First Time," *New York Times*, March 16, 2000.

45 See previous note and the Boston Metropolitan Area Planning Council Area Web site (May 1999): www.mapc.org.

46 Pisarski, *Commuting in America II*, 18; and David Rusk, *Cities Without Suburbs*

(Washington, D.C.: Woodrow Wilson Centre, 1995), 5 and following.

47 When this book was being written, the 2000 census data were only being made available. Where possible, the latest 2000 census data has been used. In many cases, only projections based on 1990 data have been available.

48 See for example David W. Chen, "Outer Suburbs Outpace City in Population Growth," *New York Times*, March 16, 2001; and Cindy Rodriguez, "City, State Take on New Cast," *Boston Globe*, March 22, 2001.

49 F. Kaid Benfield, Matthew D. Raimi, and Donald D.T. Chen, *Once There Were Greenfields: How Urban Sprawl Is Undermining America's Environment, Economy, and Social Fabric* (Washington, D.C.: Natural Resource Defense Council, 1999), 6.

50 Pisarski, *Commuting in America II*, 25–26.

51 U.S. Census Bureau, "March 1996 Current Population Survey: Income 1995–Table A: Comparison of Summary Measures of Income by Selected Characteristics: 1994 and 1995," www.census.gov.hhes/income95/in95sum.html.

52 Joel Garreau, *Edge City: Life on the New Frontier* (New York: Doubleday, 1991), 3.

53 U.S. Department of Energy, Energy Information Administration, *Buildings and Energy in the 1980's* (Washington, D.C.: Department of Energy, 1995), 3.

54 Robert E. Lang, *Office Sprawl: The Evolving Geography of Business* (Washington, D.C.: The Brookings Institution, 2000), 3.

55 Michael Polland, "The Triumph of Burbopolis," *New York Times Magazine*, April 9, 2000, 54–55.

56 Robert D. Yaro, "Growing and Governing Smart: A Case Study of the New York Region" in Bruce Katz, ed., *Reflections on Regionalism* (Washington, D.C.: Brookings Institution Press, 2000), 43.

57 See also Robert Fishman's foreword in Peter Calthrope and William Fulton, *The Regional City* (Washington, D.C.: Island Press, 2001), xv.

58 Bruce Katz, and Jennifer Bradley, "Divided We Sprawl," *Atlantic Monthly*, December 1999, 26–42.

59 Patrick Geddes, *Cities in Evolution* (London: Williams & Norgate, 1915), 48–49. See also Richard Moe and Carter Wilkie, *Changing Places: Rebuilding Community in the Age of Sprawl* (New York: Henry Holt, 1997), 47.

60 U.S. Census Bureau, *Urban and Rural Definitions*, www.census.gov/population/censusdata/urdef.txt, and *About Metropolitan Areas*, www.census.gov/population/www/estimates/aboutmetro.html.

61 See U.S. Census Bureau map gallery, www.census.gov/geo/www/mapGallery/ma_1999.pdf.

62 Lewis Mumford, *The City in History: Its Origins, Its Transformations, Its Prospects* (New York: Harcourt, Brace and World, Inc., 1961), 540–41.

S
I
X

"Charter of the New Urbanism"

Congress for the New Urbanism (1996)

Editors' Introduction

The leading current movement directed toward combating urban sprawl and creating compact, walkable neighborhoods is a professionally based movement called the New Urbanism. It emerged in the 1980s as architects and urban designers sought ways to re-create what were felt to be the best physical qualities of traditional neighborhoods and small towns – connected street grids, local shopping, community parks, rear alleys, and front porches. Initially referred to as "Traditional Neighborhood Design," the movement coalesced under the rubric of the New Urbanism in 1993 and organized itself as the Congress for the New Urbanism (CNU). Several years later, the CNU issued a charter that reaffirmed the principles articulated in the 1991 *Ahwahnee Principles*, which had been developed by the non-profit Local Government Commission with the help of leading members of the fledgling movement, and incorporated physical form approaches that had been developed in the first New Urbanist projects, particularly at the new town of Seaside, Florida. The Charter of the New Urbanism, reprinted here, outlines twenty-seven guiding principles for architecture and urban planning that focus on physical spatial structure. The principles are organized according to three interrelated spatial scales: metropolis, city, and town; neighborhood, district, and corridor; and block, street, and building.

The New Urbanism is criticized, especially in academic circles, on numerous grounds: that its traditionally inspired built forms are anti-modern and nostalgic; that its recommendations are too prescriptive and formulaic; that its emphasis on form smacks of physical determinism; that its projects are elitist because they are not particularly affordable; and that it is contributing to urban sprawl because many projects are built on greenfield sites and are of relatively low density. Nonetheless, the movement represents a coming together of many practicing professionals who, understanding the nature of the land development and housing industry, are intent on achieving urban development that is highly livable, community building, and more socially and environmentally responsible than would otherwise be built. In some ways, New Urbanism can be likened to a movement of utopian-minded former students, shocked by the realities of the "real world" and the limitations of public urban planning, and focused upon creating something better – something they thought they might be doing when they entered the field initially. And, inherent problems and faults notwithstanding, the New Urbanism is proving popular in the marketplace.

Leading practitioners of the New Urbanism, and founding members of the CNU, are Miami-based Elizabeth Plater-Zyberk and Andres Duany, Berkeley-based Peter Calthorpe, Dan Solomon of San Francisco, and Stefanos Polyzoides and Elizabeth Moule of Los Angeles. Elizabeth Plater-Zyberk and Andres Duany designed Seaside, on the Florida panhandle. Since then they have had a hand in designing numerous other communities, including the well-known Kentlands. The firm pioneered the development of form-based codes as an alternative to traditional zoning practices, developed a New Urbanist "Lexicon," and created a New Urbanist "Transect," which organizes development guidelines along a rural to urban continuum. Their own writings and writings about their work include Andres Duany and Elizabeth Plater-Zyberk, *Towns and Town-Making Principles*, edited by Alex Krieger with William Lennertz (New York: Rizzoli, 1991) and Andres Duany, Elizabeth Plater-Zyberk, and Jeff Speck, *Suburban Nation: The Rise of Sprawl and the Decline of the American Dream* (New York: North Point Press, 2000). Peter Calthorpe focuses on regional planning and has been a leading advocate for "transit-oriented development." His book *The Next American Metropolis* (Princeton, NJ: Princeton Architectural Press, 1993) promotes clustering new developments at stops along transit lines and presents guidelines for achieving walkable, transit-oriented neighborhoods. His firm has developed the concept into a plan for the Metropolitan Portland region, and several new communities have been built, most notably Orenco Station. Dan Solomon and his firm have designed numerous urban infill projects in San Francisco, the Bay Area, and elsewhere. Solomon writes eloquently about the shortcomings of modern architectural practice and design education in his book *Global City Blues* (Washington, DC: Island Press, 2003). In Britain, notable proponents of the New Urbanism include Prince Charles, who has promoted its concepts through the Prince of Wales Institute. In Europe, the most influential advocates are architects Leon and Rob Krier, who have produced many designs for new city neighborhoods.

Writings on the New Urbanism include Emily Talen, *New Urbanism and American Planning* (New York: Routledge, 2005), Peter Katz, *The New Urbanism: Toward an Architecture of Community* (New York: McGraw-Hill, 1994), and The Seaside Institute, *The Seaside Debates: A Critique of the New Urbanism*, edited by Todd W. Bressi (New York: Rizzoli, 2002).

For more information on the *Ahwahnee Principles*, see the Local Government Commission's website (http://www.lgc.org/ahwahnee/principles.html). The Congress for the New Urbanism's website provides information about New Urbanist activities and available publications (http://www.cnu.org/).

THE CONGRESS FOR THE NEW URBANISM views disinvestment in central cities, the spread of placeless sprawl, increasing separation by race and income, environmental deterioration, loss of agricultural lands and wilderness, and the erosion of society's built heritage as one interrelated community-building challenge.

WE STAND for the restoration of existing urban centers and towns within coherent metropolitan regions, the reconfiguration of sprawling suburbs into communities of real neighborhoods and diverse districts, the conservation of natural environments, and the preservation of our built legacy.

WE RECOGNIZE that physical solutions by themselves will not solve social and economic problems, but neither can economic vitality, community stability, and environmental health be sustained without a coherent and supportive physical framework.

WE ADVOCATE the restructuring of public policy and development practices to support the following principles: neighborhoods should be diverse in use and population; communities should be designed for the pedestrian and transit as well as the car; cities and towns should be shaped by physically defined and universally accessible public spaces and community institutions; urban places should be framed by architecture and landscape design that celebrate local history, climate, ecology, and building practice.

WE REPRESENT a broad-based citizenry, composed of public and private sector leaders, community activists, and multidisciplinary professionals. We are committed to reestablishing the relationship between the art of building and the making of community, through citizen-based participatory planning and design.

WE DEDICATE ourselves to reclaiming our homes, blocks, streets, parks, neighborhoods, districts, towns, cities, regions, and environment.

We assert the following principles to guide public policy, development practice, urban planning, and design:

The region: metropolis, city, and town

1 Metropolitan regions are finite places with geographic boundaries derived from topography, watersheds, coastlines, farmlands, regional parks, and river basins. The metropolis is made of multiple centers that are cities, towns, and villages, each with its own identifiable center and edges.

2 The metropolitan region is a fundamental economic unit of the contemporary world. Governmental cooperation, public policy, physical planning, and economic strategies must reflect this new reality.

3 The metropolis has a necessary and fragile relationship to its agrarian hinterland and natural landscapes. The relationship is environmental, economic, and cultural. Farmland and nature are as important to the metropolis as the garden is to the house.

4 Development patterns should not blur or eradicate the edges of the metropolis. Infill development within existing urban areas conserves environmental resources, economic investment, and social fabric, while reclaiming marginal and abandoned areas. Metropolitan regions should develop strategies to encourage such infill development over peripheral expansion.

5 Where appropriate, new development contiguous to urban boundaries should be organized as neighborhoods and districts, and be integrated with the existing urban pattern. Noncontiguous development should be organized as towns and villages with their own urban edges, and planned for a jobs/housing balance, not as bedroom suburbs.

6 The development and redevelopment of towns and cities should respect historical patterns, precedents, and boundaries.

7 Cities and towns should bring into proximity a broad spectrum of public and private uses to support a regional economy that benefits people of all incomes. Affordable housing should be distributed throughout the region to match job opportunities and to avoid concentrations of poverty.

8 The physical organization of the region should be supported by a framework of transportation alternatives. Transit, pedestrian, and bicycle systems should maximize access and mobility throughout the region while reducing dependence upon the automobile.

9 Revenues and resources can be shared more cooperatively among the municipalities and centers within regions to avoid destructive competition for tax base and to promote rational coordination of transportation, recreation, public services, housing, and community institutions.

The neighborhood, the district, and the corridor

1 The neighborhood, the district, and the corridor are the essential elements of development and redevelopment in the metropolis. They form identifiable areas that encourage citizens to take responsibility for their maintenance and evolution.

2 Neighborhoods should be compact, pedestrian-friendly, and mixed-use. Districts generally emphasize a special single use, and should follow the principles of neighborhood design when possible. Corridors are regional connectors of neighborhoods and districts; they range from boulevards and rail lines to rivers and parkways.

3 Many activities of daily living should occur within walking distance, allowing independence to those who do not drive, especially the elderly and the young. Interconnected networks of streets should be designed to encourage walking, reduce the number and length of automobile trips and conserve energy.

4 Within neighborhoods a broad range of housing types and price levels can bring people of diverse ages, races, and incomes into daily interaction strengthening the personal and civic bonds essential to an authentic community.

5 Transit corridors, when properly planned and coordinated, can help organize metropolitan structure and revitalize urban centers. In contrast, highway corridors should not displace investment from existing centers.

6 Appropriate building densities and land uses should be within walking distance of transit stops, permitting public transit to become a viable alternative to the automobile.

7 Concentrations of civic, institutional, and commercial activity should be embedded in neighborhoods and districts, not isolated in remote, single-use complexes. Schools should be sized and located to enable children to walk or bicycle to them.

8 The economic health and harmonious evolution of neighborhoods, districts, and corridors can be improved through graphic urban design codes that serve as predictable guides for change.

9 A range of parks, from tot-lots and village greens to ballfields and community gardens, should be

distributed within neighborhoods. Conservation areas and open lands should be used to define and connect different neighborhoods and districts.

The block, the street, and the building

1   A primary task of all urban architecture and landscape design is the physical definition of streets and public spaces as places of shared use.
2   Individual architectural projects should be seamlessly linked to their surroundings. This issue transcends style.
3   The revitalization of urban places depends on safety and security. The design of streets and buildings should reinforce safe environments, but not at the expense of accessibility and openness.
4   In the contemporary metropolis, development must adequately accommodate automobiles. It should do so in ways that respect the pedestrian and the form of public space.
5   Streets and squares should be safe, comfortable, and interesting to the pedestrian. Properly configured, they encourage walking and enable neighbors to know each other and protect their communities.
6   Architecture and landscape design should grow from local climate, topography, history, and building practice.
7   Civic buildings and public gathering places require important sites to reinforce community identity and the culture of democracy. They deserve distinctive form, because their role is different from that of other buildings and places that constitute the fabric of the city.
8   All buildings should provide their inhabitants with a clear sense of location, weather and time. Natural methods of heating and cooling can be more resource-efficient than mechanical systems.
9   Preservation and renewal of historic buildings, districts, and landscape affirm the continuity and evolution of urban society.

"Density in Communities, or the Most Important Factor in Building Urbanity"

from *Community Design and the Culture of Cities* (1990)

Eduardo Lozano

Editors' Introduction

Related to the sprawl debates of the past few decades, disagreements over the benefits and detriments of higher density development are just as heated. Proponents of denser and more compact cities suggest that greater densities have positive spillover effects in supporting transit, reducing car usage, enhancing local economic development, providing various environmental benefits, and promoting greater social interaction. Fear of higher density development is fostered in part by images of questionable mid-century new-town developments and poorly managed public housing. To many people, the term *density* itself frequently suggests ideas about various environmental and social problems. NIMBY (Not In My Backyard) responses to higher density residential living are common, yet often unfounded and unsubstantiated. Opponents of higher density development are typically not opposed to higher densities elsewhere – just not next to their current low-density residences. Other fears are based on the effects of social and psychological overcrowding and its impacts on behavioral freedom.

Various studies have shown, however, that overcrowding (typically a function of economics and culture) need not be necessarily equated with density (a function of geography and urban form). Other empirical research has suggested that very high densities can be reached with urban forms that are quite acceptable to most people living in the city. Different studies have shown that negative perceptions of density can be mitigated through better design, for example, the use of vegetation to screen buildings, varied building heights, richer façade detailing, and careful design of transition spaces. Many desirable neighborhoods in pre-industrial walking cities maintain densities ten times higher than the typical low-density suburban forms in use today, but are generally perceived to be of much lower density. The use of classic high-density rowhouses and low- to mid-rise building forms provides the living space, private exterior space, mixed uses, and social amenities desirable in creating livable neighborhoods.

Measures of density used by urban designers are varied, but are typically a ratio of housing units, people, or floor space over a unit of land, such as acres, hectares, or square miles. Measures of gross density include all land uses and lands within a geographic area (including public rights-of-way), while net densities usually include only the parcel lands where residential uses are located. Gross density measures are inadequate predictors of urban form or the human experience of a place, as housing units may be clustered densely on a portion of the land, or distributed evenly across it in lower densities. Measures of density are important in understanding the thresholds necessary for various services and activities to function properly in the city, such as the number of people to support a bus system versus a light rail system, or the number required in the catchment area for a supermarket.

A comparison between the urban transportation systems of Boston and Paris illustrates the effects of density.[8] The city of Boston proper, with a population of more than 600,000 in 46 square miles, has a density of 21 people per acre, whereas Paris, with a population of more than 2,200,000 over 40 square miles, has 84 people per acre – a density four times higher than that of Boston. As a result, every Parisian is within four to five blocks of one of the 279 Metro stations, where silent and clean trains run every 60 to 90 seconds. For a Bostonian, this is a dream that could never be matched, simply because four very expensive miles of subway would be needed in Boston to serve the same population that can be served with only one mile in Paris.

The relationship between density and urbanity extends beyond transportation, reaching to the viability of, and accessibility to most urban services. In the retail sector, for example, there is an increase in the number of shops and stores as residential density increases. Population density and the available income of the population living within an accessible distance determine a series of thresholds. Below a certain density, no retail stores can exist; as density increases, feasibility thresholds are reached, allowing an increasing number and variety of stores. From the point of view of the merchant, commercial feasibility is an economic consideration; for residents, easy accessibility of commercial services is not only a convenience but a social amenity.

Recognizing density thresholds is thus critical to understanding the effects of density on urbanity. In the retail sector, for example, the effects of density on the number and variety of retail stores in a commercial center can be studied through the effects of distance on shopping trips, the size of the center as an attraction to shoppers, and balance with competing commercial centers. It is also possible to sketch a series of residential density ranges and suggest an initial set of thresholds; the relationships presented below, expressed in dwelling units per acre (du/acre), correspond to the ranges commonly found in the United States; other countries have different density thresholds.

A detached single-family house normally ranges from a net density of 1 to 5 du/acre. Tighter clustering would allow a density increase of up to 8 du/acre. A semidetached two-family house ranges from 5 to 12 du/acre. A town house with party walls could range from 10 to 16 du/acre. The net density of 12 du/acre is the first urbanity threshold, since below that level it is difficult to provide community facilities in close proximity to the dwellings.[9] The tight pattern of single- and two-family houses is commonly found in many inner metropolitan areas and in most small towns of Middle America. It is a small-scale, true urban environment, catalyst of many vital communities, immortalized by Frank Capra in his films of the 1930s and 1940S. The town house (or row house) has been the basic raw material for many cities, its density allowing the generation of an urban environment with community facilities nearby; the brownstones of New York and the townhouses of Boston are among the many examples found on the eastern seaboard.

Tighter clustering with perhaps some mix of two-story flats would allow-density to increase to 20 du/acre. The net density of 20 du/acre is another threshold, since above it direct access to the ground cannot be provided from each unit, leading to a radical change in the nature of the outdoor open space, a reduction of unit identity, and a need for common parking areas.[10] Thus, the threshold of 20 du/acre is the watershed that divides the types of dwellings that can maintain a unit identity from those that are merged into larger combinations.

Low-rise apartment buildings, such as three-story walkups, have a net density that ranges from 35 to 50 du/acre. At the upper level of this range, 45 to 50 du/acre, visual intimacy can begin to be lost,[11] and a concentrated urban scale emerges. Midrise apartment buildings, six stories high, range from 65 to 75 du/acre. The upper range, 75 to 80 du/acre, is another threshold, since above this level there can be a wide variety of facilities and activities easily accessible to each dwelling, indicating that from two points of view – space and accessibility – we are already in the realm of the higher hierarchical levels of the urban environment. At the same time, the provision of parking and open space becomes an important design issue.[12]

At the top of the urban hierarchy, high-rise apartments can range from 50 to 100 du/acre. Above that range we enter the level of high-density central city buildings, with all the limitations and advantages that the core of urbanity can provide – maximum accessibility, but also limited open space, congested streets, and, in general, pressure for space. Clearly, cities that provide substitutes for the automobile in the form of good mass transit and some major open space – Central Park in New York, Luxembourg or Les Tuilleries in Paris, Hyde Park in London, Palermo in Buenos Aires – certainly offer attractive central locations.

SOCIAL AND PSYCHOLOGICAL EFFECTS OF DENSITY AND CROWDING

The pervasive aversion to urban density and the implementation of density control measures have been justified on the basis of the assumed negative effects of density on people. But how real are these effects?

It is important, first, to distinguish density from crowding. Although often confused, density and crowding are measures of different phenomena. Density is the ratio of people or dwelling units to land area. Differences in density have economic and physical implications but no clear social or psychological effects. Crowding is the ratio of people to dwelling units or rooms. Different degrees of crowding have clear social and psychological effects. Studies of urban patterns have shown that, at the neighborhood level, there is no correlation between density and crowding and that different densities have no systematic relationship to people's perception of crowding.[13] The difference between the two ratios can be shown by example: high-density and low crowding can exist in a high-rise, upper-class apartment building, where there are many dwelling units per acre but few persons per dwelling. In contrast, low-density and high crowding can be found in isolated rural shacks in a depressed region, where there are few dwellings per acre but many people per room.

The most important difference between the two concepts is that density reflects mainly physical and economic conditions, whereas crowding reflects social and psychological conditions. Thus, density is a measure of the physical (univariate) condition, involving limited use of space. In contrast, crowding is a perceived condition of limited space; it is a multivariate phenomenon due to the inter-

action of spatial, social, and personal factors and is characterized by stress.[14] As a result, density is a quantifiable index that is easy to apply universally and to measure physically and economically, whereas crowding is a subjective and highly personal experience translated into psychological stress, involving numerous factors, and impossible to apply universally. Clearly, many of the objections to density can be legitimately directed toward crowding.

The most general way to study the relationship among social variables is statistical correlation. This method will not prove that the relationship is one of cause and effect, but it will associate phenomena, highlighting areas for more conclusive research. In this respect, we must reiterate warnings about conceptualizing the relationship between the built environment and human behavior in a deterministic way, by mistakenly assuming that behavior is a direct response to environmental stimulus.[15] The effects of density have been statistically studied, and although on first impression density appears to be related to pathological behavior, more detailed analysis indicates otherwise. The small apparent effects of density on pathological behavior are reduced to insignificance when controls for social class and ethnicity are introduced.[16]

For example, a statistical analysis of census data, involving a number of public housing projects around the world,[17] correlates measures of population density with indices of social and medical pathology, as well as with the effects of intermediate variables such as income and education controlled through partial correlation. The results indicate that, although high population density is commonly associated with social disorganization, the positive correlation between density and pathology disappears when measures of social status are utilized as control variables.[18] In other words, such factors as poverty and low educational levels are at the root of social disorganization and pathological behavior. I should add that even this statistical interpretation must be qualified. There are many societies in which people with extremely low incomes and poor education lead a structured social life without such pathological behavior. The difference seems to lie in the existence of a traditional social order within the community pattern. This phenomenon is observed in the Third World, where rural migrants who lived a structured life in villages are traumatized by the breakup of traditional ties in cities, leading to social disorganization and environmental degradation, on top of economic poverty.

High density in U.S. slums is the result of the poor being forced to concentrate on expensive land around the city center in order to be near jobs and transportation, and being unable to move to other areas because of segregation barriers. Crowding is the result of the poor being forced to fit large families into small apartments because of high rents. Such concentration of poverty, with people living in crowded, deteriorating quarters, with limited access to jobs and education, is the cause of high incidences of disease, socially pathologic behavior, and the creation of a "lumpen" subculture.

Crowding, measured as the number of people per room, has been found to be highly correlated with such indices of social pathology as high mortality and juvenile delinquency since the earliest studies conducted in this field.[19] This conclusion has been supported by later studies, which strongly suggest that interpersonal pressure or crowding may be linked with pathological behavior.[20] It is important to note that studies conducted on different neighborhoods found no correlation between crowding and density and an inverse relationship between the level of crowding analysis and the importance of physical density measures at the city level.[21] But residential crowding has consistently been found to have negative consequences.[22]

Recall that crowding is a perceived condition. The perception of crowding is inversely related to one's ability to exercise behavioral freedom and to exert control over one's social and physical environment.[23] That is, crowding is experienced when the number of people in one's environment is large enough to reduce one's behavioral freedom and choice.[24] This gives rise to overmanned situations;[25] it imposes behavioral restrictions and creates social interference, leading to competition for scarce resources.[26]

Crowding is perceived when a person's demand for space exceeds the supply of space.[27] But this situation could originate in physical factors – restricted space, arrangement of space, light conditions – as much as it could originate in social factors, such as the presence of other persons felt to be competitors, since "an individual may feel crowded in the midst of strangers, but quite comfortable and secure in the presence of an equal number of friends."[28] In addition, laboratory research has shown that conditions that potentially cause crowding have no negative effects on the performance of human tasks if the physical consequences of spatial restrictions (high temperature, stuffiness, limited movement) and other environmental stresses (noise) are controlled.[29] Spatial restriction is a necessary precondition of, but is not sufficient by itself to cause, crowding stress. Thus, crowding is not objective and abstract, but subjective and personalized. The demand for space, however, originates in fairly universal needs for privacy and personal turf. Privacy does not mean withdrawal, but the ability to control visual and auditory interaction,[30] and can be defined at various levels: solitude, intimacy, anonymity in a crowd, and control of intrusion through psychological barriers.[31] But even privacy is not an absolute concept; it depends on the cultural milieu.

One of the most critical factors affecting the perception of crowding, as well as of density, is culture, which controls much of human behavior. In addition, expectations and past experiences affect one's perception of crowding. Correlation studies show that spatial restriction is not always associated with social pathology and that cultural traditions define different parameters for density and crowding. The fact that Hong Kong, with a residential density ten times higher than that of Manhattan, is a thriving city not particularly burdened with behavioral pathologies is one indication of the importance of a cultural framework. Indeed, the relationship between high neighborhood densities and social pathology is mediated by personal and cultural factors.[32]

The mediating effects of culture in spatial perception in general, and in the perception of crowding and density in particular, are probably the most important obstacle to the generalization of research findings outside of a specific environment. Different social groups have different perceptions of what constitutes trespassing in space and what constitutes permissible involvement in public and private areas, leading Hall to assert that "culture is possibly the most significant single variable in determining what constitutes stressful density," because "people brought up in different cultures live in different perceptual worlds . . . People perceive space quite differently."[33] Cultural norms mediate the perception and adjustment of interpersonal space and, thus, the sensory thresholds for residential crowding and urban density.

It has been suggested that crowding may be the result of perceived urban congestion and excessive social stimulation;[34] the inability to avoid or reduce social or visual contact[35] may cause a cognitive overload leading to stress and withdrawal.[36] Cultural norms radically change the thresholds of such perception.

However, cultural differences in the perception of crowding and density cannot be adjusted through an anthropological classification of cultures. It has been said, for example, that urban scale must be consistent with ethnic scale, since each ethnic group seems to have developed its own scale.[37] Does this mean that an Irish neighborhood must be planned in a different way than a Polish one? To what extent should subcultures be disaggregated in environments as rich as urban areas? A clue to understanding this issue, at least for planners and designers, is given by one of the oldest neighborhoods in Boston, the North End, which today is largely Italian; its tightly packed, mid-rise, relatively high-density pattern, its narrow streets and alleys seem to be ideally suited for an Italian neighborhood. However, the North End was built and settled by groups migrating from England in the eighteenth century (some of whose descendants now live in exclusive suburbs with quite different lifestyles). The explanation for this is that the original English settlers belonged to the same European urban culture from which the more recent Italians originated. Thus, a straightforward ethnic label may not account for the truly differential factors within urban subcultures, or the common elements they share.

The complexity of the relationships between built environment and human behavior has led to the formulation of various theories that go beyond the effects of crowding and density. Barker has proposed the concept of behavioral settings, in which the built environment is interpreted as affording (but not determining) behavioral opportunities;[38] in order to survive, an urban environment must be adaptable to different behaviors and to changing behaviors. One of the criticisms raised by Frampton to Modern and Postmodern architecture is that they offer limited alternatives for patterns of behavior.[39] In addition, because of cultural variations in activity, family and gender, privacy and social intercourse,[40] the same environment would be perceived and used differently by different people, according to their values, experience, and motivation.[41] In the context of the postindustrial metropolis, where cosmopolitans share urban space with locals of various cultural extractions, these concepts are critical to community designers. Sommers developed the concept of personal space, in which territoriality is a way to attain privacy through physical or symbolic barriers, and space is personalized to satisfy one's needs for identity, security, self-fulfillment, and a frame of reference.[42] The personalization of the suburb of Levittown or Le Corbusier's project at Pessac is an indication of people's preference for diversity. Territoriality, like many other urban concepts, is culture specific; the hierarchy of private to public turfs varies with different cultural parameters.

In summary, crowding stress cannot be predicted on the basis of spatial considerations alone; it is determined by a combination of environmental and personal factors acting over time. The psychological stress of crowding involves not only realization that demands for space cannot be met by the supply, but also an emotional imbalance in which a person feels infringed upon, alienated, and deprived of privacy. The size of physical space – and thus the number of people per area – is only one of the variables of crowding. The noise and light levels in a space, the number of objects and their arrangement, the social situation, the activities taking place, and the personal psychology are all factors that, together, determine the perception of crowding and the level of stress. The close, personal proximity of urban life, when seen from the vantage point of suburban life, may seem threatening since the attraction (or focus) of urban activities may not be sufficiently perceived by suburban observers. A dense urban situation may be unappealing to a person not familiar with the activities taking place in that environment and unaccustomed to the urban rituals and routines that structure – and give meaning to – dense urban life.

Crowding has an opposite: undercrowding. Undercrowding is defined as an excessive abundance of space in which an individual suffers social isolation and needs enclosure and contact with others – sometimes manifesting as agoraphobia,

Other writings on compact cities include Timothy Beatley, *Green Urbanism: Learning from European Cities* (Washington, DC: Island Press, 2000); Peter Newman and Jeffrey Kenworthy, *Sustainable Cities: Overcoming Automobile Dependence* (Washington, DC: Island Press, 1999); Katie Williams, Elizabeth Burton, and Mike Jenks, *Achieving Sustainable Urban Form* (London: E & FN Spon, 2000); Mike Jenks, Elizabeth Burton, and Katie Williams (eds), *The Compact City: A Sustainable Form?* (London: E & FN Spon, 1996); and Rod Burgess, "The Compact City Debate: A Global Perspective," in Mike Jenks and Rod Burgess (eds), *Compact Cities: Sustainable Urban Forms for Developing Countries* (London: Spon Press, 2000).

The deficiencies of today's post-industrial city reinforce the call for urgent action towards the improvement of the quality of life in the city and of its environmental impact in an ever-increasing number of publications, world conferences and research projects (particularly the WCED's Brundtland Report, 1987; the CEC's Green Paper, 1990; the UN's Earth Summit Agenda 21, 1993). Many of these reports contain sections which discuss the form and structure of a more sustainable city.

There are some who think that the city as we know it has no future, that in view of the development of ever more sophisticated communication systems there will, for many, soon be no need to work in the city centre or even in the city as is currently the case, and that the ability to work in and communicate from one's home will aggravate urban dispersal (Troy, 1996, p. 207). However, with an ever-growing majority of the globe's population living in conurbations we cannot simply abandon the city. It will continue to be the place where people work and live. Furthermore, the city is shaped by and expresses its history, the collective values and culture of its citizens which are important for their pride and the city's identity; it is a place into which vast efforts and huge sums of money have been invested which cannot simply be written off. The issue here is not the sustainable or unsustainable city; we cannot afford not to sustain the city. But it must become more readily and easily sustainable, economically as well as socially and environmentally. The all-important question is what form and structure would make the city more sustainable.

This chapter reviews the debate about the form and structure of a sustainable city. It investigates where the debate originated and compares the arguments for and against specific city forms and structures. It will become clear that the discussion is confused and inconclusive, not just because of its complexity but also because of the lack of precision in the description of urban models and a lack of focus of arguments. The call for more inter-disciplinary and multi-aspect research is therefore not surprising; what is missing is tangible evidence and convincing empirical data that one or the other urban form is, or is more, sustainable.

Nevertheless, many demands for a more sustainable city or city region are shared by those involved in the debate, and ultimately a list of sustainability criteria emerges with the help of which the potential performance of city models can be compared and evaluated.

STARTING POINT OF THE DEBATE

The basis of the sustainable city debate is the general agreement that the city we know and inhabit today causes unsustainable environmental stress, is socially stratified and functionally suboptimal, and is expensive to run. The first important warning that changes are essential to safeguard resources for future generations was given by the Brundtland Report (WCED, 1987). It was soon followed by the Green Paper on the Urban Environment published by the European Commission in Brussels (CEC, 1990), which highlights functional, social, economic and environmental problems of today's cities and puts forward objectives and directives towards a more sustainable urban environment.

At the Rio Earth Summit it was agreed that 'indicators of sustainable development need to be developed to provide solid bases for decision-making at all levels and to contribute to self-regulating sustainability of integrated environmental and development systems' (United Nations, 1993). Both the EC Green Paper and Agenda 21 have considerable significance and impact on the debate about the city because of the political weight supporting them. In the UK, they were responded to at national level by a number of guidelines and directives, the most influential of which are:

- PPG13, published in March 1994 (DoE and DoT, 1994), which introduces a major change in policy by emphasising the interrelationship between land-use planning and transport, pointing out that the need to travel can be reduced by the appropriate location of development and by encouraging forms of development which promote sustainable modes of transport, including public transport. It also encourages the maintaining of existing urban densities and, where appropriate, an increase in density.
- PPG3: *Housing* (DoE, 1992), stressing the need to bring about a maximum amount of housing within urban areas and to reduce the need to travel.
- PPG6: *Town Centres and Retail Developments* (DoE, 1993), stressing the need to integrate retail outlets into, or adjacent to, urban cores.
- PPG15: *Planning and the Historic Environment* (DoE, 1994), which refers to 'the capacity of historic towns to sustain development'.
- The UK Strategy for Sustainable Development (UK Government, 1994) in response to the Rio Earth Summit. This highlights the role of the planning system and the need to derive policy that relates land uses and transport.
- PPG1 (revised): *General Policy and Principles* (DoE, 1997), which provides a more strategic view of the role of the planning system, specifically its contribution to achieving sustainable development. It summarises other policies (land use and transport; planning for housing; importance of the town centres; rural areas; conservation of historic environment), stresses the role of design considerations in planning and contains a new section on the Citizen's Charter and propriety.

All this indicates an awakening on the part of a wide section of professionals and politicians to the understanding that the city we have – with dense cores accommodating much of the city's workplaces, retail outlets, commerce, services and amenities; sprawling and low-density, single-use suburbs; a city structure which by default generates

the need to travel and, owing to overall low population densities, car dependency and, as the result of the burning of fossil fuel, massive pollution; unattractive public transport because of low densities and underfunding; congested roads as a result of car dependency; and high energy consumption – is in the long term not sustainable owing to the destructive impact on the regional and global environment.

This understanding has generated a wide range of research projects and publications in search of sustainable urban development and living as a result of an improved form and structure for the city. Breheny's statement is symptomatic: 'if cities can be designed and managed in such a way that resource use and pollution are reduced then a major contribution to the global problem can be achieved, (Breheny, 1992a, p. 2).

THE CONCEPT OF THE COMPACT CITY

The CEC Green Paper clearly calls for a return to the compact city, certainly influenced by the fact that many historic European towns and cities have densely developed cores which are seen as ideal places to live and work (mostly by those coming to visit these places for a short time, not necessarily by their inhabitants). Such places have high population densities which, the argument goes, encourage social mix and interaction which are the major characteristics of traditional cities. The UK government largely adopts the view of the European Commission, and this causes tensions with the English ideal of suburban living.

Arguments for the compact city

The main supporters of the compact city – including the CEC (1990), Jacobs (1961), Newman and Kenworthy (1989), Elkin *et al.* (1991), Sherlock (1991), Enwicht (1992), McLaren (1992), Owens and Rickaby (1992) – believe that the compact city has environmental and energy advantages and social benefits (see Hillman, 1996, pp. 36–44; Thomas and Cousins, 1996, p. 56):

- a high degree of containment of urban development; reuse of infrastructure and of previously developed land; rejuvenation of existing urban areas and consequently urban vitality; as a result of containment and high population densities a compact city form and the conservation of the countryside;

■ affordable public transport which meets the daily needs of those without a car, the majority of the urban population; as a result, increased overall accessibility and mobility;

■ as a consequence of public transport reduced vehicular traffic volumes, related pollution and risk of death and injury in traffic; lower transport expenditure leading to less pollution; congestion spread over more roads and for shorter periods of time;

■ viability of mixed uses as a result of overall high population densities; reduced travel distances as a result of mixed uses and overall higher population densities; cycling and walking as the most energy-efficient way of accessing local facilities; less car dependency; a better environment – due to overall reduced emissions and greenhouse gases and lower consumption of fossil fuel – and consequently better health;

■ lower heating costs as a consequence of a denser urban fabric, with less energy consumption and less pollution;

■ the potential of social mix as a result of high population densities, specifically when supported by a wide range of dwelling and tenure types in the neighbourhoods;

■ concentration of local activities in communities and neighbourhoods; as a result a high-quality life, greater safety and a more vibrant environment as well as support for businessmen and services, i.e. a milieu for enhanced business and trading activities.

The main justification for the compact city is the need for the least energy-intensive patterns of activity to cope with the issues of global warming, concern about which was highlighted by the Inter-Governmental Panel on Climate Change (IPCC) in 1990, 1992 and 1995 (Hillman, 1996, p. 39).

Arguments against the compact city

There are many who insist that the case of the compact city is not proven,

that there is evidence which suggests that these claims are at the least romantic and dangerous and do not reflect the hard reality of economic demands, environmental sustainability and social expectations. The overriding problem with the compact city is that it requires us to ignore the causes and effects of decentralisation, and benefits it may bring.

Thomas and Cousins, 1996, p. 56

Or in other words, the compact city concept fails to acknowledge the poor prospects of reversing deep-seated decentralisation trends (Breheny, 1992a).

Some of the main arguments of those opposed to the compact city are:

■ that the compact city concept contradicts the profound fondness for suburban and semi-rural living in the UK (Breheny, 1992a); that at particularly high densities the advantages of concentration might change into disadvantages through congestion which would outweigh energy consumption benefits of the compact city; that the fact that telecommunication allows people to live in the country contradicts the compact city concept (Breheny, 1992a);

■ that the concept of the green city (also promoted by CEC (1990)) is in contradiction to that of the compact city (Breheny, 1992a); that open space in cities would be taken up, that as a result the city's environmental quality would suffer;

■ that the compact city policy would result in the neglect of rural communities and earlier growth centres which emerged under dispersal policy; that rural economic development would be threatened by a focus of activities within existing towns and cities (Breheny, 1992a);

■ that the compact city would cause congestion, with the increased pollution, loss of amenity space and reduction of privacy so well demonstrated in cities like Calcutta, Cairo and Rio (Knights, 1996, p. 116);

■ that in the compact city social segregation would grow as a result of the high cost of accommodation in the city centre and in the more privileged outer suburbs (van der Valk and Faludi, 1992, pp. 124–5);

■ that the scale of energy savings through concentration may be trivial in comparison to the disbenefits it causes – e.g. in terms of unpopular restrictions on movement (Breheny, 1995);

■ that optimum use of passive solar gain demands lower densities as the best energy savings are made with detached houses, semi-detached houses and bungalows; savings are less with terraced housing and less still with flats (NBA Tectonics, 1988; Breheny, 1992a);

■ that the policy of a high-density, compact city fails to take account of the uncertainty in population growth and dispersal, i.e. that the compact city would not be able to respond to the predicted increase in the number of households (in the Netherlands half a million to one million new houses are said to be needed by 2015; in the UK over the next 20 years an extra 3.3 million single households are expected to require accommodation);

■ that the power to affect local decisions and the viability of the provision of community facilities diminish with increasing scale of a compact city (McHarg, 1992, p. 153);

■ that the compact city means massive financial incentives, which are economically suspect, and a high degree of social control, which is politically unacceptable (Green, 1996, p. 151).

The relationship between transport, urban form and energy consumption has been investigated in several research projects. Results are inconclusive. There is evidence that fuel consumption per capita is highest in more rural areas but there are indications that the largest cities (e.g. London) are likely to be less efficient than medium-sized and smaller towns, presumably as a result of congestion (Bannister, 1992, p. 165). Regarding urban density, intensity and energy consumption, higher-density cities tend to exhibit lower fuel consumption rates as a result of shorter travel distances and public transport; however, 'decentralised concentration', i.e. the promotion of urban and suburban cores, might be a fuel-efficient form (Bannister, 1992, pp. 171–4).

What can be learned from the compact city arguments and counter-arguments

It is evident from this short review that research focusing on a single aspect – such as energy efficiency and transport or energy efficiency and urban form – is not likely to generate a reliable basis for the generation of concepts of a sustainable city or city region, and is not likely to come up with appropriate guidelines for planners, designers and politicians (see Breheny, 1992b). Energy issues, for instance, need to be balanced with social, economic and environmental objectives. Any improvement of one aspect of the city must be weighed against other benefits or losses.

What also needs to be taken into account is the fact that cities are all different in form and structure owing to a host of place-specific factors such as topography, history, climate, and socio-economic conditions. It cannot be expected that they should all fit the same formula when it comes to the question of a sustainable city form. After all, we are generally confronted not with the task of planning and designing new towns and cities but, rather, that of replanning and redesigning existing cities, towns and settlements to make them more readily sustainable. The claim is made that the compact city is the most energy-efficient form of city, yet no investigation is made as to how much energy would be needed to change suburban sprawl into a compact city.

Furthermore, the question whether the transformation of an existing city into a compact city is economically and socially viable is largely ignored by those promoting that concept. Some cities may have the potential to become compact without substantial financial incentives; others may never become so unless massive financial support is made available, and, in view of the tendency to expect the private sector to take over responsibility for more and more of the city's infrastructure and public realm, it is highly unlikely that such support will be forthcoming. Other cities may have the potential to be partly compact and partly decentralised. [. . .]

But there are other arguments which also need to be taken into consideration and may well influence the choice of an overall viable urban form. The city of Edinburgh, for instance, already rather compact in parts, may well have the potential to become even more compact but in pursuit of this concept nobody, I hope, would come up with the idea that Princes Street Gardens be developed, because that would destroy a unique city structure with old and new town in parallel, divided by the Gardens – quite apart from the fact that Edinburgh would lose one of its open linear spaces, the existence of which in close proximity to the highly dense old town may well render living there acceptable. By the same token, it is to be hoped that nobody would suggest, in pursuit of better environmental conditions, that a green wedge be driven through the city centre of Venice or Siena or Florence or any of the historic towns and cities with a highly dense core area.

What is suggested here is that the argument for or against a specific city form needs to take into account not only all the social, economic and environmental arguments and objectives but also the very specific structure and form of each individual city and its topographical, socio-economic and historical conditions. There may be a generally acceptable approach and there may be a shared set of objectives, but the implication of these will inevitably reflect the morphology of each individual settlement.

Then there is the problem that centrists focus their attention on the city and largely fail to discuss the relationship of the city with the countryside (with the exception of the compact city's reduced emissions as a consequence of reduced energy consumption). The city cannot exist on its own; it needs a hinterland to provide goods, food, raw materials, etc. The relationship of the city with its hinterland is therefore crucial. One point made by Susan Owens might have helped to provide additional focus for arguments, namely that the phrase 'urban sustainability' is a contradiction as 'cities will always be net consumers of resources drawing them from around the world. They are also likely to be major degraders of the environment, simply because of the relative intensity of economic and social activities in such places' (Owens, 1992, p. 79). It might therefore be more appropriate to search for structures and forms that result in a greater degree of sustainability for urban areas or regions, and investigate specifically the relationship between the city or conurbation and the countryside, rather than focus, as most of the centrists seem to do, on the sustainable city – and in this case the compact city only.

'Decentralised concentration'

Many of those opposed to the compact city support the concept of 'decentralised concentration', the concept of a multi-nucleated city or even city region in which uses concentrated in the mono-core of the compact city are dispersed into a number of smaller centres forming the nuclei of urban districts or towns or 'villages'. The concept is based on the following policies on sustainable development and urban form (Breheny, 1992a, p. 22):

- Urban containment policies should continue to be adopted, and the decentralisation process slowed down.
- Compact city proposals, in any extreme form, are unrealistic and undesirable.
- Various forms of 'decentralised concentration', based around single cities or groups of towns, may be appropriate.
- Inner cities must be rejuvenated, thus reducing further losses of population and jobs.
- Public transport must be improved both between and within all towns.
- Mixed use must be encouraged in cities, and zoning discouraged.
- People-intensive activities must be developed around public transport nodes, along the Dutch 'right business in the right place' principle.
- Urban (or regional) greening must be promoted.
- Combined heat and power (CHP) systems must be promoted in new and existing developments.

It is clear that such changes cannot be achieved in a short period of time. 'The real challenge is . . . to redesign existing urban form. Some important elements can be changed quickly (e.g. bus routes), other elements, such as railway and commercial buildings, can only be changed infrequently' (Breheny, 1992a, p. 22).

Compromise positions

We have looked at the position of the 'centrists' and 'decentrists', but there is a group of 'compromisers', as regards their views on a sustainable urban form (Breheny, 1996, pp. 13–35; Scoffham and Vale, 1996, pp. 66–73; Thomas and Cousins, 1996, pp. 53–65). They advocate a combination of the merits of centralisation, i.e. urban containment and regeneration, with benefits of the 'inevitable decentralisation' to towns and suburbs (Breheny, 1996, p. 32). The 'compromisers' propose that individual neighbourhoods should involve the community and develop a strong identity and control over local resources (Scoffham and Vale, 1996, pp. 11–12).

Though not fully explained, the call for a degree of local autonomy is based on two convictions: that the people in a neighbourhood know best what their needs and aspirations are, and that they readily take more responsibility for and ownership of their neighbourhood if they have been involved in shaping it. The call for participation has considerable consequences for the city's form and structure. For communities to become successfully involved in the shaping of their own neighbourhoods requires decentralisation of power. This in turn necessitates the decentralisation of the city form and structure because a compact city or town, unless rather small, renders a participatory approach difficult if not impossible. The decomposition of the city or city region into smaller areas such as districts and neighbourhoods makes the communities' involvement feasible and effective but necessitates a framework at city region level for the integration of all development actions in districts and neighbourhoods. Decentralisation of the city is therefore a spatial, structural, functional, social and political phenomenon which expands into transport strategies and economics and requires also a certain degree of co-ordination at regional or even wider level to make the city and city region operate effectively.

To return to the arguments of the 'decentrists' and 'compromisers', they are reinforced by the fact that urban decentralisation – a process that started with the introduction of the railway and was exacerbated by the introduction of the car as a means of mass transport – is continuous, and is driven by powerful economic forces as well as the demand for what is considered to be a high quality of living by the middle- and upper-income groups. Centralised urban forms for them exemplify a way of urban living characteristic of the lower-income groups, those without the means to escape high-density housing (see Welbank, 1996, p. 78).

In the USA, Canada, Japan and Australia suburbanisation is massive. In Europe one can observe a process of counter-urbanisation, the suburbanisation of larger cities and towns, and the growth of smaller towns and villages (Breheny, 1996, p. 21). The UK government's containment policies such as

the strategy for sustainable development (UK Government, 1994) and PPG13 on transport (DoE and DoT, 1994), based on the ECOTEC (1993) study, seem to be unable to stem this trend.

In view of this strong trend, and in view of the uncertainties of some of the qualities or perhaps deficiencies of the compact city – specifically with regard to the quality of environment provided and the quality of life in a dense environment – the old idea of the merger of the best of town and country reappears (Hooper, 1994; Lock, 1991, 1995), and so does the idea of Ebenezer Howard's concept of the 'Social City' in the form of a 'sustainable Social City' (Breheny and Rockwood, 1993).

Even the idea of social cohesion and equity as a result of high population densities in a compact city is questioned by the 'decentrists'. Welbank puts it this way:

> our society is displaying all the characteristics of lack of social cohesion. We may not like this, we may feel it to be lamentable, but until such cohesion is re-established it is unlikely that forcing people into tight physical proximity will help at all – in fact, without the pre-existence of such cohesion it would actually be destructive. The enormous voluntary exodus from our cities demonstrates that the desire for social cohesion does not override the desire to live in suburbs and low density urbanised areas, given the benefits of mobility and technology.
>
> Welbank, 1996, p. 80

There is a counter-argument to the claim that the compact city will be able to afford efficient public transport and will generate a socially more equitable community. Clean and uncongested public transport, so goes the argument, allows decentralisation because access to facilities relies on speed rather than proximity (Smyth, 1996, p. 107). If the model of the compact city is applied, the result will be a high concentration of workplaces and facilities in the compact city centre with resulting high land and property values which only high-income groups can afford. Middle-income groups will search for a place in the outer suburbs whereas the low-income groups will get stuck in a doughnut of social disadvantage, a transit zone surrounding the city centre. In essence, the compact city is socially exclusive (Smyth, 1996, p. 107). Such social stratification, according to this argument, can be avoided in a decentralised city with a series of compacted centres with smaller areas of lower-income groups located around them.

Confusion regarding the form and structure of the compact city

The concept of the compact city, as explained by the centrists, needs to be re-examined because it is still not clear what the structure and form of such a city might actually be. One's initial impression is that the compact city resembles the medieval city with the concentration of activities in a highly dense city with clear and abrupt edges to the countryside, usually in the form of a town wall (Thomas and Cousins, 1996, p. 54). But other descriptions do not seem to reinforce that impression.

'Centrist' Friends of the Earth suggest that a sustainable city 'must be of a form and scale appropriate to walking, cycling and efficient public transport, and with the compactness that encourages social interaction' (Elkin *et al.*, 1991, p. 12). This surely evokes the picture of a medieval city or perhaps Krier's '*quartiers*' (1984, pp. 70–1). But when Jenks *et al.* (1996), with reference to Haughton and Hunter (1994), summarise other suggestions of proponents of a sustainable city (p. 5) one may conclude that the compact city is not necessarily as compact and sharp-edged as the medieval city because the descriptions range from

- large concentrated centres, to
- decentralised but concentrated and compact settlements linked by public transport, to
- dispersal in self-sufficient communities (Figure 1).

Though these suggestions are listed under the heading of 'The Compact City' it remains somewhat unclear whether they refer to a compact city or alternative forms of sustainable city. The entire debate lacks clearer, more explicit descriptions of what the proponents see as the typical structure and form of a compact city, and this makes for confusion. Thomas and Cousins (1996, p. 54) also point to the lack of more explicit descriptions. They compare a number of descriptions of the properties of a compact city defined by different proponents:

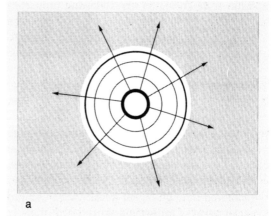

a

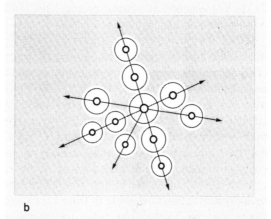

b

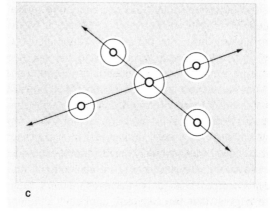

c

Figure 1 Alternative forms of sustainable cities according to Haughton and Hunter (1994):
(a) large concentrated centres
(b) decentralized but concentrated and compact settlements linked by public transport
(c) dispersal in self-sufficient communities

- Elkin *et al.* (1991) list the following properties (pp. 16 and 43): intensification of use of space in the city; higher residential densities; centralisation; compactness; integration of land uses; and some form of self-containment.
- Newman and Kenworthy (1989) list the following characteristics: more intensive land use; centralised activity; and higher densities.
- Breheny and Rockwood (1993) demand high density; mixed use; and growth encouraged within the boundaries of existing urban areas with no development beyond the city's periphery (achieved through higher-density land use and reclamation of brownfield sites).
- Owens and Rickaby (1992) list two contrasting key patterns of settlements: centralisation and decentralised concentration.

In the last case it is clear that alternative forms of sustainable city are classified; in the other cases no clear image of a compact or sustainable city's form and structure emerge as both the compact city and the city based on the 'decentralised concentration' concept may have the properties listed, though compactness may refer to the city as an entity or to parts of the city.

It becomes somewhat clearer that there are several rather different views as to what a compact city is, that the compact city is not necessarily a homogeneous phenomenon (see Burton *et al.*, 1996, p. 235). A main difference of views seems to be whether the sustainable city should be monocentric or polycentric. Another difference seems to be with regard to the extent of homogeneous compactness on the one hand and the scale or overall dimension of a compact city on the other. Some seem to suggest that a sustainable city should be compact from core to edge, with large concentrated centres like the medieval city or town – in which case the overall dimensions of such a city may have to be relatively small if good access to open space outside the city for recreational and other purposes is to be achieved. Others seem to suggest that the city may consist of different compact settlements which are decentralised but linked by public transport, in which case the open space may be in between the settlements. The distance between such settlements may vary according to their degree of independence and self-sufficiency.

It seems that the argument for and against the compact city is unnecessarily complicated by the lack of definition of what kinds of concepts people have in their minds when discussing city form. It is therefore useful to highlight what all protagonists of different city models have in common; maybe the agreed properties of a sustainable city could form the basis for the generation of models of such a city. However, before I do so, another important dimension of a sustainable city, so far missing from this discussion, needs to be included: that of the ecology of the city and its relationship to the country.

The ecological and environmental argument

The main argument of ecologists and environmentalists is that in order to achieve sustainable urban development and living it is not sufficient to solve problems of the wastage of energy and materials by behavioural changes and of pollution and congestion by introducing clean energy and promoting more energy-efficient public transport, or walking and cycling as alternative means of travel. The environmentalists argue that, in addition to all these important changes, the relationship of the city to the countryside has to change fundamentally. And this has considerable implications for the city's form and structure.

As the city is polluted and congested,

> urban society now takes refuge in the countryside in search of fresh air and natural surroundings that are denied at home. Consequently, unsustainable pressures are placed on environmentally sensitive landscapes. The advancing city has often replaced complex natural environments of woods, streams and fields with biologically sterile man-made landscapes that are neither socially useful nor visually enriching.
>
> Hough, 1989, p. 2

Hough argues further that in the city itself open spaces are not what they should be. The city, he states, incorporates vast areas of idle and unproductive land. Urban drainage, sewage disposal and other processes contribute to the pollution load. Parks are determined by horticultural science, not

differences seem to lie in the degree of compacting of the urban fabric and the degree of centralisation or decentralisation, rather than the principle. What is missing is some indication of these degrees. The decentrists put forward some models of a sustainable city applying decentralised concentration but only a general idea of the form and structure of a resulting city or city region emerges; the centrists do not sufficiently explain the structure and form of a compact city and, perhaps, much of the battle of arguments is caused by failure to provide this information.

There are some attempts to specify population densities in a compact city which range from sustainable urban net residential densities of between 225 and 300 persons per hectare (pph) (Friends of the Earth) to suggestions of optimal net residential densities of 90–120 pph to support public transport, and 300 pph to support services and facilities within walking distance (Newman and Kenworthy, 1989), but no evidence is made available as to how these figures have been derived. But there is another problem. Net densities may give a precise size of population for a limited area but do not provide a true picture of development density as they exclude other uses and all public space. High-density housing schemes, as for instance the high-rise dormitory towns of the 1960s and early 1970s in many of the UK's cities, concentrate a large number of people into a relatively small number of buildings, but these are frequently surrounded by so much open and generally unusable land that the overall population density is rather small; in such a case the net density figure may actually be misleading. Gross population densities – which indicate the population accommodated by the total area of a city or city region regardless of what the composition of uses and the size of all open space are – provide a much clearer picture of the development density of a town, city or conurbation and allow comparisons between different cities and city regions.

For these and other reasons the question of densities needs to be investigated in conjunction with the discussion of urban and residential models, without which the figures remain relatively meaningless. As Scoffham and Vale (1996, p. 66) point out, 'Density, in itself, is of little importance unless it is related to built form. Compact is meaningless unless it is related to some facts and figures.'

Though this may be generally true, densities of population become rather important when they relate, regardless of the built form, to the viability of local services and facilities, public transport and energy efficiency. It may be true that, no matter how the city itself is structured, high-density housing is in the end socially acceptable to middle- and higher-income groups only if it provides the same quality of living as the country town would (and this necessitates the investigation of dense forms of family homes with garden or other forms of private outdoor space for inner-city areas) and if there is good access to an appropriate range of local services and facilities (see Green, 1996, p. 152). These services and facilities depend, however, upon a certain threshold value of population below which they cease to be economically viable. [. . .]

CONCLUSIONS

Up to this point in the discussion the exact forms and structures that – in conjunction with, or as a result of, all the properties listed above – would render the city more sustainable remain elusive. It has become painfully evident that many of the claims in support of one or the other urban structure are not substantiated. There is no unimpaired evidence that one or the other city model would have a significantly higher or lower level of energy consumption, and investigations of the relationship between transport systems, densities and energy consumption are also largely inconclusive.

In view of the growing awareness of alternative sources of clean and renewable energy it seems only a matter of time until such sources become available on a viable economic scale. One has therefore to ask the question whether the search for a most energy-efficient city is in the long term not somewhat misguided. Reduction if not elimination of the use of fossil fuel for the generation of energy is essential not so much because this will preserve resources for future generations but because of the massive pollution that results from their use. For exactly this reason, future generations may actually not want to use these resources.

Sooner or later the quantity of energy consumed will become less relevant – on the one hand because it is clean, on the other because it is available in abundance. The major problem with car-dependent transport will then no longer be pollution but congestion, which is not solved by clean energy. Even with an abundance of clean energy, a form of urban transport is needed that uses less space; and this can be achieved by people sharing vehicles or vehicles forming trains or by efficient public transport systems. Accordingly, it is much more relevant to search for an urban form that responds to the sustainability criteria listed above, a city that is people-friendly, works efficiently and has a sustainable relationship with the regional and global hinterland. Further investigation concentrates therefore on the search for a structure that enables a high degree of mobility and access to a large variety of different services and facilities without causing congestion, a structure that allows a symbiotic relationship between city and country, a structure that enables social mix, a degree of autonomy of communities and a degree of self-sufficiency, and a structure that generates highly legible and imageable settlement forms.

REFERENCES

Bannister, D. (1992) Energy use, transport and settlement patterns, in Breheny, M.J. (ed.) *Sustainable Development and Urban Form*, Pion, London.

Breheny, M.J. (1992a) Sustainable development and urban form: an introduction, in Breheny, M.J. (ed.) *Sustainable Development and Urban Form*, Pion, London.

Breheny, M.J. (ed.) (1992b) *Sustainable Development and Urban Form*, Pion, London.

Breheny, M.J. (1995) Compact city and transport energy consumption. *Transactions of the Institute of British Geographers*, NS, 20(1), pp. 81–101.

Breheny, M.J. (1996) Centrists, decentrists and compromisers, in Jenks, M., Burton, E. and Williams, K. (eds) *The Compact City: A Sustainable Urban Form?* E & FN Spon, London.

Breheny, M.J. and Rockwood, R. (1993) Planning the sustainable city region, in Blowers, A. (ed.) *Planning for a Sustainable Environment*, Earthscan, London.

Burton, E., Williams, K. and Jenks, M. (1996) The compact city and urban sustainability, in Jenks, M., Burton, E. and Williams, K. (eds) *The Compact City: A Sustainable Urban Form?* E & FN Spon, London.

Commission of the European Communities (1990) *Green Paper on the Urban Environment*, European Commission, Brussels.

Department of the Environment (1992) *Planning Policy Guidance 3: Housing*, HMSO, London.

Department of the Environment (1993) *Planning Policy Guidance 6: Town Centres and Retail Developments*, HMSO, London.

Department of the Environment (1994) *Planning Policy Guidance 15: Planning and the Historic Environment*, HMSO, London.

Department of the Environment (1997) *Planning Policy Guidance 1 (revised): General Policy and Principles*, HMSO, London.

Department of the Environment and Department of Transport (1994) *Planning Policy Guidance 13: Transport*, HMSO, London.

ECOTEC (1993) *Reducing Transport Emissions through Planning*, HMSO, London.

Elkin, T., McLaren, D. and Hillman, M. (1991) *Reviving the City: Towards Sustainable Urban Development*, Friends of the Earth, London.

Enwicht, D. (1992) *Towards an Eco-City: Calming the Traffic*, Envirobook, Sydney.

Girardet, H. (1992) *The Gaia Atlas of Cities: New Directions for Sustainable Urban Living*, Gaia Books, London.

Green, R. (1996) Not compact cities but sustainable regions, in Jenks, M., Burton, E. and Williams, K. (eds) *The Compact City: A Sustainable Urban Form?* E & FN Spon, London.

Haughton, G. and Hunter, C. (1994) *Sustainable Cities*, Jessica Kingsley Publishers, London.

Hillman, M. (1996) In favour of the compact city, in Jenks, M., Burton, E. and Williams, K. (eds) *The Compact City: A Sustainable Urban Form?* E & FN Spon, London.

Hooper, A. (1994) Land availability and the suburban option. *Town and Country Planning*, 63(9), pp. 239–42.

Hough, M. (1989) *City Form and Natural Process*, Routledge, London and New York.

Jacobs, J. (1961) *The Death and Life of Great American Cities*, Vintage Books/Random House, New York.

Jacobs, J. (1970) *The Economy of Cities*, Vintage Books/Random House, New York.

Jenks, M., Burton, E. and Williams, K. (eds) (1996) *The Compact City: A Sustainable Urban Form?* E & FN Spon, London.

Knights, C. (1996) Economic and social issues, in Jenks, M., Burton, E. and Williams, K. (eds) *The Compact City: A Sustainable Urban Form?* E & FN Spon, London.

Krier, L. (1984) Houses, palaces, cities, in Porphyrios, D. (ed.) *Architectural Design Profile* 54.

Lock, D. (1991) Still nothing gained by overcrowding. *Town and Country Planning*, 60(11/12), pp. 337–9.

Lock, D. (1995) Room for more within city limits? *Town and Country Planning*, 64(7), pp. 173–6.

Maslow, A. (1954) *Motivation and Personality*, Harper & Row, New York.

McHarg, I.L. (1992) *Design with Nature*, John Wiley & Sons, New York.

McLaren, D. (1992) Compact or dispersed? Dilution is no solution. *Built Environment*, 18(4), pp. 268–84.

NBA Tectonics (1988) *A Study of Passive Solar Energy Estate Layout*, Department of Energy, London.

Newman, P.W.G. and Kenworthy, J.R. (1989) *Cities and Automobile Dependency: An International Source Book*, Gower Technical, Aldershot.

Owens, S. (1992) Energy, environmental sustainability and land-use planning, in Breheny, M.J. (ed.) *Sustainable Development and Urban Form*, Pion, London.

Owens, S. and Rickaby, P.A. (1992) Settlements and energy revisited, in Breheny, M.J. (ed.) The compact city, *Built Environment,* 18(4), pp. 247–52.

Riley, P. (1979) *Economic Growth, the Allotments Campaign Guide*, Friends of the Earth, London.

Salvatore, F. (1982) *The Potential Role of Vegetation in Improving the Urban Air Quality: A Study of Preventative Medicine*, York-Toronto Lung Association, Willowdale, Ontario.

Scoffham, E. and Vale, B. (1996) How compact is sustainable – how sustainable is compact?, in Jenks, M., Burton, E. and Williams, K. (eds) *The Compact City: A Sustainable Urban Form?* E & FN Spon, London.

Sherlock, H. (1991) *Cities Are Good for Us*, Paladin, London.

Smyth, H. (1996) Running the gauntlet, in Jenks, M., Burton, E. and Williams, K. (eds) *The Compact City: A Sustainable Urban Form?* E & FN Spon, London.

Thomas, L. and Cousins, W. (1996) The compact city: a successful, desirable and achievable urban form?, in Jenks, M., Burton, E. and Williams, K. (eds) *The Compact City: A Sustainable Urban Form?* E & FN Spon, London.

Troy, P.N. (1996) Environmental stress and urban policy, in Jenks, M., Burton, E. and Williams, K. (eds) *The Compact City: A Sustainable Urban Form?* E & FN Spon, London.

UK Government (1994) *Sustainable Development: The UK Strategy*, HMSO, London.

United Nations (1993) Earth Summit Agenda 21: The UN Programme of Action from Rio, United Nations, New York.

van der Valk, A. and Faludi, A. (1992) Growth regions and the future of Dutch planning doctrine, in Breheny, M.J. (ed.) *Sustainable Development and Urban Form*, Pion, London.

Wade, I. (1980) *Urban Self-Reliance in the Third World: Developing Strategies for Food and Fuel Production*, World Future Conference, Toronto.

Welbank, M. (1996) The search for a sustainable urban form, in Jenks, M., Burton, E. and Williams, K. (eds) *The Compact City: A Sustainable Urban Form?* E & FN Spon, London.

World Commission on Environment and Development (1987) *Our Common Future*, Oxford University Press, Oxford.

SIX

PART SEVEN

Elements of the Public Realm

Plate 10 Traffic congestion in Bangkok, Thailand: improvements in mass-transit infrastructure are attempting to mitigate the negative impacts of constant gridlock en-route to transforming Bangkok into a Transit Metropolis. (Photo: M. Larice)

INTRODUCTION TO PART SEVEN

The public realm – street rights-of-way, public parks, urban plazas, public buildings, and other public holdings – represents an enormous portion of a city's space. Streets alone often account for between 25 and 35 percent of all developed land. The public realm is owned by the city and it can be designed for the public. Design the public realm well and the city is likely to feel well-designed; design it poorly and the city will be less than it could be. Understandably then, the field of urban design should be and is highly concerned with the design of the public realm. This includes both the design of public spaces and the design of urban design framework plans and design guidance to shape the impacts and contributions that private development has on the public realm. Part Seven looks at public space design. The focus is primarily on plazas and streets, because these are the types of spaces urban designers are most often called upon to design. Discussion of the urban designer's roles in creating development framework plans is discussed in the readings by Jon Lang and John Punter in Part Eight.

We start with three readings that document research studies on public open spaces and how people use them – and suggest design solutions. Each study used direct empirical approaches that involved careful looking at actual places with a focus on people's everyday needs. In the first reading, three chapters from William H. Whyte's *Social Life of Small Urban Spaces*, we learn about how people use New York City's downtown urban plazas and how the amount of comfortable sitting space is the best predictor of plaza use. We also learn about intriguing urban design research methods including the use of time-lapse photography. The next reading, two chapters taken from Jan Gehl's *Life between Buildings: Using Public Space*, presents a simple and elegant theoretical framework for understanding the relationship between the quality of the physical environment and the amount and types of activities that are likely to occur there. This theory is underpinned by long study of how use of streets and public spaces in central Copenhagen has changed as the area has become progressively pedestrianized. The third reading, excerpted from Clare Cooper Marcus and Carolyn Francis's *People Places: Design Guidelines for Urban Open Space*, presents a typology of downtown plazas and their typical spatial characteristics. The next reading, taken from Randolph T. Hester's classic book *Neighborhood Space*, reminds us that beyond concerns with spatial forms, an understanding of socio-political relationships within a neighborhood can aid in community appropriate and context specific design and planning.

The next three readings focus on different dimensions of street design. We start with the concluding chapter from Allan B. Jacobs' classic book *Great Streets*, in which the most important designable qualities of streets are highlighted. Next, excerpts from David Sucher's *City Comforts: How to Build an Urban Village* provide a wealth of practical design approaches to creating pedestrian-friendly streets. We turn next to the increasingly urgent topic of how urban streets can contribute to environmental sustainability through green design approaches. Excerpts from Metro Portland's *Green Streets* handbook provide an overview of how hydrological processes work, discussion of how streets can perform water retention and pollution filtering services, and specific green street design recommendations. Finally, we turn to the issue of public transit. In a chapter taken from his book *The Transit Metropolis*, Robert Cervero debunks myths about transit in the city that have been commonly used to argue against its public provision.

"Introduction," "The Life of Plazas," "Sitting Space," and "Sun, Wind, Trees, and Water"

from *The Social Life of Small Urban Spaces* (1980)

William H. Whyte

Editors' Introduction

In working on an update of New York City's comprehensive plan in the late 1960s, William H. Whyte began pondering the design effectiveness of the city's plazas, playgrounds, and parks. Because of developers' growing desire for taller skyscrapers, the city had begun to grant density and height bonuses in exchange for public space amenities at the base of new buildings. Whyte applied for a grant to study plaza use, plaza form, general street life, playgrounds, and parks in New York and other cities. Along with a group of young, energetic research assistants, Whyte's Street Life Project researched the effectiveness and use of these public spaces over a multiyear period. The team developed innovative methods of observing and mapping physical activity in the public realm, including the use of time-lapse photography, film, unobtrusive observation, behavior mapping, questionnaires, personal interviews, and pedestrian path analysis.

In analyzing the data, many of the team's early hypotheses were both validated and refuted. Some of their key findings suggested the importance of seating supply and the adaptability of space for personal needs. Other findings about gender preferences, plaza size, sun exposure, and food vending were more surprising. Whyte's work was some of the first to recognize the impacts of the built environment on the behavioral differences of men and women; concluding that women were more careful in selecting preferred seat locations. Some of the broader conclusions from this work suggest that "what attracts people most, it would appear, is other people," and the necessity of well-used public spaces in facilitating greater civic engagement and community interaction – a function both of democracy and quality of life.

William Hollingsworth (Holly) Whyte (1917–1999) graduated from Princeton University, served in the Marine Corps and, in 1946, began his career at *Fortune Magazine* where he coincidentally came in contact with Jane Jacobs. Selling more than 2 million copies, his first literary triumph was *The Organization Man* (New York: Simon & Schuster, 1956), which chronicled the rise of corporate influence and the conformity of the middle class to corporate ideals in the mid-twentieth century. His books on urban form, public space, and design include an edited book, *The Exploding Metropolis* (New York: Doubleday, 1957), *The Last Landscape* (New York: Doubleday, 1968), and *City: Rediscovering the Center* (New York: Doubleday, 1988). A collection of Whyte's writings can be found in Albert LaFarge (ed.), *The Essential William H. Whyte* (New York: Fordham University Press, 2000).

The impacts of this book and the larger Street Life Project were numerous. In addition to the book, a short film of the same name was produced to help disseminate the results of the research to a wider audience. The Street Life Project helped the New York City Planning Commission implement new regulations and guidelines for subsequent development and design review. The observational and mapping methods of William Whyte have been adopted by a number of built environment researchers and his ideas were influential in the establishment of the Project for Public Spaces in New York City.

For additional reading on environmental behavior research and associated methods see: Edward T. Hall, *The Hidden Dimension* (New York: Doubleday, 1966); Robert Sommers, *Personal Space: The Behavioral Basis of Design* (Englewood Cliffs, NJ: Prentice-Hall, 1969); William H. Michelson, *Behavioural Research Methods in Environmental Design* (New York: John Wiley, 1975); Amos Rapoport, *Human Aspects of Urban Form: Towards a Man-Environment Approach to Form and Design* (Oxford: Pergamon, 1977); Marguerite Villeco and Michael Brill, *Environmental Design Research: Concepts, Methods and Values* (Washington, DC: National Endowment for the Arts, 1981); John Zeisel, *Inquiry by Design: Tools for Environment-Behavior Research* (Cambridge: Cambridge University Press, 1984) and *Inquiry by Design: Environment / Behavior / Neuroscience in Architecture, Interiors, Landscape, and Planning*, updated and revised edn (New York: W.W. Norton, 2006); Allan B. Jacobs, *Looking at Cities* (Cambridge, MA: MIT Press, 1985); and Jon Lang, *Creating Architectural Theory: The Role of the Behavioral Sciences in Environmental Design* (New York: Van Nostrand Reinhold, 1987). The Environmental Design Research Association also has several series of publications on this type of material.

INTRODUCTION

This book is about city spaces, why some work for people, and some do not, and what the practical lessons may be. It is a by-product of first-hand observation.

In 1970, I formed a small research group, The Street Life Project, and began looking at city spaces. At that time, direct observation had long been used for the study of people in far-off lands. It had not been used to any great extent in the U.S. city. There was much concern over urban crowding, but most of the research on the issue was done somewhere other than where it supposedly occurred. The most notable studies were of crowded animals, or of students and members of institutions responding to experimental situations – often valuable research, to be sure, but somewhat vicarious.

The Street Life Project began its study by looking at New York City parks and playgrounds and such informal recreation areas as city blocks. One of the first things that struck us was the *lack* of crowding in many of these areas. A few were jammed, but more were nearer empty than full, often in neighborhoods that ranked very high in density of people. Sheer space, obviously, was not of itself attracting children. Many streets were.

It is often assumed that children play in the street because they lack playground space. But many children play in the streets because they like to. One of the best play areas we came across was a block

on 101st Street in East Harlem. It had its problems, but it worked. The street itself was the play area. Adjoining stoops and fire escapes provided prime viewing across the street and were highly functional for mothers and older people. There were other factors at work, too, and, had we been more prescient, we could have saved ourselves a lot of time spent later looking at plazas. Though we did not know it then, this block had within it all the basic elements of a successful urban place.

As our studies took us nearer the center of New York, the imbalance in space use was even more apparent. Most of the crowding could be traced to a series of choke points – subway stations, in particular. In total, these spaces are only a fraction of downtown, but the number of people using them is so high, the experience so abysmal, that it colors our perception of the city around, out of all proportion to the space involved. The fact that there may be lots of empty space somewhere else little mitigates the discomfort. And there is a strong carry-over effect.

This affects researchers, too. We see what we expect to see, and have been so conditioned to see crowded spaces in center city that it is often difficult to see empty ones. But when we looked, there they were.

The amount of space, furthermore, was increasing. Since 1961, New York City has been giving incentive bonuses to builders who provided plazas. For each square foot of plaza, builders could add

10 square feet of commercial floor space over and above the amount normally permitted by zoning. So they did – without exception. Every new office building provided a plaza or comparable space: in total, by 1972, some 20 acres of the world's most expensive open space.

We discovered that some plazas, especially at lunchtime, attracted a lot of people. One, the plaza of the Seagram Building, was the place that helped give the city the idea for the plaza bonus. Built in 1958, this austerely elegant area had not been planned as a people's plaza, but that is what it became. On a good day, there would be a hundred and fifty people sitting, sunbathing, picnicking, and schmoozing – idly gossiping, talking "nothing talk." People also liked 77 Water Street, known as "swingers' plaza" because of the young crowd that populated it.

But on most plazas, we didn't see many people. The plazas weren't used for much except walking across. In the middle of the lunch hour on a beautiful, sunny day the number of people sitting on plazas averaged four per 1,000 square feet of space – an extraordinarily low figure for so dense a center. The tightest-knit CBD (central business district) anywhere contained a surprising amount of open space that was relatively empty and unused.

If places like Seagram's and 77 Water Street could work so well, why not the others? The city was being had. For the millions of dollars of extra space it was handing out to builders, it had every right to demand much better plazas in return.

I put the question to the chairman of the City Planning Commission, Donald Elliott. As a matter of fact, I entrapped him into spending a weekend looking at time-lapse films of plaza use and nonuse. He felt that tougher zoning was in order. If we could find out why the good plazas worked and the bad ones didn't, and come up with hard guidelines, we could have the basis of a new code. Since we could expect the proposals to be strongly contested, it would be important to document the case to a fare-thee-well.

We set to work. We began studying a cross-section of spaces – in all, 16 plazas, 3 small parks, and a number of odds and ends. I will pass over the false starts, the dead ends, and the floundering arounds, save to note that there were a lot and that the research was nowhere as tidy and sequential as it can seem in the telling. Let me also note that

the findings should have been staggeringly obvious to us had we thought of them in the first place. But we didn't. Opposite propositions were often what seemed obvious. We arrived at our eventual findings by a succession of busted hypotheses.

The research continued for some three years. I like to cite the figure because it sounds impressive. But it is calendar time. For all practical purposes, at the end of six months we had completed our basic research and arrived at our recommendations. The City, alas, had other concerns on its mind, and we found that communicating the findings was to take more time than arriving at them. We logged many hours in church basements and meeting rooms giving film and slide presentations to community groups, architects, planners, businessmen, developers, and real-estate people. We continued our research; we had to keep our findings up-to-date, for now we were disciplined by adversaries. But at length the City Planning Commission incorporated our recommendations in a proposed new open-space zoning code, and in May 1975 it was adopted by the city's Board of Estimate. As a consequence, there has been a salutary improvement in the design of new spaces and the rejuvenation of old ones.

But zoning is certainly not the ideal way to achieve the better design of spaces. It ought to be done for its own sake. For economics alone, it makes sense. An enormous expenditure of design expertise, and of travertine and steel, went into the creation of the many really bum office building plazas around the country. To what end? As this manual will detail, it is far easier, simpler to create spaces that work for people than those that do not – and a tremendous difference it can make to the life of a city.

THE LIFE OF PLAZAS

We started by studying how people use plazas. We mounted time-lapse cameras overlooking the plazas and recorded daily patterns. We talked to people to find where they came from, where they worked, how frequently they used the place and what they thought of it. But, mostly, we watched people to see what they did [Figure 1].

Most of the people who use plazas, we found, are young office workers from nearby buildings. There may be relatively few patrons from the

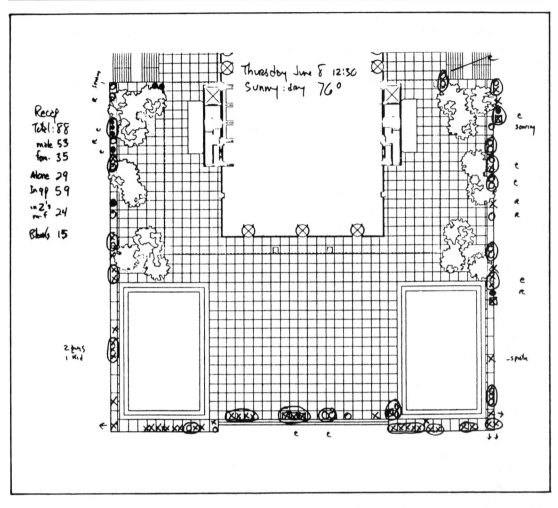

Figure 1 This is a typical sighting map. We found that one could map the location of every sitter, whether male (X), female (O), alone, or with others (XO), in about five minutes, little more time than a simple head count would take.

plaza's own building; as some secretaries confide, they'd just as soon put a little distance between themselves and the boss. But commuter distances are usually short; for most plazas, the effective market radius is about three blocks. Small parks, like Paley and Greenacre in New York, tend to have more assorted patrons throughout the day – upper-income older people, people coming from a distance. But office workers still predominate, the bulk from nearby.

This uncomplicated demography underscores an elemental point about good urban spaces: supply creates demand. A good new space builds a new constituency. It stimulates people into new habits – al fresco lunches – and provides new paths to and from work, new places to pause. It does all this very

quickly. In Chicago's Loop, there were no such amenities not so long ago. Now, the plaza of the First National Bank has thoroughly changed the midday way of life for thousands of people. A success like this in no way surfeits demand for spaces; it indicates how great the unrealized potential is.

The best-used plazas are sociable places, with a higher proportion of couples than you find in less-used places, more people in groups, more people meeting people, or exchanging goodbyes. At five of the most-used plazas in New York, the proportion of people in groups runs about 45 percent; in five of the least used, 32 percent. A high proportion of people in groups is an index of selectivity. When people go to a place in twos or threes or rendezvous there, it is most often because they have decided to. Nor are these sociable places less congenial to the individual. In absolute numbers, they attract more individuals than do less-used spaces. If you are alone, a lively place can be the best place to be.

The most-used places also tend to have a higher than average proportion of women. The male-female ratio of a plaza basically reflects the composition of the work force, which varies from area to area – in midtown New York it runs about 60 percent male, 40 percent female. Women are more discriminating than men as to where they will sit, more sensitive to annoyances, and women spend more time casting the various possibilities. If a plaza has a markedly lower than average proportion of women, something is wrong. Where there is a higher than average proportion of women, the plaza is probably a good one and has been chosen as such.

The rhythms of plaza life are much alike from place to place. In the morning hours, patronage will be sporadic. A hot-dog vendor setting up his cart at the corner, elderly pedestrians pausing for a rest, a delivery messenger or two, a shoeshine man, some tourists, perhaps an odd type, like a scavenger woman with shopping bags. If there is any construction work in the vicinity, hard hats will appear shortly after 11:00 A.M. with beer cans and sandwiches. Things will start to liven up. Around noon, the main clientele begins to arrive. Soon, activity will be near peak and will stay there until a little before 2:00 P.M. Some 80 percent of the total hours of use will be concentrated in these two hours. In mid and late afternoon, use is again sporadic. If there's a special event, such as a jazz concert, the flow going home will be tapped, with people staying as late as 6:00 or 6:30 P.M. Ordinarily, however, plazas go dead by 6:00 and stay that way until the next morning.

During peak hours the number of people on a plaza will vary considerably according to seasons and weather. The way people distribute themselves over the space, however, will be fairly consistent, with some sectors getting heavy use day in and day out, others much less. In our sightings we find it easy to map every person, but the patterns are regular enough that you could count the number in only one sector, then multiply by a given factor, and come within a percent or so of the total number of people at the plaza.

Off-peak use often gives the best clues to people's preferences. When a place is jammed, a person sits where he can. This may or may not be where he most wants to. After the main crowd has left, the choices can be significant. Some parts of the plaza become quite empty; others continue to be used. At Seagram's, a rear ledge under the trees is moderately, but steadily, occupied when other ledges are empty; it seems the most uncrowded of places, but on a cumulative basis it is the best-used part of Seagram's.

Men show a tendency to take the front-row seats, and, if there is a kind of gate, men will be the guardians of it. Women tend to favor places slightly secluded. If there are double-sided benches parallel to a street, the inner side will usually have a high proportion of women; the outer, of men.

Of the men up front, the most conspicuous are girl watchers. They work at it, and so demonstratively as to suggest that their chief interest may not really be the girls so much as the show of watching them. Generally, the watchers line up quite close together, in groups of three to five. If they are construction workers, they will be very demonstrative, much given to whistling, laughing, direct salutations. This is also true of most girl watchers in New York's financial area. In midtown, they are more inhibited, playing it coolly, with a good bit of sniggering and smirking, as if the girls were not measuring up. It is all machismo, however, whether uptown or downtown. Not once have we ever seen a girl watcher pick up a girl, or attempt to.

Few others will either. Plazas are not ideal places for striking up acquaintances, and even on the most sociable of them, there is not much mingling. When strangers are in proximity, the nearest thing to an exchange is what Erving Goffman has called civil inattention. If there are, say, two smashing blondes on a ledge, the men nearby will usually put on an elaborate show of disregard. Watch closely, however, and you will see them give themselves away with covert glances, involuntary primping of the hair, tugs at the earlobe.

Lovers are to be found on plazas. But not where you would expect them. When we first started interviewing, people told us we'd find lovers in the rear places (pot smokers, too). But they weren't usually there. They would be out front. The most fervent embracing we've recorded on film has

children, instead of, for example, where there are only toys. In residential areas and in city spaces, comparable behavior among adults can be observed. If given a choice between walking on a deserted or a lively street, most people in most situations will choose the lively street. If the choice is between sitting in a private backyard or in a semiprivate front yard with a view of the street, people will often choose the front of the house where there is more to see.

In Scandinavia an old proverb tells it all: "people come where people are."

A series of investigations illustrates in more detail the interest in being in contact with others. Investigations of children's play habits in residential areas[4] show that children stay and play primarily where the most activity is occurring or in places where there is the greatest chance of something happening.

Both in areas with single-family houses and in apartment house surroundings, children tend to play more on the streets, in parking areas, and near the entrances of dwellings than in the play areas designed for that purpose but located in backyards of single-family houses or on the sunny side of multistory buildings, where there are neither traffic nor people to look at.

Corresponding trends can be found regarding where people choose to sit in public spaces. Benches that provide a good view of surrounding activities are used more than benches with less or no view of others.

An investigation of Tivoli Garden in Copenhagen,[5] carried out by the architect John Lyle, shows that the most used benches are along the garden's main path, where there is a good view of the particularly active areas, while the least used benches are found in the quiet areas of the park. In various places, benches are arranged back to back, so that one of the benches faces a path while the other "turns its back." In these instances it is always the benches facing the path that are used.

Comparable results have been found in investigations of seating in a number of squares in central Copenhagen. Benches with a view of the most trafficked pedestrian routes are used most, while benches oriented toward the planted areas of the squares are used less frequently.[6]

At sidewalk cafés, as well, the life on the sidewalk in front of the café is the prime attraction.

Almost without exception café chairs throughout the world are oriented toward the most active area nearby. Sidewalks are, not unexpectedly, the very reason for creating sidewalk cafés.

The opportunity to see, hear, and meet others can also be shown to be one of the most important attractions in city centers and on pedestrian streets. This is illustrated by an attraction analysis carried out on Strøget, the main pedestrian street in central Copenhagen, by a study group from the School of Architecture at the Royal Danish Academy of Fine Arts.[7] The analysis was based on an investigation of where pedestrians stopped on the walking street and what they stopped to look at.

Fewest stops were noted in front of banks, offices, showrooms, and dull exhibits of, for example, cash registers, office furniture, porcelain, or hair curlers. Conversely, a great number of stops were noted in front of shops and exhibits that had a direct relationship to other people and to the surrounding social environment, such as newspaper kiosks, photography exhibits, film stills outside movie theaters, clothing stores, and toy stores.

Even greater interest was shown in the various human activities that went on in the street space itself. All forms of human activity appeared to be of major interest in this connection.

Considerable interest was observed in both the ordinary, everyday events that take place on a street – children at play, newlyweds on their way from the photographers, or merely people walking by – and in the more unusual instance – the artist with his easel, the street musician with his guitar, street painters in action, and other large and small events.

It was obvious that human activities, being able to see other people in action, constituted the area's main attraction.

The street painters collected a large crowd as long as their work was in progress, but when they left the area, pedestrians walked over the paintings without hesitation. The same was true of music. Music blaring out on the street from loudspeakers in front of record shops elicited no reaction, but the moment live musicians began to play or sing, there was an instantaneous show of lively interest.

The attention paid to people and human activities was also illustrated by observations made in connection with the expansion of a department store in the area. While excavation and pouring of foundations were in progress, it was possible to see into the building site through two gates facing the pedestrian street. Throughout this period more people stopped to watch the work in progress on the building site than was the case for stops in front of all the department store's fifteen display windows together.

In this case, too, it was the workers and their work, not the building site itself, that was the object of interest. This was demonstrated further during lunch breaks and after quitting time – when no workers were on the site, practically nobody stopped to look.

A summary of observations and investigations shows that people and human activity are the greatest object of attention and interest. Even the modest form of contact of merely seeing and hearing or being near to others is apparently more rewarding and more in demand than the majority of other attractions offered in the public spaces of cities and residential areas.

Life in buildings and between buildings seems in nearly all situations to rank as more essential and more relevant than the spaces and buildings themselves.

NOTES

1 Gehl, Ingrid. *Bo-miljø* (Living Environment-Psychological Aspects of Housing). Danish Building Research Institute, report 71. Copenhagen: Teknisk Forlag, 1971.

2 Ibid.

3 Gehl, Jan. *Attraktioner på Strøget.* Kunstakademiets Arkitektskole. Studyreport. Copenhagen, 1969; Gehl, Jan. "Mennesker til fods" (Pedestrians). *Arkitekten* (Danish) 70, no. 20 (1968): 429–46; Jacobs, Jane. *The Death and Life of Great American Cities.* New York: Random House, 1961; Whyte, William H. *The Social Life of Small Urban Spaces.* Washington D.C.: Conservation Foundation, 1980.

4 Kjærsdam, Finn. *Haveboligområdets fællesareal.* Parts 1 and 2. Part 1 published by: Den kongelige Veterinær og Landbohøjskole, Copenhagen, 1974. Part 2 by: Aalborg Universitetscenter, ISP, Aalborg, 1976; Morville, Jeanne. *Planlægning af børns udemiljø i etageboligområder* (Planning for Children in Multistory Housing Areas). Danish Building Research Institute, report 11. Copenhagen: Teknisk Forlag, 1969.

5 Lyle, John. "Tivoli Gardens." *Landscape* (Spring/Summer 1969): 5–22.

6 Gehl, Jan. *Attraktioner på Strøget.* Kunstakademiets Arkitektskole. Studyreport. Copenhagen, 1969; Gehl, Jan. "Mennesker til fods" (Pedestrians). *Arkitekten* (Danish) 70, no. 20 (1968): 429–46; Kao, Louise. "Hvor sidder man på Kongens Nytorv?" (Sitting Preferences on Kongens Nytorv). *Arkitekten* (Danish) 70, no. 20 (1968): 445.

7 Gehl, Jan. *Attraktioner på Strøget.* Kunstakademiets Arkitektskole. Studyreport. Copenhagen, 1969; Gehl, Jan. "Mennesker til fods" (Pedestrians). *Arkitekten* (Danish) 70, no. 20 (1968): 429–46

"Urban Plazas"

from *People Places: Design Guidelines for Urban Open Space* (1998)

Clare Cooper Marcus and Carolyn Francis

Editors' Introduction

Cities intent on improving the quality of development have found it useful to develop design guidelines that speak to community aspirations and give direction to developers and their landscape architects and architects. Design guidelines take different forms. Some are highly prescriptive, stating exactly what should be done in terms of dimension, materials, and/or style. Others are performance-based, describing general desired outcomes but leaving the specifics up to individual designers.

A number of design guidelines for public open spaces have recently been developed by cities and non-profit groups concerned with creating places that work for people, most notably the Project for Public Space's *How to Turn a Place Around* (2000). Many of these guidelines draw on the pioneering work done by landscape architect Clare Cooper Marcus, who has spent years observing how people use public spaces and asking people how they feel about the spaces they use. In this reading from *People Places: Design Guidelines for Urban Open Space*, co-authored with Carolyn Francis, the role of urban plazas is defined, a classification of different plaza types based on their purposes is presented, and the important spatial characteristics of each plaza type are articulated. The typology does not prescribe what or where physical elements should be put within particular plazas, but rather speaks to the role each plaza type plays in supporting urban public life. The intent of the authors is to provide a basis upon which cities might develop their own local guidelines for specific plaza types.

Clare Cooper Marcus is a Professor at the University of California, Berkeley, College of Environmental Design. She taught social factors in design to several generations of landscape architects. Her other writings that articulate design guidelines based on research findings include *Housing as if People Mattered: Site Design Guidelines for Medium Density Family Housing*, with Wendy Sarkissian (Berkeley, CA: University of California Press, 1988) and *Healing Gardens, Therapeutic Benefits and Design Recommendations*, with Marni Barnes (New York: John Wiley, 1999).

IS THERE A ROLE FOR THE URBAN PLAZA?

In *Dreaming of Urban Plazas*, Robert Jensen points out that we are frequently disappointed in the urban places we create because we imagine that they will be like the quintessential urban spaces of Siena or Barcelona.

It is indicative that we call them plazas or sometimes piazzas. Our own English word "place" won't do. "Place" is derived from the Latin word "platea" meaning an open space or broadened street as in the Spanish "plaza" and the Italian "piazza" ... The word is at once too common and too diverse in its meaning to designate what we want in an urban center downtown. So we turn to Spanish and Italian. That is what we want.

(Jensen 1979, 52)

But not necessarily what we get.

The modern plaza is not the piazza of days gone by, yet it does have some relevant contextual and functional parallels. Is it farfetched to consider the corporate sky-scraper the modern equivalent of the medieval cathedral, each symbolizing, for its era, the seat of power? The public outdoor space next to each is, or was, crowded at certain times of the day because that particular building function attracted people. In each case, the primary people generator (cathedral and corporate office tower) has, or had, a vested interest in the appearance of the space and in how it is used. What is undoubtedly different is that the contemporary office plaza has a very limited range of uses compared with those of the medieval piazza. Indeed, according to observation studies of modern plaza use, sitting, standing, walking, and their combination with eating, reading, watching, and listening account for more than 90 percent of all use.

It is not surprising that many Americans spend their vacations in Europe. Part of the attraction is that the hearts of many European cities are dedicated to pedestrian movement. To a resident of Los Angeles – where a planning report referred to the pedestrian as "the largest single obstacle to free traffic movement" (Rudofsky 1969, 106) – strolling through the streets and plazas of a French, German, or Italian city must indeed seem like being in another world.

Rudofsky suggested in *Streets for People* that just as we look back in amazement to an era when people were willing to share the streets with trains, so "future generations will marvel at the obtuseness of people who thought nothing of sharing the street with motor cars" (1969, 341). Although Americans have less of a tradition than do Europeans for strolling, promenading, or frequenting outdoor cafes, studies of street life in U.S. cities indicate that more and more people are recreating in downtown outdoor space. Whyte found a 30 percent increase between 1972 and 1973 in the number of people sitting in plazas and small parks in Manhattan; he found between 1973 and 1974 an additional 20 percent increase. He concluded that more people are getting into the habit of sitting in plazas and that

with each new plaza the clientele grows (1974, 27). Outdoor eating has also become more popular: "There is more picnicking in the parks and plazas and on the library steps at lunch, and throughout the day so much street corner munching of hot dogs and knishes it is a wonder every New Yorker is not fat" (Whyte 1974, 28). Also on the increase are public displays of affection, smiling, street entertainment, crazy characters, shmoozing (groups engaged in sidewalk gossiping), and impromptu sidewalk "conferences" among businesspersons.

Gehl (1987) reported that as the total area of pedestrian streets and squares in Copenhagen tripled between 1968 and 1986, the number of people standing or sitting in those areas tripled also, while the total city population remained the same. Thus, even in northern Europe, with no particular tradition for street life, public outdoor activities are on the rise.

Studies of the U.S. West Coast confirm the same trends in San Francisco and Seattle. Enhancing these trends are the economic situation, encouraging more people to bring lunches from home; the demographic trend of more people living alone and perhaps seeking relaxed conversations and companionship during the lunch hour; and the stress of office environments.

Downtown open spaces are also being used more by the people who live there, not just by the lunch-hour crowd. In San Francisco, Union Square is an outdoor living room for elderly people who live in downtown hotels, a place for lunch-hour relaxation for nearby shop and office workers, and a weekend neighborhood park for Chinatown residents.

DEFINITION

According to J.B. Jackson (1985), a plaza is an urban form that draws people together for passive enjoyment. Kevin Lynch (1981, 443) suggested that

> the plaza is intended as an activity focus, at the heart of some intensive urban area. Typically, it will be paved, enclosed by high-density structures, and surrounded by streets, or in contact with them. It contains features meant to attract groups of people and to facilitate meetings.

For our purposes, a plaza is defined as a mostly hard-surfaced, outdoor public space from which cars are excluded. Its main function is as a place for strolling, sitting, eating, and watching the world go by. Unlike a sidewalk, it is a place in its own right rather than a space to pass through. Although there may be trees, flowers, or ground cover in evidence, the predominant ground surface is hard; if grass and planted areas exceed the amount of hard surface, we define the space as a park rather than a plaza.

A TYPOLOGY OF DOWNTOWN PLAZAS

The purpose of the following typology is to make sense of the categories of downtown open space in U.S. cities. Although this typology was developed in San Francisco, we believe it can be applied to most cities. It may be used as a basis (1) for understanding the variety of spaces described in this chapter as urban plazas, (2) for categorizing plaza spaces in a specific city, and (3) for developing local guidelines for specific plaza types.

One could categorize downtown spaces in many ways: by size, use, relationship to street, style, predominant function, architectural form, location, and so on. But because this book is concerned with the interplay of form and use, our classification is based on a mix of form and use, moving from the smallest to the largest in size. We describe six broad categories of plazas plus subcategories of each. [. . .]

The street plaza

A street plaza is a small portion of public open space immediately adjacent to the sidewalk and closely connected to the street. It sometimes is a widening of the sidewalk proper or an extension of it under an arcade. Such spaces are generally used for brief periods of sitting, waiting, and watching, and they tend to be used more by men than by women.

■ *The seating edge:* A seating-height wall or stepped edge to a sidewalk.
■ *The widened sidewalk:* A widened portion of the sidewalk furnished with seating blocks, steps, or bollards. Used primarily for viewing passersby.
■ *The bus-waiting place:* A portion of the sidewalk at a bus stop, sometimes furnished with a bench, shelter, kiosk, or litter container.

■ *The pedestrian link:* An outdoor passage or alley that connects two blocks or, sometimes, two plazas. Sometimes it is wide enough for planting; other times it is just a passageway between buildings. It is used almost exclusively for walking.
■ *The corner sun pocket:* A building footprint that is designed to open up a small plaza where two streets meet and where there is access to sun during the peak lunchtime period. It is used for sitting, viewing, eating lunch, and the like.
■ *The arcade plaza:* A sidewalk that is widened by means of an extension under a building overhang. It is sometimes furnished with chairs or benches.

The corporate foyer

The corporate foyer is part of a new, generally high-rise building complex. Its main function is to provide an elegant entry and image for its corporate sponsor. It is usually privately owned but accessible to the public. It is sometimes locked after business hours.

■ *The decorative porch:* A small decorative entry, sometimes planted or supplied with seating or a water feature. It often is too narrow or shaded to encourage much use.
■ *The impressive forecourt:* A larger entry plaza, often finished in expensive materials (marble, travertine) and sometimes designed to discourage any use but passing through.
■ *The stage set:* A very large corporate plaza flanked by an impressive tall building that it helps frame. The plaza often is detailed so as to discourage use by "undesirables" or to minimize its use for sitting or eating. It is primarily a stage set with building as a backdrop.

The urban oasis

The urban oasis is a type of plaza that is more heavily planted, has a garden or park image, and is partially secluded from the street. Its location and design deliberately set this place apart from the noise and activity of the city For example, Banerjee and Loukaitou-Sideris (1992, 93–94) quote advertising for Citicorp Plaza in Los Angeles: ". . . a park-like setting where wide benches and shade trees provide a quiet respite from a busy day," or, in the words of the architect, a place where greenery was intended to hide from the users the "horrifying buildings and the city outside and create a truly refreshing difference in the daily life of the city and a high contrast to its buildings pavements and sidewalks." The urban oasis is often popular for lunchtime eating, reading, and socializing, and is the one category that tends to attract more women than men or, at least, equal proportions of each. The urban oasis has a quiet, reflective quality.

- *The outdoor lunch plaza:* A plaza separated from the street by a level change or a pierced wall and furnished for comfortable lunchtime use. It is often attractively planted, provided with more than adequate seating, and sometimes incorporates a café or take-out restaurant.
- *The garden oasis:* A small plaza, often enclosed and secluded from the street, whose high density and variety of planting conveys a garden image. The garden oasis sometimes includes flower planters and a water feature and usually supplies a variety of seating possibilities. It is popular for lunchtime use and as a quiet respite from city activities.
- *The roof garden:* A rooftop area developed as a garden setting for sitting, walking, and viewing. Its access is sometimes poorly signed, perhaps to discourage heavy use.

The transit foyer

The transit foyer is a plaza type of space created for easy access in and out of heavily used public transit terminals. Although the detailing may not encourage any activity but passing through, the captive audience of transit users sometimes draws street entertainers, vendors, and people watchers.

- *The subway entry place:* A place for passing through, waiting, meeting, and watching. It sometimes becomes a favorite hangout for a particular group (e.g., teens) who can reach this place by public transit.
- *The bus terminal:* Where many city bus lines converge and many commuters arrive and leave the city center each day. It is primarily a space

to move through, but it sometimes attracts vendors of newspapers, flowers, light snacks, and the like.

The street as plaza – pedestrian and transit malls

When a street is closed to traffic, it has the potential to take on the role of a plaza – that is, to become a place where people stroll, sit, eat, and watch the activity going on around them. Most often located in the traditional downtown area, pedestrian malls comprise a number of continuous blocks along a shopping street that have been relegated entirely or primarily to use by people on foot. In most cases a pedestrian mall has modified paving, either eliminating or narrowing the pre-existing roadway; increased planting; and some level of street furniture. It may or may not include amenities such as food sources, vendors, entertainment, or public art – as relevant here as in more standard plazas.

- *Traditional pedestrian mall:* A street completely closed to traffic with permanent bollards, curbs, or other design detailing.
- *Mixed mall:* A pedestrian mall that allows limited use by automobiles, perhaps only during certain hours, and typically on a constricted roadbed.
- *Transit mall:* A pedestrian mall that incorporates public transit, allowing for buses, shuttles, or other transit, but not for private automobiles. The most common type of pedestrian mall being developed in the United States in the 1980s and 1990s.

The grand public place

The grand public place comes closest to our image of the old-world town square or piazza. When located near a diversity of land uses (office, retail, warehouse, transit) it tends to attract users from a greater distance and in greater variety (by age, gender, ethnicity) than do other plazas. Such an area is often big and flexible enough to host brown-bag lunch crowds; outdoor cafes; passers-through; and occasional concerts, art shows, exhibits, and rallies. It is usually publicly owned and is often considered the heart of the city – the place where an annual Christmas tree might be erected or guests are taken for a visit.

- *The city plaza:* An area predominantly hard surfaced, centrally located, and highly visible. It is often the setting for programmed events such as concerts, performances, and political rallies.
- *The city square:* A centrally located, often historic place where major thoroughfares intersect. Unlike many other kinds of plazas, it is not attached to a particular building; rather, it often encompasses one or more complete city blocks and is usually bounded by streets on all four sides. Hardscape and planting are often finely balanced, so that this place could be considered midway between a plaza and a park. Sometimes it contains a major monument, statue, or fountain. It attracts a variety of users and activities. However, sometimes, because of its central location and high land value, the city square has been redesigned to incorporate underground parking.

This typology is not necessarily exhaustive; rather, it is presented as a starting point for thinking about downtown plaza spaces. [. . .] guidelines can apply to any or all of these types. Some cities (for example, San Francisco) have developed their own categories or guidelines for each. We have not attempted to do this because we hope that our recommendations can be utilized or modified for settings that differ by region, climate, and culture.

NOTES

With special thanks to Eva Liebermann, Tridib Banerjee, and Anastasia Loukaitou-Sideris for input to the revised edition.

REFERENCES

Banerjee, Tridib, and Anastasia Loukaitou-Sideris. 1992. *Private Production of Downtown Public Open Spaces: Experiences of Los Angeles and San Francisco.* Los Angeles: University of Southern California, School of Urban and Regional Planning.

Gehl, Jan. 1987. Life between buildings: Using public space. New York: Van Nostrand Reinhold.

Jackson, J.B. 1985. Vernacular space. *Texas Architect.* 35(2): 58–61.

Jensen, Robert. 1979. Dreaming of urban plazas. In *Urban Open Spaces*, L. Taylor (ed.). New York: Rizzoli.

Lynch, Kevin. 1981. *Good City Form.* Cambridge, MA: MIT Press.

Rudofsky, Bernard. 1969. *Streets for People: A Primer for Americans.* Garden City, N.Y.: Anchor Press/Doubleday.

Whyte, William. 1974. The best street life in the world. *New York Magazine*, July 15, 26–33.

"Neighborhood Space"

from *Neighborhood Space* (1975)

Randolph Hester

Editors' Introduction

As the twenty-first century gets under way, it might be expected that making fine, satisfying neighborhoods would be a well-understood activity. Generally (if not universally) admired residential quarters have been created throughout history, providing plentiful lessons for contemporary neighborhood design. Each of us can fall into easy reverie over the everyday places we like to visit and where other people live, work and socialize with comfort and ease. Great neighborhoods abound in our travels, yet often seem distant memories in the current places where many of us live. The emphasis in contemporary development on investment returns and speculative profit-making has resulted in very different physical outcomes compared to the pedestrian-oriented traditional neighborhoods of the pre-industrial era. Issues of gigantism, the privatization of public space, incomplete services, zoned-out land uses, and placelessness litter the critiques of the contemporary development scene. These critiques have induced improvements in some places, but these tend to be rare. With the waning effectiveness of "as-of-right" permitting and comprehensive citywide planning (particularly in larger and harder-to-manage American cities), many municipalities have become increasingly drawn to neighborhood planning and discretionary decision-making as a means of directing improvements in livability, services, and place-making at a more manageable scale. Recently, neighborhood planning has been made more effective by growing levels of participation from local residents, enhanced oversight with respect to new development, and greater responsiveness by officials to neighborhood issues.

Various definitions and concepts of *neighborhood* have been identified over the years – many seemingly synonymous with the idea of community. A key distinction between these two terms, however, is the required geographic nature of neighborhoods – whereas communities can exist outside place-based settings. Prior to analyzing the concept of *Neighborhood Space*, in his book of the same name, Randolph Hester reviews the evolution and history of neighborhood theory. He defines neighborhood space as "that public outdoor territory close to home, which, because of the residents' collective responsibility, familiar association, and frequent shared use, is considered to be their own." Hester is writing for a design audience here, introducing designers to new concepts of urban analysis, public participation and neighborhood planning. Throughout the writing, he remains focused on user-responsive design and place-making. He advocates a highly participatory and grassroots type of planning here, long before it became a popular method of ensuring plan ownership and local relevancy in current practice.

Randolph Hester is a Professor of Landscape Architecture and Environmental Planning at the University of California, Berkeley, where he teaches neighborhood design, sacred landscapes, community design participation, and environmental justice. He is a founder of the research movement to apply sociological principles to environmental design. Much of his recent research is focused on environmental issues and cultural landscapes in Taiwan. Other books by Randolph Hester include: *Community Goal Setting*, written with Frank J. Smith and Ali Smith (Stroudsburg, PA: Hutchinson Ross, 1996, original 1982), *Planning Neighborhood Space with People* (New York: Van Nostrand Reinhold, 1982), *The Meaning of Gardens: Idea, Place and*

Action, edited with Mark Francis (Cambridge, MA: MIT Press, 1990), *Community Design Primer* (Mendocino, CA: Ridge Times Press, 1990), *A Theory for Building Community*, written with S. Chang (Taipei: Yungliou Press, 1999), and *Democratic Design in the Pacific Rim*, written with C. Kweskin (Mendocino, CA: Ridge Times Press, 1999).

For classic and contemporary literature on neighborhoods, neighborhood design, and planning see: Clarence Perry, 'The Neighborhood Unit,' in *Regional Study of New York and its Environs, Volume 7* (New York: Sage, 1929); American Public Health Association, *Planning the Neighborhood: Standards for Healthful Housing* (Chicago, IL: Public Administration Service, 1948); Herbert J. Gans, *The Urban Villagers: Group and Class in the Life of Italian Americans* (New York: Free Press, 1962); Suzanne Keller, *The Urban Neighborhood: A Sociological Perspective* (New York: Random House, 1968); Milton Kotler, *Neighborhood Government* (Indianapolis, IN: Bobbs-Merrill, 1969); Tridib Banerjee and William C. Baer, *Beyond the Neighborhood Unit: Residential Environments and Public Policy* (New York: Plenum Press, 1984); William M. Rohe and Lauren B. Gates, *Planning with Neighborhoods* (Chapel Hill, NC: University of North Carolina Press, 1985); C. Silver, "Neighborhood Planning in Historical Perspective," *Journal of the American Planning Association* (vol. 51, no. 2, pp. 161–174, 1985); William R. Moorish and Catherine R. Brown, *Planning to Stay: Learning to See the Physical Features of your Neighborhood* (Minneapolis, MN: Milkweed, 1994, 2000); Sidney Brower, *Good Neighborhoods: A Study of In-Town and Suburban Residential Environments* (Westport, CT: Praeger, 2000); Andres Duany and Elizabeth Plater-Zyberk, "The Neighborhood, the District, and the Corridor," in Peter Katz and Vincent Scully, Jr. (eds), *The New Urbanism: Toward an Architecture of Community* (New York: McGraw-Hill, 1993); Ali Madanipour, "How Relevant Is 'Planning by Neighborhoods' Today?" *Town Planning Review* (vol. 72, no. 2, pp. 171–191, 2001); and Urban Design Associates, *The Architectural Pattern Book: A Tool for Building Great Neighborhoods* (New York: W.W. Norton, 2004).

NEIGHBORHOOD SPACE

Important components of a neighborhood

There has been much discussion of what a neighborhood is.[1] Physical planners, designers, and geographers have concentrated on the physical dimensions of a neighborhood, describing it in terms of boundaries and areas. The work of Robert E. Park and E.W. Burgess stressed the physical features of a neighborhood: land use, density, street patterns, "natural" boundaries, condition of dwelling units, and amount of open space. Since 1915, when Robert E. Park introduced the idea of neighborhood as an ecological concept with planning implications,[2] physical planners and designers have dealt with the idea of neighborhood in a rather simple manner: in particular, they have considered the physical aspects without giving much thought to the social aspects of a neighborhood. Although flavored with social idealism, the work of the best-known planners and designers – Clarence Stein, Walter Gropius, Le Corbusier, and Frank Lloyd Wright – reflects a physical concept of neighborhood by defining it in terms of desirable features, such as:

- A neighborhood focal point – usually the elementary school with its recreation area; (ideally centered in a park with not more than a half-mile walk through the park to the kindergarten or the elementary school without crossing traffic).
- Peripheral access road with connections to a free-flowing parkway.
- Elimination of rectangular residential islands surrounded by concrete rivers, with their streaming traffic.
- Safe residential streets laid out in short loops and culs-de-sac.
- Every home abutting planned park and recreation spaces.
- Neighborhood facilities – an auditorium, parklets, parks, and recreation courts, including: tot lots, baseball and basketball, "golden age" areas, and nationality games such as bocci and bowling on the green.
- A neighborhood shopping center.[3]

On the other hand, social planners and sociologists have stressed with equal singularity the social dimensions of a neighborhood. They view the neighborhood in terms of its symbolic and cultural aspects, and emphasize shared activities and experiences, the resulting social groupings, and common values and loyalties. The physical environment is taken for granted. P.H. Mann has concluded that the physical aspects of a neighborhood are sterile, unrelated to the social aspects, and therefore not useful.

A few planning theoreticians have pled for a unified definition that combined both the social and the physical aspects of a neighborhood. In England, Ruth Glass recognized both an area with physical characteristics and a territorial group with primary social interaction. Terrence Lee proposed that the urban neighborhood be defined as a socio-spatial schema, a definition that most clearly combined the social and physical components of neighborhood into a unified schema. The work of Glass and Lee signaled the beginning of a serious effort to define "neighborhood" more comprehensively from the users' points of view. Many planners, designers, and social scientists have since tried to define neighborhood by relating human behavior and geography, land development and social predictions, and city planning and social change.

At the same time another group of sociologists contended that the neighborhood did not need to be defined; rather, it needed to be discarded as a planning unit. This group claimed that neighborhood planning was security-seeking nostalgia, a desire to reconstruct the simplicity of the days when the New England town meeting symbolized the urbanites' rootedness in local territory. R.R. Issacs and Louis Wirth, among others, sought to show that man's behavior was no longer oriented to the local area but to the city, nation, and world. Therefore, a definition of the neighborhood was irrelevant.

Throughout the debates surrounding the various definitions of neighborhood, the application of the neighborhood unit to planning continued. The concept reached an apex of popularity among planners after World War II, but the results were disillusioning, and the concept fell out of favor. Many of the theories were impracticable. They suffered in practice not only because they were applied in a simplistic manner that smacked of physical determinism, but also because they did not describe

neighborhood from the users' point of view. In addition, the theoretical definitions largely ignored the political aspects of local communities. This became painfully clear in the 1960s when planning proposal after planning proposal was opposed by neighborhood groups.

What is needed, then, is a practical definition that embraces the interest of residents in their neighborhoods, that recognizes the social, spatial, and political aspects of a neighborhood, that is user defined, that speaks to the issue of the importance of local relationships, and that is useful in planning.

Milton Kotler has proposed such a definition: "The neighborhood is a political settlement of small territory and familiar association, whose absolute property is its capacity for deliberative democracy."[4] A sixth grader has said: "A neighborhood is when you can get a hundred and fifty people to protest at City Hall against a highway proposed to go through a residential area." Kotler's recognition of the political nature of neighborhood is particularly important in light of recent demands by citizens to have an active role in shaping their neighborhoods and in light of their willingness to go to great lengths to control their neighborhoods. But what does he mean by "small territory" and "familiar association"? These concepts formed the basis of many of the theories of neighborhood discussed earlier, but few planners were able to combine them in a socio-spatial schema.[5] That Kotler accepts both concepts within his political context makes his definition extremely useful. Kotler's definition implies that the size of the small territory depends upon the political issue. If the location of a thoroughfare will affect several thousand people in an area of several square miles, that is a small territory. If the location of a minipark will affect only a block, that is a small territory. The idea that a neighborhood changes in the minds of the residents according to the nature of the political crisis or issue is basic to a functional definition, but it does not deal with the nonpolitical aspects of daily human activity and interaction, which are just as important to many residents. A small child was not concerned about any political issue when she remarked that "a neighborhood is when your friend lives on the same block." There are many people who would define a neighborhood similarly, without political overtones. It is appropriate, then, to view the small territory as that area close to home which, because of frequent use and familiarity, is considered one's own. This aspect of Kotler's definition is particularly relevant to heavy users of the area close to home, the young, the old, the poor, the homebound, and the deviant. It still suggests that neighborhood boundaries are not fixed but vary from person to person, depending on life-cycle stage, life style, ethnicity, and personal preference. Planners, when referring to this dynamic quality, say that the small territory "roves." For a child, it might be centered around home, school, a friend's home, and a park; for a father, around home and work.[6]

Inasmuch as these personal patterns overlap with others within the territory close to home, familiar associations develop. These associations may be with people using the same facilities or seeking the same services, with the "place" itself, or with events that have occurred in the place, all of which contribute to a sense of familiarity and ownership.[7] Such shared spaces can be mapped, and neighborhood boundaries can be drawn around these commonly used places.

Andrew Jackson said: "But you must remember, my fellow citizens, that eternal vigilance by the people is the price of liberty, and that you must pay the price if you wish to secure the blessing."[8] To secure the blessing and guarantee themselves a quality life at home, residents have begun to show greater interest in their neighborhoods. By defining a neighborhood as a political settlement, Kotler drives to the heart of this matter. For years planners clung to the idea that the neighborhood planning unit fostered grass-roots democracy and local loyalties. Kotler suggests that the opposite is true. He contends that, because of local concerns and shared problems, local loyalties and common values develop. These are expressed through grassroots democracy, and a feeling of neighborhood results.[9] As people become more determined that a good environment begins at home, this sense of neighborhood increases, which makes local relationships correspondingly more important. In this way a neighborhood is defined by the residents, not the planners, and is expressed in the political actions taken by the residents. This suggests acceptance of what has been called a "collective responsibility," which is not dictated from a higher authority but arises from the people of an area because of shared values, use patterns, and common problems. This suggests that the planned

"neighborhood should be small enough to encourage participation of all families in common neighborhood concerns."[10]

In contrast to many definitions of neighborhood, the idea of a neighborhood as a political settlement is also extremely useful in planning social and physical change. It is increasingly apparent that residents want to control their neighborhoods; therefore, change at the neighborhood level must be consistent with the values and goals of the residents, or it will not occur.[11]

In summary, Milton Kotler's "political settlement of small territory and familiar association, whose absolute property is its capacity for deliberative democracy" provides an umbrella definition that is both relevant to the people who make up a neighborhood and useful to the designers who must plan change for a quality neighborhood environment. By including the concepts of small territory and familiar association, the definition echoes the social and spatial concepts of neighborhood that were central to the earlier literature. Kotler asserts that the political view of the neighborhood supersedes the importance of either the spatial or social aspects, and it is this emphasis in Kotler's definition that speaks most directly to the renewed interest of laymen in their neighborhoods. But most importantly, the definition provides the context for discussing an approach to socially suitable neighborhood design.

Neighborhood space

Having defined neighborhood in political terms, we shall now examine the concept of neighborhood space. Neighborhood space is that territory close to home, including houses, churches, businesses, and parks, which, because of the residents' collective responsibility, familiar association, and frequent shared use, is considered to be their "own." But in this discussion, the concept of neighborhood space will be limited to that *pubic*[12] *outdoor* territory close to home which, because of the residents' collective responsibility, familiar association, and frequent shared use, is considered to be their own. For designers seeking to design socially suitable neighborhoods, two questions must be answered. First, the concept of "residents' own" must be clarified. Second, designers need to know how to *delineate* those spaces which the residents consider their own.

The concept of "own" refers to a sense of collective symbolic ownership, not legal, individual, property ownership. It is similar to the sense of "turf," which urban street gangs demonstrate in their guarding of certain city areas; no member of the gang owns the area legally, but the entire gang, as a group, has a sense of owning it symbolically. On the other hand, a resident may actually have legal ownership of his home, but it is less likely that it will be symbolically owned by the remaining residents. But if there is a vacant lot in the neighborhood, the private ownership of which is unclear to the residents, it is frequently used as their "own."

In other words, public and ambiguously owned private spaces lend themselves to collective symbolic ownership more than clearly privately owned properties. Included in this definition are such spaces as parks, street corners, storefronts, alleys, rooftops, sidewalks, schoolyards, playgrounds, parking lots, streets, paths, junkyards, yards, front porches, streams, abandoned lots, undeveloped lots, secret niches, plazas, churchyards, trash dumps, woods, bus stops, front steps, gardens, outdoor cafés, phone booths, forbidden places, favorite places, utility spaces, floodplains, ponds, greenways, conservation easements, beautification areas, and transportation corridor spaces.

In addition to being publicly or ambiguously owned, these outdoor spaces have other characteristics that increase the residents' sense of collective symbolic ownership. Residents frequently share the use of these spaces, and they can be involved collectively in acquiring, planning, and changing these spaces. As frequency of use and intensity of involvement increase, collective symbolic ownership increases. Symbolic ownership also appears to increase as the users perceive that the space meets their special needs, as the space increases in value as a status object to outsiders, especially outsiders of high status, and as the space increases in value to one's peer group. All these factors contribute to a sense of collective ownership of outdoor spaces.

Designers also need to know how to delineate those spaces which the residents collectively consider their own and which therefore constitute their neighborhood. A good way to start is to designate the "natural" boundaries, such as streams, undeveloped lands, major transportation corridors, social class changes and historic districts. Although the area designated by such general boundaries is usually larger than the territory that residents consider their own, the designer can use a number of techniques to delineate the smaller area after he defines the natural boundaries. He can attend political meetings in the area to determine what the issues are and where the concerned people live, review the minutes of neighborhood organization meetings, talk to political leaders, or review newspapers for issues in the area. Another technique is to observe outdoor spaces in the area and to ask the people who use those spaces why they use them and where they live. This approach should provide an overview of the important spaces and of the users; both the spaces and the home location of the present users can then be mapped. When these points describe new boundaries within the "natural" boundaries, the areas can be considered to constitute a neighborhood. Where these areas overlap most frequently should describe the neighborhood focal points. The designer may consider the people living in that delineated area as the potential users or may expand the boundary if the neighborhood space he is planning can accommodate a larger use area.

Use of neighborhood space

Generally, public outdoor space near the home is used for work and leisure activities, political gatherings, educational projects, and movement from place to place within or through the neighborhood. How the space is used obviously depends on the activity occurring there: work (delivering mail, repairing the automobile, keeping shop), leisure activities (taking a walk, playing football, jogging, bicycling, hanging out, sitting on the front porch, swinging, playing checkers, throwing a frisbee, dancing), political gatherings (protesting a city policy, planning a park, closing a street), educational projects (showing a class the effects of soil erosion, identifying trees), and movement from place to place (walking to work, driving to the grocery). Although it is easy to predict these theoretical uses of neighborhood space, the actual specific use is much more difficult to predict, as a number of recent studies of neighborhood designs show [Figure 1].

ACTIVITY / SOCIAL ECOLOGY CHARACTERISTICS	Single Person, Structured or Unstructured Setting–Place Specific	Single Person Moving from Place to Place	Small Group, Unstructured Setting Place Specific	Small Group Moving from Place to Place	Small Group, Structured Setting Place Specific	Medium Size Group, Unstructured Setting–Place Specific	Medium Size Group Moving from Place to Place	Medium Size Group, Structured Setting–Place Specific	Large Group, Structured or Un-Structured Setting–Place Specific
Working	Y	Y →	Y	Y →					
Leisure	YN	YN →	YN	YN →	YN	YN	YN →	YN	Y
Political						Y	Y →	Y	Y
Education			N						Y
Moving		YN →		YN →					

KEY: Diagrams indicate general spatial arrangements

Y = Activity that usually requires special props
N = Activity that usually does not require special props
YN = Activity that includes sub-activities usually requiring props
→ = Movement

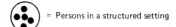 = Persons in unstructured setting

= Persons in a structured setting

Figure 1 Generalized Neighborhood Space Use: Activity Social – Ecology Potentials: The different activities that occur in neighborhood spaces require different types of spaces, depending on the characteristics of the social interaction of each activity.

In their study of children's behavior in a Turnkey III housing development in a southern city, Henry Sanoff and John Dickerson found that much more activity occurred in the streets and sidewalks than in the central playground and open field designated for activity:

The site plan provided a "large open space" for multiple activities, yet it accounted for less than 3.0 percent of all the people observed . . . the predominant activity in the field was ball play where young children and adolescents were the major participants.[13]

Similarly, it would have been difficult to predict that the most activity would occur in the cul-de-sacs, but "the major pattern of activity clusters occurred in the five cul-de-sacs."

Young children were frequently observed cycling, playing ball, and engaged in random play there. Sanoff and Dickerson explain that this clustering of activity is due to traffic patterns and the design of the streets:

The high observed densities in the cul-de-sacs occurred primarily because they were relatively free from vehicular traffic that would interrupt the predominant street activities of ball playing. The cul-de-sacs were so distributed throughout the site that they served as meeting places since they were easily identified and described by children.[14]

Even though there was a lot of activity in the cul-de-sacs, it was all of the same type:

> While the observed density was high for the "cul-de-sac," the diversity (density divided by activity types) was rather low, suggesting that the spatial cues and boundary conditions limited the range of activities. Though the "cul-de-sac" is a street, its very shape encourages more organized group activities than the street where the activities are more predominantly cycling and walking.[15]

The design of the spaces had resulted in a use pattern that the designers had not expected. Although more space was available in the field, the specific qualities of the streets and cul-de-sacs made them more attractive to the residents.

Another study of the use of neighborhood space in low-income housing projects, by the Committee on Housing Research and Development at the University of Illinois at Urbana, also demonstrates the difficulty the designer has had in predicting the specific use of neighborhood space. The study found that the designers had misjudged the use of the circulation routes through the site. Central sidewalks designed to be the major pedestrian collectors were rarely used. The residents were observed walking parallel to the sidewalk through small courts or green spaces; they often took shortcuts through the existing playground. The study concluded that the mistakes could have been prevented by more socially sensitive design:

> Assumptions like these result from lack of information on how people behave in specific situations. Another example was seen in the design of toddler's play areas at Orton Keyes Courts. These play areas were observed to be utilized less than the drying courts. In fact, the dwelling courts contained more activity than the institutionalized play areas on site. The same was true for Fairgrounds Valley, where the dwelling units were arranged around parking / service courts; the parking courts proved to be very popular play areas among the children.[16]

In spite of the difficulty in predicting the use of neighborhood space, a number of studies have begun piecing together the puzzle of how people use the space near their homes. A study done by the Department of Planning in Baltimore, Maryland, indicates that neighborhood space is used for outdoor leisure time activities in two distinct patterns: in home-based areas (dispersed throughout the neighborhood) and in recreational facilities (concentrated in areas such as playgrounds). Of these, the home-based spaces account for the major portion of recreational time.[17] In addition, those spaces which can be used while retaining visual access to the home tend to be used more. This is corroborated by the Bangs and Mahler study of rowhouse open space use in Baltimore County. They found that "most people will not regularly use a local open space if it is further than 400 feet away from their homes," but that if good pedestrian and visual access exists, use will not drop off abruptly as distance increases [Figure 2].[18]

In a similar study, Pamela Dinkel found that use of open space within the neighborhood was higher in a black than in a white neighborhood, and that use of open space outside the neighborhood was consistently higher in the white neighborhood. Park use in general was considerably higher in the black neighborhood. There seemed also to be significant differences in the informal neighboring and the resulting use of space, which the Dinkel study indicates is more dependent upon the socioeconomic status of neighbors than any other variable.[19] Low-income blacks interacted with their neighbors more frequently near home, resulting in greater use of home-based spaces. The Baltimore Department of Planning Study similarly found a significant difference in the use of open space, depending upon socioeconomic class: "In the low-income sites studied, most outdoor leisure activities occurred in the front of the house; on porches, steps, sidewalks, or street."[20] "Conversely, in the upper-middle-income sites studied, socializing and playing most often took place in the controlled setting of private yards."[21]

These studies do not offer a complete picture of how neighborhood space is used, but indicate that there are a number of important factors operating at the macro level which influence the use of the neighborhood space:

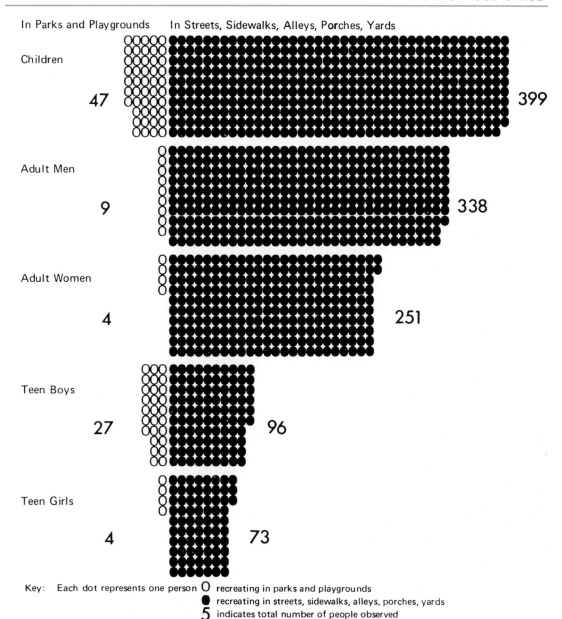

Key: Each dot represents one person O recreating in parks and playgrounds
● recreating in streets, sidewalks, alleys, porches, yards
5 indicates total number of people observed

AVERAGE NUMBER OF PEOPLE OBSERVED RECREATING PER DRIVING CENSUS
IN VARIOUS NEIGHBORHOOD SETTINGS

Figure 2 A study in Baltimore, Maryland, indicates that home-based activities rather than park activities account for the major part of recreation time. Source: Department of Planning and the Department of Housing and Urban Development, *Baltimore Community Renewal Program Interim Report: A Progress Report on the First Year of a Two Year Program Study* (Baltimore, MD: May 1972, p. 149).

1 The qualities and quantity of the space.
2 The social makeup of the potential users –
   especially socioeconomic class and life-cycle
   stage, but also sex, ethnicity, and region.
3 The psychological factors influencing personal
   preference.
4 The accessibility of local versus nonlocal spaces,
   facilities, and services.

The same idea has been stated by Suzanne Keller:

> The concentrated use of local services and facilities varies widely according to the economic and cultural characteristics of the residents, the types of facilities and their adequacy, the accessibility of nonlocal facilities and their adequacy, and the degree of isolation of the area, economically, ecologically, and symbolically.[22]

The important point is that the use of neighborhood space depends on many factors other than the design of the space and varies significantly from neighborhood to neighborhood.[23]

In addition to the factors just mentioned that influence the use of neighborhood space, there has been a significant general increase in the use of neighborhood open space for leisure activities.[24] As well as serving for circulation space and children's play space, those places close to home are increasingly perceived as work space, political gathering space, and leisure activity space for all age groups. Given the recent trends toward qualitative life styles, energy and resource conservation, and grass-roots governments, an increase in home dependence and in neighborhood space utilization should be expected. Some planners indicate that there already has been a change.

To summarize, it appears easy to describe generally how neighborhood space is used: for work activities, leisure activities, political gatherings, educational projects, and movement from place to place. It is more difficult to predict the specific use of the outdoor space near one's home. Recent studies indicate that the use of neighborhood space is influenced by important factors other than the planning and design of the space.

Although these factors are easily recognized, the actual use of neighborhood space is unique to each neighborhood. There is also a general increase in the use of neighborhood space for leisure activities.

Importance of increased user expectations

Renewal of interest in the neighborhood will certainly have an effect on the design of neighborhood space. In fact, it already has: urban renewal, highways, and large-scale development projects have been halted or changed because the residents opposed the impact that such projects would have on their neighborhood environments and their daily lives.[25] Citizens' groups have reacted against the growth-at-any-expense ethic, which allowed widespread destruction of urban neighborhoods. The Temple University renewal and expansion plan was adamantly opposed by neighborhood residents. After months of debate, a compromise plan, radically different from the original scheme, was accepted. The Inner Belt proposed through central Boston was opposed by Cambridge residents, who feared that their neighborhoods would be destroyed by the freeway; the Inner Belt was stopped. In Boulder, Colorado, when faced with uncontrolled growth that threatened the quality of life throughout their neighborhoods, citizens forced a reevaluation of the amount of land in industrial and high-density zoning. The result has been to discourage large-scale development throughout the city.

The common theme of these citizens' activities is the demand that planners do a better job of creating quality neighborhoods by designing spaces that meet the needs of the users, spaces that are not rubber stamps of other neighborhood spaces, spaces that are consistent with the values and desires of the neighborhood and not necessarily those of the planners. This theme has been manifested in several ways that are important to designers. Residents are willing to exert veto power over development plans that would negatively affect their neighborhoods. Even if the veto is not used, citizens are increasingly exerting their power to control the planning and design of local space. Neighborhood groups expect to play a major role in planning and designing their communities, and they expect the designer to produce socially suitable plans that take into account the unique needs of the residents by considering social variations and group idiosyncrasies.

At the same time that designers have been called on to produce socially suitable plans, the misuse and lack of use of neighborhood facilities designed in the past point to the previous difficulties that designers have had in assessing the needs of users. This has been particularly true when the users differed from the designer in terms of socioeconomic status, life-cycle stage, and ethnicity. In fact, designers have been guilty of a kind of physical design determinism, ignoring social factors and

planning authority can't do that now with streets – which seems the case – why will it be able to do it with rail? What will likely be ignored – as is typical with enormous capital projects of any kind – is the fine-grain finish work to make it comfortable. In fact the hot debate over rail vs. buses vs. cars vs. heavy rail vs. elevated monorail vs. light rail typically proceeds with little attention to how things are to be done.

I fear that such debates over the wisdom of building rail systems will take the place of truly discussing the built environment and deter us from dealing with the thousands of small things we can do now and in a few years to build better cities. We will plan to buy a fire truck while Rome burns.

Traffic calming

Along with contextualism in the design of buildings, the most significant new idea in city planning in the last thirty years is traffic calming.

Traffic calming is a set of street design techniques. It involves a variety of small modifications to street geometry and dimensions to accommodate the automobile and yet give the pedestrian psychological precedence.

These techniques assume – as Buckminster Fuller did – that we should reform the environment, not personal behavior. Rather than modifying human behavior, Fuller suggested that it is far easier to get people to act differently by redesigning their environment.

Traffic calming also recognizes that the car will not wither away. It is too popular and indeed too sensible to disappear. Here is an instrument that gives people personal autonomy over their own lives, their own daily to and fro. People struggle for freedom and the car is a very real means to use it. The car will not disappear without authoritarian rule.

But too many cars lead to uncomfortable cities. The person as driver overwhelms the very same person as walker. The real icon of America should be Janus, the god with two perspectives: one the driver, the other the walker. Some form of peaceful coexistence between our personas as driver and walker must be found.

Consider the annoying and dangerous phenomenon of speeding. Ninety years after the first speeding regulations, and who knows how many speeding tickets later, many of us still exceed the posted limits. We do it because the roads are designed to allow us to do so. There is a natural speed for any given road configuration. Many roads are marked 30 miles per hour and yet are designed to be driven comfortably at 55 mph because of sight lines, lane width, and shallow curves. Design will win out. More police will not. Redesign the roads to make better use of our natural inclination to drive as quickly – or as slowly – as the road design itself suggests.

Traffic calming seeks to find such design solutions. Its theme is to moderate vehicles' speed, give more physical space to the walker, and reclaim some of the walker's space. Its goal is not to make driving impossible but to slow it down within cities to a more human pace.

We illustrate with a variety of techniques the basic principle of changing driver behavior though design. But the subject is so new, so complex, and so uncharted that the reader is forewarned that this is only the start of the possibilities.

Traffic calming through design should not be read to mean that human behavior is of no importance. For example, driver education could be broadened to include respect for pedestrians and the reasons for traffic-calming design. Many drivers are frustrated by traffic-calming devices because they do not understand the reason behind them – i.e., traffic calming aims not to anger drivers but to calm them.

Consider curbs

The basic tool of traffic calming is the placement of the curb. Walkers may cross it; drivers may not. It defines their respective realms.

A rural lane – and perhaps even a low-volume city street – does not need a curb because the amount of traffic is trifling. There are not enough vehicles to create a sense of threat to the walker.

But a city is defined by greater traffic, and the curb provides order; it is a physical barrier. To drive over the curb onto the sidewalk is uncomfortable for the driver, hard on the car, and an outrage to social niceties.

The curb – apparently an insignificant line – nevertheless has a large part in defining how we experience the city. Moving it one way or the other

Figure 1

changes the relative size of the pedestrian/car realms and their consequent sense of power.

"Bulb" the corner for more pedestrian space

The struggle for urban space between walker and driver turns on a detail: the placement of the curb.

Moving the curb out into the roadway creates a "curb bulb." It is a superior and more comfortable spot for the walker crossing the street. It widens the sidewalk. It reduces the distance pedestrians must cross and makes them more prominent and visible to drivers. It takes back road from the driver and gives it to the walker [Figure 1].

Though largely made of concrete, the urban environment is really very plastic. Reshaping the curb line has a large effect on our use of the street.

Build streets on a grid

Why even mention such an obvious thing as the continuous street grid? Probably because the suburban systems of cul-de-sacs became the predominant pattern of platting after World War II and the grid is not the automatic solution in subdivision. It is now a choice; but it is a choice not often made because of simple economics. Platting raw land

based upon a system of cul-de-sacs creates more lots and hence more profit than does a rectangular street grid.

To clarify further, it is not critical that the grid be rectangular, i.e., a literal ninety-degree grid. The important thing is that the streets be continuous and create continuous thoroughfares. What one wants to avoid are dead ends, which tend to concentrate traffic on arterials rather than diffuse it through a broader network.

Additionally, a grid system allows a complex hierarchy of streets: arterials, collectors, and feeders, each one with different amounts of traffic.

Dedicated bike paths are difficult to establish and can never hope to satisfy the demand for bike routes. Restriping the existing street space to create bike lanes provides a greater opportunity. But continuous side streets, for example, as part of a grid make an excellent path for cyclists. They already exist and are a less trafficked and continuous route from here to there. These low-traffic feeders can, with very little investment, serve as bicycle routes.

Make blocks short

The short block (less than 240 feet or so) is a traffic-calming device of the first order. Short blocks mean more intersections. More intersections mean more places where cars must stop, thus lowering average auto speed. Short blocks also create more opportunities for walkers to cross the street. The short block is more interesting for walkers. A journey seems quicker, livelier, and more eventful when punctuated by crossing streets.

There's an economic attraction to short blocks: more corners. From a real estate value perspective, the corner is the best place to be: it has frontage on two streets, hence more visibility. Its cornerness also provides greater flexibility for site planning, which is the very first and most important part of designing a building. The corner at the intersection of a city's most heavily traveled thoroughfares provides the greatest access and visibility; it is traditionally the very best place in a retail district, its central place, and is known in the trade as "the one hundred percent corner." It represents the highest possible value, and everything declines away from that spot.

The value of corners is recognized for dwellings, too, as a house on the corner has more light and air than a midblock site and is typically more valuable.

Obviously it's a bit late, in most cities, to form a street grid with short blocks; most of our cities were platted (divided up into streets, blocks, and lots) long ago. Furthermore, it is not very easy to cut a new street through an existing block; one needs to condemn and purchase private property. We live with the glories and the mistakes of our forebears. Yet we are still expanding into the suburbs, building edge cities, and that gives us a chance to plat with short blocks.

Adding streets may be problematic; however, eliminating a street is often a cinch. The streets are a commons, and the "tragedy of the commons" is that no one cares for it as their own. Owners of adjacent property may request that the public's interest in a street be relinquished. This vacation – what a smoothly soothing term – of street right-of-way is a neat and low-cost way to increase the size of an urban property, but often at high cost to the streetscape.

Vacations were an urban planning fad in the 1950s and 1960s, which saw the development of the superblock. Here, two or more blocks plus the adjacent right-of-way are combined into one building site. The practice contributes to a city scaled for cars and is a grave error, but it is still being carried out as large institutions covet "free" land in the public right-of-way and local governments lend their approval to that transfer.

Use shortcuts to create a grid

One naturally thinks that a sidewalk must be next to a road. But why not build sidewalks without streets?

For example, there may be a good reason to develop a new plat with some cul-de-sacs, or it may be already built. Such a pattern prohibits through traffic and creates quiet dead-end streets.

But it also creates few side streets, and all the traffic is channeled to a few arterials. One side of a cul-de-sac doesn't connect with another one. The parent with a stroller, the jogger, the child on a bike, or even the casual perambulator must detour to a busy street to go just a block.

Figure 2

The solution is to connect two cul-de-sacs (or two long parallel streets, for that matter) with bike-throughs. These can be installed when the plat is designed, but since they take so little space they would be a relatively inexpensive element in retrofitting an existing plat as well.

Here is a practical comfort test for the design of a new plat: children should be able to visit friends, get to school, and run to the store without having to walk or ride on a busy arterial. Some jurisdictions now require that site plans for new multi-acre developments show walking-time contours to specific places, such as a transit stop.

The bike-through shown [in Figure 2] goes through a block of houses and connects two streets. It is a way for children to travel with less danger from cars and it creates urban spaces where cars cannot go, shifting the psychological feel of a neighborhood by creating car-free spaces.

But don't let drivers everywhere on the grid

Interrupt the grid every so often. Block some intersections so cars cannot pass. When the street

Figure 3

the noble impulse behind the pedestrian mall. But it doesn't always work.

It isn't real, for one thing, and it's not something you can do many places. It's really inconvenient, and while it may be wonderful in theory, comfortable cities are built in practice. Behind the pedestrian-only mall is a theme-park vision of a city, something fascinating and quaint and worthy of a visit, but not something one might use every day. The reality is that we have personal vehicles. The task is not to ban them (impossible) but to calm them (readily done). Like the Colt .45 of the frontier, traffic calming is the equalizer of the auto age.

Mixing cars and pedestrians:

- Increases the eyes on the street – some pedestrian malls look pretty lonely.
- Maintains or even increases on-street parking spaces.
- Is convenient, will be used, and thus creates sustainable places.

Traffic-calming structures can be casual

Public works departments have authority over what happens in the street right-of-way (sidewalk and roadbed). They have a legitimate responsibility for public safety and for assuring improvements in that space.

But sometimes their requirements are set so high that nothing gets done. Very informal traffic-calming devices, such as planting areas at the edge of streets, appear to work well – certainly at the very least as an experiment to determine how particular traffic devices will work.

Traffic calming must be seen as a systemic approach, and with literally thousands of discrete intersections and conditions in a city, there is need for empirical experimentation.

Decrease the turning radius

The various codes of city planners and city engineers, with their standards and requirements, form a "genetic code," a set of instructions to direct future growth. These infinitesimal elements of a city's street engineering code have a large influence on

grid is broken, the speedy and sometimes annoying flow of traffic through a neighborhood is disrupted. But adding a bollard (a post which prevents vehicles from entering an area) to a curb cut allows fire trucks and ambulances to move rapidly when needed [Figure 3].

Breaking the grid may appear to contradict the idea of building streets on a grid, and to some degree it does. But both patterns should exist. We start with the presumption of continuous routing – i.e., the grid – and then vary it with devices for traffic calming and visual diversity. The resolution is balance, and viewing streets as a system.

And let cars and people mix

After all the emphasis these days on pedestrian-friendly streets it may seem counterintuitive to suggest that cars and people should mix. Wouldn't it be more civil and humane to create a place just for walkers? Take a busy commercial street, ban the cars, and leave it for people afoot. Such was

Figure 4

Figure 5

our behavior as drivers and pedestrians. One key code standard is the "turning radius" of the curb at intersections.

This turning radius is determined by the placement of the curb and is the size of the circle that will fit in the corner. The smaller the circle's radius, the sharper the turn. The sharper the turn, the slower one must drive [Figure 4].

Typically, in a residential subdivision, the standard will be twenty-five feet. But by reducing this radius to fifteen feet, the engineer still allows free auto movement but signals to the driver that a slower speed is appropriate. In addition, the narrower curve places the pedestrian closer to the goal: the other side of the street.

Slow traffic with circles

The traffic circle is simply a curb placed in the middle of an intersection – a deliberate obstruction in a stream of traffic, which forces drivers to slow down. It creates a quieter, calmer, and more "residential" environment [Figure 5].

Once again, there are no silver bullets that solve all problems. The traffic circle is not manna. Some drivers loathe them and will change their route to avoid them, which may serve to decrease traffic on one block, only to divert it to another, increasing traffic there.

Use traffic circles only as part of a systems approach to neighborhood traffic. Their surface is a good place for neighborhood decoration.

Consider roundabouts

Roundabouts are grown-up traffic circles. Their purpose is not so much to slow down traffic but to allow movement through intersections without having to stop. They are quite common in parts of Europe and are starting to be used in the United States.

The most frustrating aspect of urban traffic is not so much its overall speed but its stop-and-go nature. A slow but steady pace would achieve the same overall time from origin to destination but without the mental aggravation of continual acceleration, braking, stopping, and then accelerating again.

Raise the crosswalk

Let the pedestrian cross without stepping down onto the street bed. Raise the pedestrian crossing so that it is level with the sidewalk.

Figure 6

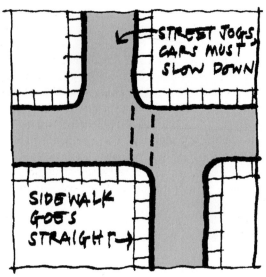

Figure 7

The extra six inches of height makes the walker more visible to drivers, particularly if one uses a pavement of contrasting texture and color. The change of grade is also a long-wave speed hump, which forces the driver to slow down to avoid an unpleasant bump [Figure 6].

Where entire intersections are raised so that traffic going in both directions will slow down, these intersections are known as "sleeping policemen" because of their ongoing deterrence effect.

Curve roads to narrow sight lines

Traffic engineers are taught to hate curves in roads. Here is inertia, the most basic law of physics, manifesting itself. The path of least resistance is the path one is already upon; the curve is a detour and takes extra force to negotiate. A body at rest remains at rest; a body in motion remains in motion. The engineers want to let the car act like a stone in outer space, continuing on its path without diversion.

Curves slow down cars. Curves cause drivers to use their brakes. As the traffic engineer sees it, curves are to be straightened at any chance so that the traffic may flow more smoothly and hence more rapidly.

It is for precisely that reason that we should let the road curve. It makes drivers use their brakes. *It makes them slow down.*

It creates a slower and more tranquil environment when the car and the walker have to share the same space in cities. In fact, the new revolution of traffic calming will ask us to add curves in some situations.

Narrow the street

Some streets are wider than they need to be for the traffic on them now or in the future. The excess room in the roadway can be given over to pedestrians or plants. The narrower street will signal to the driver to slow down as well.

Jog to calm traffic

The street jogs, the car must slow; the sidewalk doesn't and the walker may proceed, giving symbolic precedence and greater ease to the walker. Whether this street pattern was planned or was an accident of platting has long been forgotten. But it works and is an example of a serendipitous detail [Figure 7].

Consider vehicle size in street design

City streets are designed for the size and speed of the vehicles that are expected to drive upon them.

A city car need not be able to go 110 miles per hour. In fact, 30 mph is sufficient for a great number of trips. Twenty-five to thirty mph is a typical speed limit on most nonarterial streets. A vehicle much smaller and slower than our current cars –

gutters, storm sewers or other elements of artificial drainage.

As the natural landscape is paved over, a chain of events takes place that typically ends in degraded water bodies. This chain begins with alterations to the hydrologic cycle and the way water is transported and stored.

Urbanization changes a watershed's response to precipitation. The most common effects are reduced infiltration and decreased travel time, which significantly increase peak discharges and runoff. Runoff is determined primarily by the amount of precipitation and by infiltration characteristics related to soil type, soil moisture, antecedent rainfall, land surface cover type, impervious surfaces and surface retention. Travel time is primarily determined by slope, length of flow path, depth of flow and roughness of the flow surface. Peak discharges are based on the relationship of these parameters and on the total drainage area of the watershed, the location of development (which includes encroachment into the floodplain and loss of wetlands), the effect of any flood-control structures or other human-constructed storage facilities, and the distribution of rainfall during a given event.

Assessment of the amount, timing and duration of stream flow events can be visually presented in a hydrograph. A hydrograph is a plot comparing the rate of runoff against time for a point on a channel or hillside [Figure 1].

Impervious land as an indicator of ecological health

Research indicates that when impervious area in a watershed reaches 10 percent, stream ecosystems begin to show evidence of degradation, and coverage more than 30 percent is associated with severe, practically irreversible degradation. Degradation occurs in five ways:

1 Rainwater is no longer trapped by vegetation to the same extent.
2 Much of the rainwater is prevented from moving into the soil, where it recharges groundwater, and stream baseflows are thereby reduced.
3 Because the larger volume of rainwater cannot infiltrate into the soil, more rainwater runs off, creating greater flows more frequently. This enlarges the stream channel, causing bank erosion and associated reduction of habitat and other stream values.
4 Runoff flowing across impervious surfaces collects and concentrates pollutants from cars, roadways, rooftops, lawns, etc. ("nonpoint" sources), significantly increasing pollution in stream and other waterbodies.
5 Impervious surfaces retain and reflect heat, causing increases in ambient air and water temperatures. Increased water temperatures negatively impact aquatic life and oxygen content of waterbodies.

The three basic tenets of reducing imperviousness – retaining the natural landscape, minimizing pavement and promoting natural infiltration to the soil – are simple concepts that can be understood by most people.

Trees and the hydrologic cycle

Street trees provide many stormwater management and water quality functions, including: interception, throughfall, evapotranspiration and conveyance attenuation.

Interception

Interception is the precipitation that falls on the surfaces of trees, including foliage surfaces, bark and branches. Intercepted water is either evaporated directly to the atmosphere, absorbed by the canopy surfaces or transmitted to the ground surface via stem flow. Interception of rainfall by street trees reduces the volume and velocity of water flow, and temporarily stores rainfall in the canopy layer. Leaves, needles, twigs, branches, trunks and bark all capture, absorb and transpire rainfall. Branching structure, foliage and bark texture and canopy density determine how much rainfall is intercepted. Other factors that influence the amount of rainfall intercepted by the tree canopy include relative humidity, temperature, wind speed and the characterization and magnitude of rainfall events. For example, during periods of low temperature, the amount of evaporation from the tree

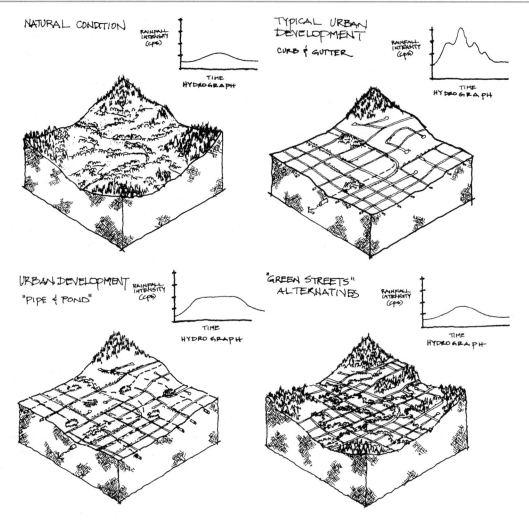

Figure 1 The effect of stormwater management system alternatives on the rate of stormwater discharge to streams.

surfaces is reduced. During light rainfall, coniferous trees have greater water retention ability than deciduous trees. If there are several overlapping canopies and a thick litter layer, there may have to be up to one inch of rainfall before water will reach the soil.

The March 2002 report from the Center for Urban Forest Research, *Western Washington and Oregon Community Tree Guide: Benefits, Costs and Strategic Planting*, used a numerical simulation model to estimate annual rainfall interception of three common street tree species of small, medium and large stature. The model includes rainfall intercepted, as well as the amount of throughfall and stem flow. The study found that the average annual benefits of a tree increase with age and that larger trees produce more water quality savings. The

average stormwater management cost savings of a large tree (red oak) was $15.25 per year through the first 40 years of the tree life. By the 40th year, the tree intercepts more than 1,100 gallons of stormwater, providing a cost savings of more than $30. Therefore, designing a street that encourages large long-lived tree species with broad canopies and maintaining those trees through a long life provides a tangible cost benefit to a stormwater management agency.

Throughfall

Throughfall is the portion of precipitation that reaches the ground directly through gaps in the vegetation canopy and drips from leaves, twigs and stems. Leaf surface area, canopy density and branching structure determine how much through-fall there will be during a normal storm event and the rate at which the throughfall will occur. The throughfall of stormwater in street trees generally occurs at a slower rate, or is attenuated, than storm-water falling directly on or draining to an impervi-ous surface, increasing the likelihood of infiltration into the ground. Excessively heavy rainfall and wind will increase the amount of throughfall.

Evapotranspiration

Evapotranspiration is the process by which plants release moisture in the form of water vapor. A single tree can transpire up to 100 gallons of water in a day; one acre of vegetation can transpire up to 1,600 gallons of water on a sunny summer day. Evapotranspiration rates are fairly well known for large stands of forests, but are rarely calculated for individual trees in urban settings. Transpiration ratios are used to calculate water use efficiency, which in arid regions are typically rather high. Trees that can capture and evapotranspire large amounts of water can significantly reduce the amount of rainfall that becomes stormwater runoff.

Conveyance attenuation

During periods of heavy rainfall, when water is being conveyed through biofiltration swales, filter strips or linear detention facilities, trees can act as check dams, slowing down water and increasing the potential for infiltration. Trees are particularly effective at conveyance attenuation where grades exceed 2 percent.

Shade and temperature reduction

Urban areas have a high percentage of impervious surfaces that absorb and radiate heat, causing the urban heat island effect. Although the urban heat island effect is commonly used to describe an increase in air temperatures, it has far reaching impacts on water quality. Cool rainfall collects on hot impervious surfaces, is warmed up and conveyed by a system of curbs to stormwater drains. These drains are connected to a series of underground pipes that discharge the warm polluted water to local streams and rivers. This increase in water temper-ature removes the amount of dissolved oxygen in the stream, making it difficult for fish and aquatic organisms to survive. Trees control temperature by absorbing and reflecting solar radiation, controlling windflow, and intercepting, transpiring and evap-orating rainfall. The amount of temperature reduc-tion and shade depends upon the species of tree. Temperature measurements in Sacramento, California have shown that neighborhoods with considerable tree cover are three to six degrees cooler than newer neighborhoods lacking tree cover. By shading impervious street surfaces, street trees help reduce the temperature of stormwater runoff.

Phytoremediation

The process of using plants to remove contamina-tion from soil and water is called phytoremediation. Many plants have the ability to absorb excess nutrients, filter out sediments and break down pollutants commonly found in stormwater runoff. Plants with deep roots improve these functions by breaking up soils and keeping them permeable. Planting areas can be designed to encourage plants to grow deep roots or an extensive root system.

Tree roots absorb and transpire water, and some tree roots can stabilize waterborne pollu-tants from runoff. Pollutants partially controlled by trees include nitrogen oxides, sulfur dioxides, carbon monoxide, carbon dioxide, ozone and small particulates less than 10 microns in size. Up to 47 percent of surface pollutants can be removed in the first 15 minutes of a storm event, including pesti-cides, fertilizers and biologically derived materials and litter. In order to prevent these pollutants from entering our streams and rivers via the conventional piped stormwater system, areas for infiltration and treatment should be created. Pro-viding pervious surfaces that capture stormwater runoff increases opportunities for pollutant removal and attenuation of flow velocity.

Scientists have found that many plants can stabilize toxins by absorbing them and storing them in their tissues. Plants can be used to clean up metals, pesticides, solvents, explosives, crude oil, polyaromatic hydrocarbons and landfill. Plants actually need metals such as zinc and copper for growth and most do not distinguish between these and other metals. Some plants enlist soil bacteria to detoxify organic compounds. Phytoremediation of toxic heavy metals reduces the overall volume of contaminated material.

Phytoremediation research shows that there are six biological processes at work: phytoextraction, rhizofiltration and phytostabilization, which use plants to stabilize or remove metals from soil and/or groundwater; and phytodegradation, rhizodegradation and phytovolatization, which use plants to stabilize, degrade and/or remove organic contaminants. These biological processes remove, metabolize, neutralize, sequester, filter and contain toxins, metals, sediment, minerals, salts and pollutants that would otherwise move through a watershed from hydrologic and other forces.

Earth as a sponge and soil type

Infiltration rates of soils vary widely and are affected by subsurface permeability as well as soil intake rates. Soils are classified into four hydrologic soil groups according to their minimum infiltration rate, which is calculated from bare soils after prolonged wetting. Most urban areas are only partially covered by impervious surfaces; however the soil remains an important factor in runoff estimates. Urbanization has a greater affect on runoff in watersheds with soils having high infiltration rates (i.e., sands and gravels) than in watershed predominantly of silt and clays, which generally have low infiltration rates. Also, development typically results in the removal of topsoil leaving generally heavily compacted soils that have a reduced pollutant treatment capacity.

Stream corridor processes and habitat functions

Buffers are required to protect stream systems from the impacts of development for water quality and flood protection purposes. The primary goal of buffering is to maintain the areas adjacent to the stream system in native vegetation. This vegetation can be grass, shrub or riparian forest depending on the character of the watershed in a given stream reach. This vegetation provides food and cover to a variety of insects, animals and fish. Because of their linear nature, stream corridors function as migration zones for mammals, amphibians and insects moving between upland and lowland communities or within a community. Streamside vegetation also can provide shade to lower stream temperatures. Maintaining native vegetation also buffers the stream corridor from storm water impacts by slowing and filtering runoff and protecting soils from erosive stream flow.

Buffers are not synonymous with corridors. Buffers are political boundaries at set widths while a stream corridor width is natural and can vary along its length. One important component of a corridor that contributes to this variation is the floodplain. Floodplains are integral parts of the stream system that store and convey periodic high flows. Some extend less than 100 feet from the stream and others extend beyond this set buffer width. Periodic flood flows deposit sediment on the floodplain and rejuvenate native vegetation. Confluences (places where streams come together) are particularly diverse because they are subject to dynamic flows from several streams.

In urbanizing areas, natural stream corridor systems are typically overlaid with a geometric street grid. If not properly designed, the stream crossings that result from the intersections of these natural and built systems can negatively impact the natural vegetation buffering the stream and the timing and extent of low, bankfull and flood flows. This, in turn, reduces habitat functions of the stream channel and floodplains within the stream corridor. Excessive numbers of stream crossings can fracture a continuous corridor and reduce the availability of food and cover, the shading quality, and inhibit the ability of fish and other species to migrate through the stream and stream corridor system.

The role of biofiltration and swales

Biofiltration is a term adopted for processes in which stormwater receives treatment through physical, chemical or biological interaction with vegetation and the soil surface. Pollutant removal occurs by settling, infiltration, plant uptake and/or ion exchange with soil particles. The processes include (1) sheet flow over a broad vegetated "filter strip," (2) small created wetlands or infiltration basins or (3) flow at some depth through a vegetated channel, or "swale."

A "swale" is a vegetated channel that looks similar to, but shallower and wider than, a ditch. Swales are designed and maintained to transport shallow depths of runoff slowly over vegetation. The slow movement of runoff allows sediments to settle out and particulate to be filtered and degraded through biological activity (microsomes). Research in the early 1980s along Washington state highways and in Florida demonstrated the ability of biofiltration swales to remove solids and metals effectively from stormwater runoff.

DESIGN SOLUTIONS

In developing a stormwater management and stream protection system, the goal should be to incorporate the natural watershed functions described within this chapter in order to reduce the negative impacts of development on water quality and stream habitat. The following objectives have been identified to support this goal:

1 Minimize the generation of storm runoff by reducing the amount of impervious surface within the right of way.
2 Manage the runoff volume by infiltrating wherever possible.
3 Provide detention, retention, infiltration and/or water quality benefits as close to their source as possible by incorporating these functions into the overall right of way design.
4 Protecting stream corridors with buffer areas and design crossings to assure the minimum of impact on not only the stream channel but to the stream corridor.

Ideally the design solutions illustrated herein would be designed in concert with the roadway eventually forming a Green Streets framework for the community. This should occur with new development such as in the yet undeveloped areas. Within existing development where roads are already built and hemmed in by adjacent development the only option is generally retrofit. Retrofitting, though generally requiring compromises to an optimal system, is nonetheless important as an incremental improvement.

A: THE WHOLE PICTURE – A WATERSHED STRATEGY

It is essential that a connection be made between the health of a stream and the overall health of the watershed. Furthermore, the effectiveness of design solutions that improve runoff quality at a point downstream may be irrelevant if the water quality of runoff from upstream locations is not improved. Streams can be most effectively protected from water quality and habitat degradation caused by stormwater if complementary efforts are made at a number of different levels. The following is a watershed protection approach that applies eight tools to protect or restore aquatic resources. Not all of these relate directly to road network and right of way design but it is important to put the tools that apply to rights of way within a broader context.

Land-use planning

This is the single-most important tool requiring consideration of the effect of land-use change on water resources, obtaining consensus on the water resource goals for the entire watershed and developing a land-use pattern at the sub-watershed scale that can meet these goals. Acceptable and effective land-use planning techniques also must be selected to reduce or shift future impervious cover; including: *Watershed-based zoning*: involves defining existing watershed conditions, measuring current and potential future impervious cover and modifying master plans and zoning to shift location and density of future development into management categories – "restorable," "non-supportive," "sensitive" and "impacted." *Overlay zoning*: mapped districts that place special restrictions or specific development criteria onto existing zoning provisions.

Land conservation

In order to sustain the integrity of aquatic uses and to maintain desired human uses, five types of land may need to be conserved in the sub-watershed: critical habitats, aquatic corridors, hydrologic reserve areas (sustaining a stream's regime), water hazards that pose a risk of potential pollution spills and cultural/historic areas that are important to identity of a place. Land conservation strategies should consider the value of the existing aquatic resources within the watershed and regional ecosystem. Cultural values (e.g., fisheries, recreation, aesthetics, etc.) should be overlaid with the ecological values to develop a conservation strategy. Techniques for conserving land include outright acquisition, conservation easements, landowner and public sector stewardship, and setback of water pollution hazards.

Aquatic buffers

A buffer physically protects the stream or wetland from future encroachment and serves as a riparian right of way with benefits to habitat and flood protection. Buffers also are used to protect natural floodplain processes that are necessary for ecosystem health. Tree cover, periodic flooding, sediment and organic transport are all essential features/functions of aquatic buffers. Maintaining these processes and functions are the most important goals performed by buffers. Therefore, all human impacts to watershed hydrology should be mitigated outside the aquatic buffer areas.

Better site design

Individual development or road building projects can reduce the amount of impervious cover they create and increase the natural areas they conserve. A key planning technique is "cluster development" that creates protected open spaces that have environmental and market benefits. Techniques such as "green" parking lots and a rooftop runoff management strategy are site specific techniques.

Erosion and sediment control

A combination of clearing restrictions, erosion prevention and sediment controls, coupled with a diligent plan review and strict construction enforcement are needed to mitigate the impacts of erosion and sediment discharge into streams. In addition, new techniques for trapping and coagulating sediment from construction sites are in the offing as technology improves.

Stormwater treatment practices

The general goals for treatment practices include: maintaining infiltration and groundwater recharge and quality, reducing stormwater pollutant loads, protecting stream channels from increased quantity and frequency of flow, preventing increased overbank flooding and safely conveying extreme floods.

Non-stormwater discharges

This tool concerns itself with how wastewater and other non-stormwater flows are treated and discharged in a watershed. Key program elements consist of inspections of private septic systems, repair or replacement of failing sanitary sewers and spill prevention.

Watershed stewardship programs

Once a watershed is developed, communities still need to invest in ongoing watershed stewardship in order to increase public understanding and awareness.

B: REDUCING RUNOFF ON-SITE – PRIVATE PROPERTY DESIGN SOLUTIONS

On-site stormwater treatment is a key element in an overall stream protection strategy. This section briefly illustrates techniques available.

Buildings

One way to reduce the runoff from buildings is to minimize the overall land area occupied by buildings. The solution is creating narrower and taller buildings. Understandably, this strategy is subject to other community goals. The old-fashioned rain barrel that collects roof runoff can be effective in reducing and slowing runoff by collecting rainwater for future use or by slowly allowing it to seep out. Another technique is an underground dry well connected to the roof's downspout that slowly infiltrates water. Concerns of nuisance ponding and basement flooding can be solved by slightly sloping stormwater away from the building. These concerns, however, are only significant within 10 to 15 feet from the building's foundation.

Parking lots

An overall strategy for reducing parking lot runoff is reducing the actual parking areas by using shared parking facilities and structured parking. Structural design solutions include stormwater bioretention areas, vegetated swales and filter strips that can be integrated into landscape areas and traffic islands. If designed properly, these areas would be both functional and attractive. The use of porous pavements also is appropriate for parking lots where travel speeds are significantly lower than roadways. An option that can add visual interest to the parking lot and increase permeability is to vary the surface materials between travel lane and parking stall.

Infiltration basins

Generally considered to be outside of the right of way, infiltration basins have the potential of working in conjunction with other design solutions. Depending upon location, they can either be part of a "treatment train" or collect runoff directly from the street or parcel. Infiltration basins generally come in wet and dry types.

A "wet pond," also known as a "constructed wetland," is a permanent pool of water that detains and treats runoff. The primary treatment method is settling with the secondary method being bioremediation by the plants along the edge of the pond. The simplest form is a rectangular basin with a deeper forebay that traps floatables and larger settleable solids. The remainder of the pool is more shallow with wetland vegetation. The inclusion of islands and jetties create longer flow paths and, hence, higher treatment performance, longer edges for ecological interactions, and greater diversity of habitats for all organisms. The islands can be a benefit specifically to waterfowl, which use them as protective and nesting areas. Wet ponds can be designed to be an attractive amenity for the community.

Dry infiltration basins, also known as "extended detention ponds," store water during storms for a short period of time and then either discharge to adjacent water sources or allowed to infiltrate. To infiltrate, the bottom of the basin is designed with permeable material. Dry ponds can be incorporated into the community as functional open space during dry periods and water features during inundation.

C: DESIGN SOLUTIONS WITHIN A RIGHT OF WAY

Choosing the appropriate design solution or combination of design solutions for public rights of way is accomplished by identifying the desired functions and evaluating the site and watershed conditions. The primary functions include runoff reduction, detention, retention, conveyance and water quality mitigation. Retention is commonly achieved through infiltration. It is preferred because it reduces the amount of stormwater runoff while providing water quality benefits. More than any other function, its application is limited by natural site conditions such as soil, slope and groundwater. Not all land in a metro area is suitable for infiltration solutions. In areas that are suitable, maintenance is essential to preserve function. Because appropriate areas are limited, the first step in selection should be to determine where, within the right of way, it is possible to infiltrate stormwater. The extent to which infiltration is possible will determine the location and size of infiltration areas.

Stormwater and water quality functions of other solutions to be used in the right of way, in conjunction with infiltration designs, can then be determined. The next step should be to determine where it is possible to reduce impervious surfaces. This will affect the stormwater volume requiring treatment by the solution and accordingly the size and function of the solution.

Table 1 Green streets design solutions	Runoff Reduction	Detention	Retention	Conveyance	Water quality
Street trees	X	X			X
Reduced imperviousness	X				
Permeating or eliminating curb and gutter	X				X
Vegetative filter strips				X	X
Swales		X		X	X
Linear detention basin		X		X	X
Infiltration trench		X	X		X
Infiltration basin*		X	X		X

typically located outside of the right of way.

Finally, the design of the selected solution should respond to the right of way site conditions and intensity of vehicular and pedestrian use within the right of way. To facilitate the selection process, we have grouped the green streets design solutions into five functional categories: (1) runoff prevention; (2) detention; (3) retention; (4) conveyance; and, (5) water quality [Table 1].

Many solutions combine several of these functions. Within the context of the Green Streets project, runoff reduction is primarily accomplished by substituting impervious surfaces with pervious surfaces wherever practicable, by retaining or planting large street trees with broad canopies and by removing curbs and gutters. Detention is temporary storage. It attenuates peak runoff rates by slowing the rate at which runoff enters the stream system but does not significantly affect the volume of runoff reaching the stream. Green streets design solutions that provide this function include low-gradient swales, swales with check dams and linear detention basins. Retention, or permanent storage, reduces both the volume and rate of runoff by decreasing the amount of runoff in the system through infiltration or evapotransporation. Green streets design solutions that provide retention include infiltration trenches and linear infiltration

basins. Conveyance describes the movement of stormwater from one location to another. Vegetative conveyance mechanisms such as swales are used within green streets to provide water quality and detention benefits while conveying stormwater. Water quality benefits provided by the green streets design solutions include the removal of suspended solids, oils and grease, nutrients, metals and bacteria. This is accomplished by slowing the flow of runoff by either holding it in a basin or passing it through vegetation, aggregate, or soil allowing sediments to come out of suspension. Green streets water quality design solutions include filter strips, sediment trenches and swales.

Matching design solutions to regional street designs

The intensity of human land use also affects the selection of stormwater and water quality design solutions. Available right of way widths, traffic volumes and speeds, preferred bicycle lane location, and density of adjacent development will influence the extent to which permeable surfaces or curbless elements can be incorporated into the right of way design. For this reason, green streets design solutions can be adjusted to accommodate these uses. Green streets design solutions are intended to integrate stormwater management into the overall design of the right of way. However, limited right of way widths and land-use intensity may necessitate the use of traditionally engineered solutions, such as sand filters, separators and catch basin inserts.

Maintenance tradeoffs of green streets design solutions

Frequency of required maintenance for innovative solutions may be unknown. More intensive maintenance or monitoring may be required for the first year or two until regular requirements are clearly understood. More intense maintenance will be required to ensure the successful establishment of filter strip and swale vegetation. Once established, maintenance will only be required two to three times a year. Periodic irrigation may be required to keep vegetation alive in extreme drought. Timing of street sweeping and debris removal is more critical for green streets solutions. Ideally the entire street and stormwater system will be cleared immediately before the first fall storms to minimize the volume of fine sediment contributed to the filter systems.

A note about street sweeping

Improvements in street sweeping technology have caused a recent reevaluation of their effectiveness. New studies show that conventional mechanical broom and vacuum-assisted wet sweepers reduce nonpoint pollution by 5 to 30 percent; and nutrient content by 0 to 15 percent, but that newer dry vacuum sweepers can reduce nonpoint pollution by 35 to 80 percent; and nutrients by 15 to 40 percent for those areas that can be swept. While actual reductions in stormwater pollutants have not yet been established, information on the reductions in finer sediment particles that carry a significant portion of the stormwater pollutant load is available. Recent estimates are that the new vacuum-assisted dry sweeper might achieve a 50–88 percent overall reduction in the annual sediment loading for a residential street, depending on sweeping frequency.

Street sweepers that can show a significant level of sediment removal efficiency may prove to be more cost-effective than certain stormwater treatment practices, especially in more urbanized areas with higher areas of paving. It is important to note that although street sweeping improves the quality of the runoff, it has no effect on the quantity.

Street trees

Street trees perform many functions and provide many benefits to urban residents. Some of the recognized benefits of trees are their role in providing shade and moderating the climate, conserving energy and water and creating an aesthetically pleasing streetscape.

There has been considerable research on the role that trees play in improving air quality, conserving energy, preventing soil erosion, replenishing moisture in soil and groundwater and absorbing and transpiring rainfall; however very little research has been conducted in urban areas about the ability of street trees to perform these functions. The information that we do have about trees in their native habitat indicates that street trees may be effective stormwater management tools.

The stormwater benefits that street trees may be able to provide to urban areas include runoff reduction and detention, achieved through interception and evapotranspiration; conveyance attenuation; and water quality mitigation, achieved through a reduction in stormwater runoff temperature, and absorption and stabilization of pollutants from street runoff. All street trees perform these functions; particular species may perform them better than others depending on various characteristics such as:

- persistent foliage
- resistance to pollutants (both air and water)
- canopy spread
- tolerance to poor soils
- longevity
- root pattern
- growth rate
- bark texture
- drought tolerance
- foliage texture
- tolerance to saturated soils

■ branching structure
■ canopy density.

Other aspects that have an influence on how street trees perform include resistance to exposure (wind, ice and heat) and resistance to disease and pest infestations.

Guiding principles

Principles for existing and new urban development integrate both neighborhood-scale designs and regional strategies that reduce the impact of streets on streams and rivers. These principles help achieve other goals such as improving the diversity of the urban forest, improving habitat for a variety of wildlife species and creating natural connections to valuable resource areas. Local codes, ordinances, street tree lists and planting specifications should be consulted before applying the following principles. Street tree selection is a site-specific action, and it is recommended that an inventory of street trees and site characteristics be conducted prior to selection. In sensitive natural resource areas natives may be required. In order to protect urban trees from disease infestations, no single species of tree should account for more than 10 percent of the city's tree population.

Preserve existing trees

Existing trees should be preserved, protected and properly cared for in order to live long, healthy lives. Existing trees and vegetation are adapted to local conditions, which can reduce both short- and long-term maintenance costs, and are well established, which can reduce infrastructure and energy costs. Current development practices remove most vegetation and the associated nutrient rich layers of soil, which increases soil erosion, compaction and surface water runoff. The largest single killer of urban trees is soil compaction. Large areas around these trees (beyond the drip line) should be protected from soil compaction so that drainage is not compromised and root systems are not damaged.

The green streets approach encourages the protection of existing vegetation by finding opportunities to incorporate them into parks and open space, individual lots and along streets. Where possible, groves of trees and large trees should be protected to retain the biological activity and nutrient availability in the soil. Street trees can be thought of as an extension of parks and open space areas, and can be fully incorporated into development plans. These connections are valuable to wildlife and contribute to the biodiversity of the ecosystem.

Plant large trees with wide canopies

When using street trees as a stormwater management tool, the primary goal is to maximize canopy coverage and to create opportunities for canopy overlap to shade impervious surfaces and provide increased interception area. In order to achieve this, a minimum 8-foot planting strip is recommended to allow for larger trees that have wide-spreading canopies. Where constraints limit large trees, stormwater bioswales that contain a variety of small trees and shrubs can create a diverse habitat for a variety of wildlife species, while achieving desired stormwater goals.

Where trees cannot be protected, trees that attain considerable canopy spread should be planted. If streets are designed properly, large trees can be planted without causing damage to infrastructure, such as sidewalks and underground pipes. This is easier to achieve in new developments where large well-designed planter strips can be incorporated into the street right of way during preliminary planning stages.

[. . .]

Heavy equipment should not be used near areas where trees are going to be planted so that soil does not get compacted. Planting areas that have been compacted or disturbed need to be re-aerated and amended with organic material to encourage the re-establishment of biological activity. Nutrient rich and oxygen rich soils are important to the long-term health of the tree and promote healthy and often vigorous growth. Biological activity in the soil retains soil structure and improves drainage.

In areas where planter strips [are constrained], or where tree wells are required, structural soils can be used to provide areas for large tree roots. Structural soils can be used along the length of the planter strip and beneath sidewalks to encourage

by government sponsors (e.g., regarding timetables, fares, and routing), the lowest-cost provider delivers the service. Thus, the public sector retains control over how services are configured, leaving it to market forces to determine at what price.

In successful transit metropolises, profiteering is not discouraged as long as the broader public good is promoted. Tokyo's rail-served suburbs are a product of real estate speculation that has enriched many large conglomerates yet also produced a strong, transit-oriented built form. For the past three decades, Japanese railway companies have branched into real estate, retailing, construction, and bus operations. The practice of value capture is alive and well in Tokyo's suburbs, with some companies reaping more than 30 percent returns on investment from joint railway and new town projects. Society as a whole has gained from the efficient codevelopment of transit services and suburban settlements. A spirit of entrepreneurship also pervades Mexico City's transit sector. Thanks to thousands of independent paratransit owner-operators, efficient minibus services link outlying neighborhoods with Mexico City's regional metro system. Driven by the profit motive, minibus operators fill the market niches left unserved by the region's metro service.

Giving transit priority

Many transit metropolises go the extra distance in making transit time-competitive with the private automobile. Ottawa and Zurich give clear preference to high-occupancy vehicles in the use of scarce road space, under their respective "transit first" programs. Signal preemptions allow Zurich's trams and Ottawa's buses to move briskly along surface streets, with stops largely limited to picking up and dropping off customers. Copenhagen and Zurich have reassigned large amounts of downtown streets to trams, buses, and nonmotorized transportation. Over the past decade, both cities have managed to absorb the growth in downtown-destined travel without expanding road capacity.

Small is beautiful

A number of this book's cases have shown that large-scale megaprojects are not necessarily the only means of creating a successful transit metropolis. Small, incremental steps matter – provided they are guided by some general vision of where the region is headed. Much of the successes in Curitiba, Karlsruhe, and Zurich are attributable to the cumulative effects of many modest, low-cost, but fast-turnaround actions. Ingenuity and a willingness to experiment and take risks have also played a role. Examples abound across case-study cities: Curitiba's trinary road system and raised boarding tubes; Karlsruhe's dual-mode, track-sharing technologies; Adelaide's adaptation of O-Bahn to a linear parkway with restricted right-of-ways; and Zurich's reappropriation of road space and resignalization of traffic lights, transforming tramways into something akin to grade-separated light rail services. Reasons for "small is beautiful" approaches have varied – from a desire to establish political credibility (Curitiba) to a need to preserve the traditional core city (Zurich) to a spirit of entrepreneurship and innovation (Karlsruhe).

Cities such as Ottawa and Curitiba show that low-cost systems do not necessarily equate to low-quality services. Nor do large-scale transit projects have to be budget busters. Ottawa, Adelaide, and Curitiba have designed bus-based guideways that match the service features of many underground metro systems at a fraction of the cost. The integration of mainline and feeder services within the same vehicle has virtually eliminated transfers along some corridors. Curitiba's inventive system of boarding tubes and transfer stations has made transferring between tangential and radial bus lines seemingly effortless.

Urban design: cities for people and places

An overarching design philosophy of most transit metropolises is that cities are for people, not cars. High-quality transit is viewed as consonant with this philosophy. In Copenhagen, Munich, and Curitiba, parts of the historical cores have been given over to pedestrians and cyclists. In all of the European cases, trams and light rail vehicles blend nicely with auto-free zones, moving at a pace and operating at a scale that is compatible with walking and cycling.

Urban design is every bit as important outside central cities. In the suburbs of Copenhagen and

Stockholm, transit stations are treated as community hubs. Rail stops and the civic spaces that surround them are often town gathering spots. They are the places where people congregate during national holidays, community celebrations, and public protests. Many civic squares do double duty as farmers' markets and venues for open-air concerts. Nearby retail shops, grocery stores, newsstands, and movie houses benefit from having transit customers delivered to their front steps. Street furniture, greenery, urban art, and water fountains add comfort and visual aesthetics. Through conscious design, transit is both physically and symbolically at the community's core.

It is tempting to label design features such as pedestrian-ways and public squares "amenities." Providing functional spaces for pedestrians and cyclists is, however, no more of an amenity than providing parking and freeway on-ramps for cars. They are basic provisions, not amenities.

Auto equalizers

Many transit metropolises have matched provisions for pedestrians, cyclists, and transit users with restraints on motoring and auto ownership. In Singapore, Tokyo, and Stockholm, this has mainly taken the form of punitive pricing: steep surcharges on gasoline and automobile purchases, hefty vehicle import duties, and expensive central city parking. Singapore has done more than any city to pass on true social costs to auto motoring. It was the first city to introduce road pricing on a large scale – initially through an area licensing scheme that charged motorists to enter downtown zones during peak hours, and more recently using a sophisticated combination of radio frequency, optical detection, imaging, and smart card technologies to instantaneously pass on variable charges. Singapore's vehicle quota system, which indexes the allowable number of vehicle registrations for anyone year to ambient traffic conditions, is also one of a kind.

In other transit metropolises, automobile restraints have been achieved through regulations and physical design strategies. Tokyo's garaging requirements have long held the city's car population in check. Mexico City's alternating ban on car usage, depending on vehicles' license plate numbers, has curbed auto motoring as a pollution reduction strategy. Munich, Zurich, and Curitiba have slowed down, redistributed, and sometimes deterred automobile travel, especially in residential neighborhoods, by narrowing roads, designing in speed tables, and necking down intersections. Parking management is also crucial to rationalizing the use of central city space. Limited parking encourages transit riding and frees up more central city space for pedestrians, cyclists, and transit vehicles. When heading downtown, most residents of medium-size cities like Zurich and Ottawa either leave their cars in their garages or at peripheral park-and-ride lots.

Transportation planners often refer to these measures as "auto disincentives," though this is a misnomer. More accurately, they are "auto equalizers" – they seek to level the playing field by removing many of the built-in biases that encourage and indeed sometimes reward auto motoring.

Hierarchical and integrated transit

Many of this book's successful transit metropolises feature well-designed, hierarchical forms of transit services. By carefully integrating services – high-capacity mainline services, intermediate connectors, and community-scale feeders – many origin-destination combinations can be efficiently served by transit offerings. Trams are critical intermediate carriers in Zurich, Munich, Karlsruhe, and Melbourne. In Munich, Curitiba, and Ottawa, limited-stop buses provide essential tangential, crosstown connections. In many cases, integration also extends to fares. Zurich, Munich, Copenhagen, and Karlsruhe reward frequent travelers through various discount arrangements. Unified tariffs and ticketing allow passengers to transfer across and modes without having to pay an extra fare.

Flexibility

For some of the smaller transit metropolises reviewed in this book, bus-based technologies not only reduce capital outlays but also provide important flexibility advantages. Flexibility is a fundamental trait of adaptive transit cities. As settlement patterns evolve and unfold, rubber-tire services – be they minibuses, conventional buses, or bi-articulated buses – can easily adjust to shifting patterns of travel. Ottawa's cost-effective mix of peak-hour express services and off-peak timed-transfer feeder services epitomizes the inherent malleability of bus-based transit. Curitiba's joint running of limited-stop, direct-line services and frequent-stop, high-capacity services along structural axes is also only possible with a bus-based system. Moreover, experiences in Ottawa, Curitiba, and Adelaide underscore the staging advantages of busways over rail lines. In all three cities, busways have opened in phases, prior to the completion of entire projects, because regular surface streets were available in place of missing links. In Curitiba and Ottawa, busways are also used by fire trucks, ambulances, and emergency vehicles.

Necessity is the mother of invention

The technological advances reviewed in this book were not a result of new gadgetry seeking an application, but rather a real-world need spawning an innovation. Munich's call-a-bus form of "smart paratransit" was launched because of the need to provide flexible and efficient one-to-many connections between S-Bahn stations and their hinterlands. Munich's real-time displays of park-and-ride information at the entrances to multistory garages has similarly filled a market need – to expedite transit commuting by reducing the time spent finding a parking space. Other technologies that have filled a market gap and materially benefited transit users include Ottawa's real-time, signpost-based passenger information system, Zurich's citywide dynamic traffic signalization system, and Karlsruhe's versatile dual-voltage trains that run on both tramway tracks and intercity train tracks.

Serendipity

Of course, not all of the outcomes reviewed in this book are the result of deliberate actions and far-sighted planning. Some are beneficiaries of good timing and good fortune. Curitiba's ability to build trinary roads along structural axes stemmed in part from the generous supply of right-of-ways that were preserved following the 1943 Agache Plan. Part of the reason Ottawa and Adelaide were able to build

busways at relatively low costs was the availability of linear riverside parkways. Ottawa's heritage of national capital planning and land preservation provided the right-of-ways for western portions of the city's busway as well as natural buffer strips that insulated surrounding neighborhoods. And, in many cases, transit investments occurred during an upswing in a particular region's economy, ensuring a close correspondence between transportation and urban growth.

DEBUNKING MYTHS

The twelve case studies reviewed in this book also serve to dispel at least six myths about transit and the city.

Myth 1: Except for downtown destinations, only poor people ride transit

Statistics for three cities alone will hopefully bury this myth. In Stockholm, one of the wealthiest cities in one of Europe's wealthiest countries, well over 60 percent of commute trips to downtown and more than a third of commutes to suburban job sites are by public transportation. Zurich averages more transit trips per capita than virtually any city in the world while also holding honors as one of the world's most affluent cities, with purportedly the highest commercial real estate prices anywhere. Curitiba has both the highest per capita income and the highest transit modal split of any Brazilian city.

Transit and affluence are not inherently at odds. To the contrary, prosperous urban environments and good transit are mutually supporting.

Transit's stigma as a means of transport for the underclass and marginalized populations stems in large part from its legacy of poor-quality services. In affluent societies, poor services generally attract poor people. Ramping up the quality of service will polish transit's image faster than anything.

In Melbourne and Zurich, tramways have been partly credited with luring well-to-do households back to central city neighborhoods. There is also ample evidence that middle-class households are willing to give up a second car when living in transit-supportive central city environs.

Myth 2: Public transit is inherently a public sector enterprise

This view is largely rooted in the belief that transit engenders economies of scale and is therefore a natural monopoly.[1] Defenders of the status quo argue against competition under the guise of "cream skimming" – profit-seeking operators will concentrate only on lucrative services, leaving the money-losers to public operators. The reality is that there is often little cream, or profits, to skim from public operators, and in most places where competition is encouraged (such as Stockholm, Copenhagen, and Curitiba), governments still control the supply, quality, and price of services. Once standards are set, the lowest bidder is awarded the service. The customer is blind to the fact that a private interest owns the transit vehicle and a non-government employee is driving it. What matters to the customer is service quality and value for price.

The majority of transit metropolises profiled in this book – Stockholm, Copenhagen, Singapore, Munich, Curitiba, Karlsruhe, and Adelaide – have warmly embraced competition as a means of producing cost-effective services. In all of these places, governments own and control the non-rolling-stock assets – guideways, railway tracks, land, and buildings, like maintenance facilities. Competitive tendering is used primarily to reduce costs. In the case of Tokyo, the private sector's role is far greater – many private intrametropolitan railway companies own all assets outright. Public oversight in Tokyo is largely restricted to ensuring that public safety standards are met.

Too often, "public transit" is literally interpreted to mean publicly owned and operated transit. "Public" simply means the service is available to the general populace, not that there is a public sector provider. As demonstrated by the world's transit metropolises, the business of designing and providing efficient transit services is not inherently a public sector enterprise – it involves private participation as well, to varying degrees.

Myth 3: Transit always loses money

Experiences in Singapore, Tokyo, and Curitiba prove otherwise. In 1995–96, the Singapore Mass Rapid Transit, Limited (MRT) – the owner-operator of the city's metro system and a company whose shares are sold on the Singapore stock exchange – collected passenger revenues that exceeded operating costs and expenses for debt service. All of the city's private bus companies today operate in the black. Most of Tokyo's private railway companies earn modest profits from rail services and more or less break even from operating ancillary bus services. As noted, their windfalls come from real estate development around rail stops. In Curitiba, private bus companies earn enough profit to replace rolling stock, on average, every three years, making the city's bus fleet among the newest in the world.

Transit need not be a deficit-riddled business. The key to attracting riders and generating profits is to provide high-quality service. The key to providing high-quality service is to match transit supply to the cityscape – in short, to become a transit metropolis.

Myth 4: Bus transit is incapable of shaping urban form and attracting high-rise development around stops

Besides buses being stigmatized as a second-class form of conveyance, the conventional wisdom holds that buses repel development because of their negative by-products: diesel toxins that spew from tailpipes. Experiences around busway stops in Ottawa and Curitiba should put this myth to rest. In both cities, some of the priciest condominiums anchor sites adjacent to busway stops. Retail and office developers have also flocked to busway corridors in both cities.

Good-quality service – whether vehicles are propelled by electricity or fossil fuels, or whether they roll on steel wheels or pneumatic tires – will spawn compact development. It is the accessibility premium that attracts real estate development, not the type of transit equipment. In fact, compared to freeways and even railway corridors, busways produce relatively low ambient noise levels. Its inherent flexibility advantages and superior adaptability to spread-out patterns of development make bus transit – especially when combined with dedicated busways – a potentially stronger shaper of growth patterns than rail transit in some settings.

Table 1 Knowledge versus action and associated terms

Understanding cities	Designing cities
Past/present	Present/future
What was, what is	What should be
Descriptive	Prescriptive
Substantive	Normative
Research, reflection, knowledge	Action, projection
Urban science, urbanism	Urban design

Closer to home, Kevin Lynch's work is a good case in point illustrating the tensions between the two conceptual poles: Lynch researched people's mental images and constructs of cities and analyzed the history, evolution, and meaning of places in order to seek better ways to design cities. However, while in *The Image of the City* (1960), substantive information is separated from prescriptive or normative advice, in *A Theory of Good City Form* (1981), the two are closely interwoven. Similarly, as Christopher Alexander and his team (Alexander *et al.* 1977) rampage through existing cities they deem "good," they collect, sort out, and discard bits and pieces that they believe will constitute patterns or elements for designing new cities. However, they are essentially not interested in describing critically existing environments per se. On the architectural side of urban design, the brothers Krier (*Rational Architecture* 1978) have peaked [sic] into a prototypical medieval town for identifying the antidote to the ills of modern design theories. Their American followers, architects and town planners Duany and Plater-Zyberk (Knack 1989), have found their norms in the late nineteenth-century American small town, which, after some study, they have then modified and spiced up with garden city and city beautiful theories to establish their own theory of design.

The attractiveness of the normative stand is obvious: it provides unmitigated guidance for designers in their everyday endeavors. Yet its limitations are serious: in the light of the wholesomeness and complexity inherent to design, all normative theories eventually run into difficulties and often fail outright. Further, it is disturbing to find that many normative theories use research to justify or substantiate a priori beliefs when, in fact, the reverse should take place, and research results should be interpreted to *develop* theories.

In order to enter the next generation of urban design theory, urban designers will need to pay more attention to the substantive side of research and to refrain from making quick prescriptive inferences from such research. They will need to separate conceptually the art of description from that of prescription and to devise clean and honest ways of evaluating existing or past situations (for opposing views on this matter, see Jarvis 1980 and Oxman 1987). This is not to say that description or substantive work is "true"; that is, free of values and interpretation. Description is just as subjective – dependent on who is doing it – as prescription. As the art of seeing, hearing, smelling, feeling, and knowing, description can only reflect the capabilities and sensitivities of the researcher (Relph 1984). But if descriptive activity is just as morally bounded as prescription, and if it tells what is right or wrong subjectively, it does nonetheless stop short of venturing into what should be done.

For the design and planning professions to mature properly, time must be taken to focus on substantive information. Some scholars have even advocated the need to describe solely without seeking explanation because they see explanation (the "why" attached to the "what") as yet another incentive not to grasp fully the object or phenomenon being described (Relph 1984). Whatever the case may be, substantive approaches will force designers and planners to engage personally in the information at hand, to interpret it, and to apply it to the specific context of their activities.

The gap between knowledge and action is not an easy one to bridge. It requires careful synthesis. As the perennial example of substantive information, the use of historical studies provides a case in point: today work in history is fashionable and touched upon by many urban designers, yet the dialectic between practice and historical knowledge remains elusive at best, and so far seemingly capricious and idiosyncratic. Careful assessment must precede jumping to practical conclusions.

For these reasons, this article focuses on substantive research and theories. A companion part to this article remains to be written which would map the scope and breadth of normative theory in urban design. Some of this work has been done by French urbanist and philosopher Françoise Choay.

Choay has framed an epistemology of urban design as a normative, prescriptive field in two seminal books that, although they include Anglo-Saxon literature, are only available in French. One, *Urbanisme, utopies et réalités* (1965), is an anthology of key texts on urban design since the nineteenth century. The second, *La régle et le modèle* (1980), posits two fundamental texts defining an explicit, autonomous conceptual framework "to conceive and produce new spaces": Alberti's *De re aedificatoria* (1988), first published in 1452, which, according to Choay, proposes rules for urban design, and Thomas More's *Utopia* (1989), first published in 1516 as a model for urban design.

Others have started to assemble normative theories of urban design, notably Gosling and Maitland (1984), Jarvis (1980), and recently, Geoffrey Broadbent (1990). Broadbent's latest book paints a broad yet condensed chronological picture of "emerging *concepts* in urban *space* design" (my emphases of Broadbent's book title). It promises to encourage future critical assessment of the significance and effectiveness of normative theories of and paradigms in urban design.

Concentrations of inquiry

Substantive research and theories can first be classified by their area of concentration, according to specific views and aspects of the city on which they are focusing. Establishing different concentrations of inquiry is to accept that there are several different lenses through which the design and the making of the city can be viewed and that, in consequence, no single approach to design may suffice. As pedestrian as this realization may appear to, for instance, engineers or physicians who are used to studying their problems from many different angles, it is a challenging proposal to the urban designer accustomed to thinking about singular, "correct" approaches. Nine concentrations of inquiry have been identified: urban history studies, picturesque studies, image studies, environment-behavior studies, place studies, material culture studies, typology-morphology studies, space-morphology studies, and nature-ecology studies. The definition and contents of these areas constitutes the second part of this article.

Research strategies

The specific research strategies that can be used to develop knowledge are, again, several. One quickly discovers that the choice of a research strategy unveils the true philosophical basis of the research itself. The first research strategy is termed the *literary approach*: it emanates from the humanistic fields – literature and history being the most prominent ones – and it relies on literature searches, references and reviews, and archival work of all kinds, as well as personal accounts of given situations. The intent of the literary approach is to relate a story of a given set of events.

Second is the *phenomenological approach*, which projects a holistic view of the world, everything being related to everything, and whose practice depends entirely on the researcher's total experience of an event. It is similar to the artist's approach because it is both learned and intuitive, synthetic and wholesome, or eidetic (signifying that it uses specific examples of behavior, experience, and meaning to render descriptive generalizations about the world and human living [Seamon 1987, 16]). Phenomenologists describe events with all their feelings, senses, and knowledge. They usually refuse to explain the "why" of their findings because they see explanation rooted in interpretation and misinterpretation – leading quickly to abuse of information. Phenomenologists therefore oppose the third research strategy, *positivism*, which portrays the value of description in explanation.

Positivism maintains that knowledge is based on natural phenomena to be verified by empirical science. Positivism implies certainty of cause and effect. It is the tool of the sciences, which are based on the reduction of wholes and on systems of interconnected parts.

While most attempts to describe built environments have used literary or positivistic approaches, phenomenological approaches have recently flourished, according to Seamon, because of a practical crisis in the design fields, where nonholistic approaches have led to partially successful environments, and because of a philosophical crisis in the sciences due to the limitations of positivist thinking. Recently also, however, there have been attempts to reconcile positivism and phenomenology and to see them as complementary (Hardie *et al.* 1989; Seamon 1987).

Modes of inquiry

Specific modes of inquiry are identified to distinguish further between the various research strategies used. Two modes seem to prevail. One is the *historical-descriptive* mode, in which the research is based primarily on accounts of historical evidence – whether on site or via historical documents, plans, drawings, painting, archives, or analyses of the topic. The historical-descriptive mode is generally not used for theory-building purposes, but focuses on highlighting specific events and things. Literary and phenomenological research strategies usually use this mode of inquiry.

The other mode is *empirical-inductive*, where the research is set to observe a given phenomenon or to collect information on it, which is then described via an analysis of the information gathered. Through induction, the explanations of the phenomenon may be generalized upon to develop a theory. ("Empirical" means relying on experience and observation alone, often without due regard for system or theory, or capable of being verified by observation or experiment. "Empiricism" is the theory that all knowledge originates in experience or the practice of relying upon observation and experiment; it is especially used in the natural sciences.) This mode prevails in positivistic research but can be found in phenomenological work as well.

A third mode is *theoretical-deductive*, in which a theory is developed on the basis of past knowledge, which is then tested via research. Used primarily in quantitative research (Carter 1976), this mode is rarely found in the design fields, because they encompass problems that are either too complex or, as Horst Rittel has termed them, too wicked to be approached quantitatively (Rittel and Webber 1972). In such cases, this mode seems to lead to truisms (e.g., all grid plans are the result of a planned approach to making cities) or then to problems that have limited significance to the design of whole environments (e.g., economic theories related to real estate taxation, theories of land use allocation, housing choices, and so on).

Research focus: object/subject

A third screen needs to be applied to areas of inquiry, the research focus. Most research in this country focuses on the study of *people* in the environment. This *subject* orientation enlarged in the 1960s when research became seen as a necessary addition to the practice of planning and design. The primacy of subject-oriented research can be explained as a reaction to "old guard" designers' earlier focus on the physical components of the environment. Theirs was an orientation toward the *object* – a second possible research focus – which became increasingly suspicious as theories of good health, safety, and welfare relying on the need for clean, airy environments continued to bring unsatisfactory results. The ultimate blow to the object orientation of physical planners was the failure of urban renewal schemes, which proved that poverty, not environment, was the primary reason for epidemics, crime, and ethically questionable lifestyles. That good environments can do little to alleviate the basic state of poverty was a hard lesson to learn after four decades of work. From then on, research on the object qualities of the environment became unpopular, and a single focus on people in the environment prevailed with, for instance, sociologist Herbert Gans (1969) as its greatest advocate.

Later, some researchers urged concentration on the interaction between people and environments as a specific phenomenon that could explain well the nature of our environments (see Rapoport 1977; G. Moore *et al.* 1985). Today, the field of person-environment relations, or environment-behavior research, is at least present in planning and design. At the same time, it is under heavy inside and outside criticism largely for neglecting the "environment" part of the person-environment couplet. A return to the study of the object has been advocated by many, especially architects influenced by theorists like Rossi (1964, 1982), who has gone so far to argue the autonomy of architecture as a discipline that is separate from the sciences and the arts. More modest postures favoring a return to object-oriented research, with complementary rather than primacy over subject research, have been argued as well by, for instance, geographer M.P. Conzen (1978), environmental psychologists D. Canter (1977) and J. Sime (1986) and architect L. Groat (Moudon 1987). This trend corresponds also to a rising interest in the study of vernacular environments as the physical evidence of people's long-standing interactive relationship with their surroundings.

Vernacular environments thus offer attractive prospects: many are unusual physical objects, yet not the objects of a few planners and designers, but those traditions and customs that are an intrinsic part of culture. Indeed this "culturally ground object" can uncover the deep relationship between people and environments.

Research ethos: etic/emic

Finally, research needs to be screened for its ethos – this term is selected to depict the "heart" of the research. Two categories of ethos come to mind: the *etic* and *emic* ethos. Borrowed from anthropology, these terms were first popularized in design circles by Amos Rapoport (1977). They come from *phonetic* – related to the written language – and *phonemic* – related to the spoken language. The difference can be further grasped by comparing the two French terms, *la langue* and *la parole*, the first being language as a structured system of sound (or signs to be studied for its internal logic, and the second, a no less structured, yet only practiced system of sounds. Applied to studies of people and culture, etic and emic relate to the nature of the source of the information gathered – etic in the case of the informant being the researcher, the person who will use the information, and emic in the case of the informant being the person observed.

Environment and behavior studies were the first to seek to bring an emic orientation to the design professions: they unearthed information about the uses of environments directly from the users, without relying on the opinion of design and planning professionals. However, the actual methods used in person-environment studies can be more or less emic. For instance, unstructured interviews, oral histories, and self-studied methods of all sorts are straightforwardly emic. But observations of behaviors, although emic in their intent, are, in the true sense of the term, etic because they are done by professionals. Rapoport has called these research approaches "derived" etic and has opposed them to "imposed" etic approaches, which he condemns as mere fabrications of the researcher's mind.

The importance of getting emically significant information about the environment cannot be understated. Lynch's (1960) studies of people's images of cities popularized the need for an emic ethos in the information necessary to the planning and urban design professions. These studies complemented earlier works in parallel areas of anthropology and sociology: the Lynds' critical description of people's lives in *Middletown* (1929); W. Lloyd Warner's *Yankee City* (1963); Herbert Gans's controversial *The Levittowners* (1967); E.T. Hall's compilation of an environment both limited and enhanced by our physiological beings, in *The Silent Language* ([1959] 1980) and *The Hidden Dimension* (1966); and Robert Sommer's *Personal Space* (1969). All opened the door to an enormous field of yet untapped information.

AREAS OF CONCENTRATION

The concerns of urban designers and the nature of the decisions they make are necessarily wide-ranging. The interdisciplinary nature of urban design is likely to remain, and it is doubtful that the field will ever become a discipline with its own teachings separate from the established architecture, landscape, and planning professions. But if primarily architectural research (in, for instance, building science, architectural styles, or programming) and urban planning research (in employment, transportation, and housing demand) are only tangentially relevant to urban design, general socioeconomic issues relating to the environment always loom near the foreground of urban design concerns. In this sense, all social science research pertaining to the environment is of interest. Similarly, all information concerning urban space and form will be useful. Yet a search for breadth must nonetheless be constrained for the sake of practicality. The literature surveyed focuses on the products of urban design or the human relationship with the built environment and related open spaces. The city, and more generally, the landscape as modified by people; its physical form and characteristics; the forces that shape it; the ways in which it is designed, produced, managed, used, and changed – all are central to a search for work that informs urban design. This essentially humanistic view of urban design justifies, at least within the confines of this article, further exclusions – to wit, literature on development and real estate finance, marketing, economic theory, and urban political theory that, unfortunately, relate only marginally at this point to the powers of urban designers.

The literature assembled according to these criteria has then been subjected to various classification exercises in an effort to identify salient areas of relevant inquiry. Thus the classification proposed emphasizes the types of questions posed by the different research, and groups the different works on the basis of the similarity of their quests rather than on the particularities of the methods used. The classification also offers a conceptual framework that is simple enough for both students and practitioners to remember and to work with over time.

Nine areas of concentration are proposed to encompass research useful to urban design. The list of areas should be seen as open-ended. Further, individual researchers can belong to one or more areas of inquiry, depending on the scope of their particular works. Some of these concentrations will be readily accepted as mainstream urban design. But some will raise eyebrows and need further discussion. The following pages review the nature and coverage of each area of concentration. Included is a tentative critique of each area's current status with respect to the level of its development and its current place in building an epistemology for urban design.

Urban history studies

The study of urban history has expanded remarkably over the past two decades to include now significant information for the practicing urban designer. This area's early dependence on art history and its traditional emphasis on "pedigreed" environments (Kostof 1986), on their formal and stylistic characteristics, is gone. Studies of places inhabited by ordinary people, explanations of why and how they inhabit them, have become the focus of an increasing number of scholarly works. Women, special needs groups, and the lower economic echelons of our social class structure are now an integral part of urban historical research. The history of the Anglo-Saxon suburb occupies an important place in historical studies today as suburbs constitute a substantial part of contemporary cities. Further, while Western influences continue to prevail, the overreliance on the European experience is waning, especially with Asian, Islamic, and other cultures embarking into internationally recognized scholarly endeavors.

Classical work on the history of urban form has come from design and planning historians, to include S.E. Rasmussen (1967), A.E.J. Morris (1972), and John Reps (1965), and from historical geographers such as Gerald Burke (1971), Frederick Hiorns (1956), Robert Dickinson (1961), Marcel Poëte (1967), and Henri Lavedan (1941). On the architectural side, there are Norma Evenson (1973, 1979), Spiro Kostof (1991), Norman Johnston (1983), Mark Girouard (1985), and Leonardo Benevolo (1980). Lewis Mumford (1961) remains a powerful critic, although his influence is diminishing with the emergence of more detailed research on various aspects of his writings. But the classical understanding of the history of the city is being enriched and also challenged by the growing explorations of ordinary landscapes, as in the works of Sam Bass Warner (1962, 1968), J.B. Jackson (1984), David Lowenthal and Marcus Binney (1981), Reyner Banham (1971), and recently, John Stilgoe (1982), Edward Relph (1987), and Michael Conzen (1980, 1990). James Vance (1977, 1990) emerges as a wide-ranging scholar of the processes shaping the physical construct of the urban environment. Considering the social history of environments also adds reality to historical forms that in the work of Bernard Rudofsky (1969), Alan Artibise and Paul-André Linteau (1984), Roy Lubove (1967), Anthony Sutcliffe (1984), Dolores Hayden (1981), and Gwendolyn Wright (1981), for instance, come alive in the descriptions of people's everyday struggles to shape their surroundings. How cities have actually been built is another subject of increasing interest – with Joseph Konvitz (1985), David Friedman (1988), and Mark Weiss (1987) standing out as promising contributors.

Work in history continues to be primarily etic and based on literary research (Dyoz 1968). However, derived etic research is beginning to dominate social history. Similarly, phenomenological approaches are increasingly taken – Relph's and J.B. Jackson's work being some of the best received by urban designers. Historians in this category can be object- or subject-oriented, or they can deal with the interaction between people and the physical environment.

The many new publications on increasingly varied subjects related to urban history exercise a growing influence on design and planning professionals. Correspondingly, a few historians are willing to venture into discussing the implications of historical experience for the present – for instance, Joseph Konvitz (1985), Robert Fishman (1977, 1987), Richard Sennett (1969), and Kenneth Jackson (1985; Jackson and Schultz 1972). Conversely, design-oriented scholars are reaching out into history in an attempt to develop theory – as for instance, Dolores Hayden (1984), Peter Rowe (1991), and Geoffrey Broadbent (1990).

The emerging richness of the field warrants further classification and analysis to help the urban designer to select the appropriate works and to uncover more than can be recognized in this article. Work in historical geography and urban preservation is worth reviewing as it includes critical inventories of urban environments. Similarly, historical guidebooks of cities, as well as contemporary guides emphasizing a city's history (Wurman 1971, 1972; Lyndon 1982) yield material that adds to historical knowledge of particular cities. Finally, journalistic criticism is an area that parallels history in its evaluative approach to existing environments and needs to be explored. While such criticism used to be limited to the isolated, yet powerful works of a few – for example, Jane Jacobs (1961), Hans Blumenfeld (1979), Ada Louise Huxtable (1970), Robert Venturi and Denise Scott Brown (Venturi *et al.* 1977), and recently, Joel Garreau (1991) – several publications have emerged that begin to provide a vehicle for systematic and continued critiques of implemented ideas (for instance, *Places*, *the Harvard Architectural Review*, and others).

Picturesque studies

Picturesque studies of the urban landscape were the foundations and the keystone of urban design until the late 1960s. Today they keep a prominent position in both education and practice, and they offer some of the widely read introductory texts in urban design. These studies are running personal commentaries of the attributes of the physical environment. Authors identify and describe both verbally and graphically what they think are "good" environments. Such good environments are analyzed for their relevance to contemporary urban design problems.

Object-oriented, these works emphasize the visual aspect of the environment, which is seen as a stage set or a prop of human action. Gordon Cullen's *Concise Townscape* (1961) remains one of the most memorable contributions to urban design in the picturesque style. Cullen caught the fancy of both architects and planners disturbed by the technical, barren aspects of modernism. He helped them to formulate the scope of urban design as an interdisciplinary activity requiring both architectural and planning skills.

Precursors of the picturesque genre include Camillo Sitte ([1889] 1980) and Raymond Unwin (1909), both of whom have recently regained popularity in urban design. While the postwar work of Thomas Sharp (1946) on English villages has yet to be rediscovered by urban designers, Paul Spreiregen's *Urban Design: The Architecture of Towns and Cities* (1965) remains a standard introductory urban design text today. Also prominent are the writings of Edmund Bacon (1976) and Lawrence Halprin (1966, 1972).

The term "picturesque" is not widely recognized to encompass the works of Sitte, Cullen, Bacon, or Halprin. It has been used in this capacity by Panerai *et al.* (1980) in an effort to capture the emphasis on the pictorial component of the environment that characterizes works in this category. Robert Oxman (1987) used Cullen's own words and called the work "townscape analysis."

For all their popularity, however, picturesque studies are unevenly "practiced," and there have not been publications following this research and thinking mode in several years. Developments in the intellectual context of urban design have lessened the forcefulness of the original picturesque argument. First, if these studies are etic and phenomenological in nature – stands that remain in good currency in contemporary planning and design discourse – they do not espouse these philosophical beliefs in a conscious manner. Rather, they appear to assume a naïve "good-professional-knows-it-all" posture that has been rightfully questioned since the early 1970s. Simply put, they lack the literary references of more recent phenomenological writings such as Relph's or J.B. Jackson's. And they lack the theoretical and philosophical underpinnings of a Norberg-Schultz. It follows that picturesque studies do not fare well either with social science approaches in planning and design research: their unabashed etic stand is unacceptable on this score, and the idiosyncratic swinging between highly personal descriptions and specific prescriptions puts these works in an old-fashioned league.

Finally, whereas picturesque studies were innovative in their early consideration of vernacular landscapes, they have been superseded recently by the several bona fide historical works of scholars such as Thomas Schlereth (1985b), Dell Upton (Upton and Vlach 1986), John Stilgoe (1982), R.W. Brunskill (1981, 1982), Stefan Muthesius (1982), and others. Thus picturesque studies maintain a high profile for the beginning student of urban design but do not sustain well more rigorous and deep investigation.

Image studies

Image studies include a significant amount of work on how people visualize, conceptualize, and eventually understand the city. This category would not exist without Kevin Lynch's *The Image of the City* (1960), whose influence was paramount in launching subsequent research. In fact, many planners and designers see image studies as the main contribution of urban design to the design fields. The Lynchian approach is sometimes understood as continuing the picturesque tradition because of its focus on how the urban environment is perceived visually. Yet the posture of image studies is reversed from that of picturesque studies: it is the people's image of the environment that is sought out, not the professional observer's. Thus image studies are intrinsically emic and subject oriented. Lynch had been influenced by the works of E.T. Hall ([1959] 1980, 1966), Rudolph Arnheim (1954, 1966), and Gyorgy Kepes (1944, 1965, 1966). As a student, he was part of Kepes's MIT group of environmental thinkers who sought to create and understand environmental art – art in space and art as space, so to speak.

Image studies are witnesses to the growing influence of the social sciences on design since the 1960s. They focus on the physiological, psychological, and social dimensions of environments as they are used and experienced by people, and on how those aspects do or should shape design and design solutions. The importance given in these studies to the lay person's view of the surrounding environment has transformed urban design activity: not only are Lynch's five elements used (and, according to Lynch himself, abused [Lynch in Rodwin and Hollister 1984]), but questionnaires, surveys, and group meetings are now standard fare backing up the majority of complex design processes. Among the many studies looking to verify and to expand on Lynch's findings, the ones that brought systematic comparisons (and oppositions) between the professional's and the lay person's views, were his own student's, Donald Appleyard (Appleyard *et al.* 1964; Appleyard 1976). Working closely with psychologist Kenneth H. Craig, Appleyard's group at the University of California at Berkeley trained many students to research people and environments as a sound basis for urban design. Robin Moore (R. Moore 1986) and Mark Francis (Francis *et al.* 1984; Francis and Hester 1990) are products of Lynch's and Appleyard's programs and are now themselves eminent contributors in this arena. The scientific basis of their work has in effect closed the loop linking image studies and environment-behavior studies, and these researchers are now commonly associated with this latter area of concentration.

Environment-behavior studies

The study of relations between people and their surroundings is an interdisciplinary field whose history has yet to be documented fully. Stemming from work done since the turn of the century in environmental psychology and sociology, these studies have grown rapidly since the 1960s, supported by a variety of federally sponsored laws in such areas as community mental health, energy conservation, environmental protection, and programs directed at special needs populations, children, the elderly, the physically impaired, and others.

In the 1960s, the design and planning professions turned to sociology and environmental psychology as sources of valuable information in this new emic realm of research on the environment. Since then, person-environment relations has become a bona fide part of the architectural profession, covering research on how people use, like, or simply behave in given environments. The field also rapidly spread to urban design as Amos Rapoport, Kevin Lynch, and Donald Appleyard began to investigate the human dimension of neighborhoods, urban districts, and cities at large.

Environment-behavior research, as it is increasingly called today, has until recently been almost totally positivistic. Actually, its original influence on design was due to its science-based approach, which was deemed more serious, reliable, and rational than the then-traditional intuitive, often highly personal, design process. The introduction of the social sciences to planning and design was part of a broader trend of interest in multidisciplinary activity, itself the product of system-thinking developed by the military during World War II. In England, the influences of both modernism and the systems approach divided architectural schools of the postwar period into two groups: one at the Bartlett, where Llewelyn Davis was to assume a multidisciplinary approach to design, and the other at Cambridge University, with Martin and March, which was to focus on space, urban form, and land use (Hillier 1986).

In the United States in the early 1960s, the University of California at Berkeley was first to create a College of Environmental Design, thus expanding the professions of architecture and planning to the general design of environments, including industrial design. In the new curriculum at Berkeley, "user studies" (meant to collect information on people expected to use the facilities to be designed) and "design methods" involving the coordination of different interests and expertise (from the user to the investor) ranked high on the list of important courses that students were to take.

Although environment-behavior studies have recently suffered some setback at least in architecture (their development is perceived to have taken away from design – or is it Design?), they are in fact well entrenched in design thinking. People like Amos Rapoport (1977, 1982, 1990), Robert Gutman (1972), Michael Brill (Villeco and Brill 1981), Sandra Howell (G. Moore et al. 1985), Jon Lang (1987), Karen Franck (Franck and Ahrentzen 1989), Clare Cooper Marcus (1975; Marcus and Sarkissian 1986), and Oscar Newman (1972, 1980) remain important figures in education and practice

nationally. The Environmental Design Research Association (EDRA) celebrated its twentieth year with many of its members holding appointments in schools of design around the country (Hardie et al. 1989). The term "environmental design research" has been proposed to cover those studies that relate specifically to design and to eliminate the polarity and actual conflicts that the couplet environment-behavior engenders (Villeco and Brill 1981).

Influential figures contribute to the field: I. Altman (1986; Altman and Wohlwill 1976–85), D. Canter (1977), L. Festinger (1989), D. Stokols and I. Altman (1987), and J.F. Wohlwill (1981, 1985), among others. Principal authors directly related to issues of urban design include: Amos Rapoport (1977, 1982, 1990) on residential environments, city, and settlement; Donald Appleyard (1976, 1981) on city and streets; W.H. Whyte (1980) on urban open spaces and city; Jack Nasar (1988) on environmental aesthetics; Robin Moore (1986) on children and environments; Mark Francis (Francis et al. 1984) on urban open space; William Michelson (1970, 1977) on neighborhoods; Clare Cooper Marcus (1975; Marcus and Sarkissian 1986) on residential environments; both Jan Gehl (1987) and Roderick Lawrence (1987) on streets and residential environments; Oscar Newman (1972, 1980) on residential environments; and S. and R. Kaplan (1978) on open spaces. Further, if most of the studies conducted in this area relate to ordinary environments some deal with differences in values and preferences between professional designers and lay people (Cant 1977; Nasar 1988).

The broad, multidisciplinary nature of the field makes information retrieval somewhat difficult. There are many organizations sponsoring and publishing research (G. Moore et al. 1985), and many journals that have yet to provide comprehensive indexes. However the School of Architecture at the University of Wisconsin in Milwaukee, has published a handy bibliography for use by their doctoral students (G. Moore et al. 1987). Useful surveys of the field are also being produced (Altman and Wohlwill 1976–85; G. Moore et al. 1985; Stokols and Altman 1987; Zube and Moore 1987). As an interesting aside, G. Moore et al. (1987) include J.B. Jackson and other geographers as part of environmental design research. But in our classification,

Hillier, Bill. 1986. Urban morphology: The UK experience, a personal view. In *A propos de la morphologie urbaine*, Françoise Choay and Pierre Merlin, eds. Tome 2. *Communications*. Laboratoire "Théorie des mutations urbaines en pays développés," Université de Paris VIII. Paris: Institut d'urbanisme de l'Académie de Paris, E.N.P.C. March.

Hillier, Bill, and Julienne Hanson. 1984. *The social logic of space*. Cambridge: Cambridge University Press.

Hiorns, Frederick R. 1956. *Town-building in history: An outline review of conditions, influences, ideas, and methods affecting 'planned' towns through five thousand years*. London: George G. Harrap.

Hiss, Tony. 1990. *The experience of place*. New York: Knopf.

Hough, Michael. 1984. *City form and natural process: Towards a new urban vernacular*. Beckenham, Kent: Croom Helm.

Hughes, J. Donald. 1975. *Ecology in ancient civilization*. Albuquerque: University of New Mexico Press.

Huxtable, Ada Louise. 1970. *Will they ever finish Bruckner Boulevard?* New York: Macmillan.

Jackson, J.B. 1984. *Discovering the vernacular landscape*. New Haven, Ct.: Yale University Press.

——. 1980. *The necessity for ruins and other topics*. Amherst: University of Massachusetts Press.

Jackson, Kenneth. 1985. *Crabgrass frontier: The suburbanization of the United States*. New York: Oxford University Press.

Jackson, Kenneth, and Stanley Schultz, eds. 1972. *Cities in American history*. New York: Knopf.

Jacobs, Allan. 1985. *Looking at cities*. Cambridge, Ma.: Harvard University Press.

——. 1978. *Making city planning work*. Chicago: American Society of Planning Officials.

Jacobs, Jane. 1961. *The death and life of great American cities*. New York: Random House.

Jakle, John A. 1987. *The visual elements of landscape*. Amherst: University of Massachusetts Press.

Jarvis, R.K. 1980. Urban environments as visual art or as social settings? *Town Planning Review* 51, 1: 50–65.

Johnston, Norman. 1983. *Cities in the round*. Seattle: University of Washington Press.

Kaplan, Stephen, and Rachel Kaplan. 1978. *Humanscape: Environments for people*. North Scituate, Ma.: Duxbury.

Kepes, Gyorgy. 1966. *Sign, image, symbol*. New York: G. Braziller.

——. 1965. *The nature and art of motion*. New York: G. Braziller.

——. 1944. *Language of vision*. Chicago: P. Theobald.

Knack, Ruth Eckdish. 1989. Repent, ye sinners, repent. *Planning* 55, 8: 4–13.

Konvitz, Joseph. 1985. *The urban millennium: The city-building process from the early middle ages to the present*. Carbondale: Southern Illinois University Press.

Kostof, Spiro. 1991. *The city shaped: Urban patterns and meanings through history*. Boston: Bulfinch Press/ Little, Brown.

——. 1986. Cities and turfs. *Design Book Review* 10 (Fall): 9–10, 37–39.

Lang, Jon. 1987. *Creating architectural theory: The role of the behavioral sciences in environmental design*. New York: Van Nostrand Reinhold.

Lavedan, Henri. 1941. *Histoire de l'urbanisme: Renaissance et temps modernes*. Paris: Laurens.

Lawrence, Roderick. 1987. *Housing, dwellings and homes: Design theory, research and practice*. New York: Wiley.

Lerup, Lars. 1977. *Building the unfinished: Architecture and human action*. Beverly Hills, Ca.: Sage.

Lewis, Peirce F. 1975. Common houses, cultural spoor. *Landscape* 19, 2: 1–22.

Lowenthal, David, and Marcus Binney, eds. 1981. *Our past before us?* London: Temple Smith.

Lubove, Roy. 1967. The urbanization process: An approach to historical research. *Journal of the American Institute of Planners* 33: 33–39.

Lyle, John T. 1985. *Design for human ecosystem: Landscape, land use and natural resources*. New York: Van Nostrand Reinhold.

Lynch, Kevin. 1981. *A theory of good city form*. Cambridge, Ma.: MIT Press.

——. 1972. *What time is this place?* Cambridge, Ma.: MIT Press.

——. 1960. *The image of the city*. Cambridge, Ma.: MIT Press.

Lynch, Kevin, and Lloyd Rodwin. 1958. A theory of urban form. *Journal of the American Institute of Planners* 24: 201–14.

Lynd, R.S., and H.M. Lynd. 1929. *Middletown: A study in contemporary American culture*. London: Constable.

Lyndon, Donlyn. 1982. *The city observed: Boston*. New York: Random House.

McHarg, Ian. 1971. *Design with nature*. Garden City, N.Y.: Doubleday.

March, Lionel. 1977. *Architecture of form*. Cambridge, Ma.: MIT Press.

Marcus, Clare Cooper. 1975. *Easter Hill Village: Some social implications of design*. New York: Free Press.

Marcus, Clare Cooper, and Wendy Sarkissian. 1986. *Housing as if people mattered: Site design guidelines for medium-density family housing*. Berkeley: University of California Press.

Maretto, Paolo. 1986. *La casa veneziana nella storia della città, dalle origini all'ottocento*. Venice: Marsilio Editori.

Martin, Leslie, and Lionel March, eds. 1972. *Urban space and structures*. Cambridge: Cambridge University Press.

Michelson, William. 1977. *Environmental choice, human behavior, and residential satisfaction*. New York: Oxford University Press.

———. 1970. *Man and his environment*. Reading, Ma.: Addison-Wesley.

Mitchell, William J. 1990. *The logic of architecture, design, computation, and cognition*. Cambridge, Ma.: MIT Press.

Moll, Gary, and Sara Ebenreck, eds. 1989. *Shading our cities: A resource guide for urban and community forests*. Washington, D.C.: Island Press.

Moneo, Raphael. 1978. On typology. *Oppositions* 13 (Summer): 23–45.

Moore, Charles W., William J. Mitchell, and William Turnbull, Jr. 1988. *The poetics of gardens*. Cambridge, Ma.: MIT Press.

Moore, Gary T., and the Faculty of the Ph.D Program. 1987. *Resources in environment-behavior studies*. Milwaukee: School of Architecture and Urban Planning, University of Wisconsin.

Moore, Gary T., D. Paul Tuttle, and Sandra C. Howell, eds. 1985. *Environmental design research directions, process and prospects*. New York: Praeger Special Studies.

Moore, Robin. 1986. *Childhood domain: Play and place in child development*. London: Croom Helm.

More, Thomas, Sir, Saint (1478–1535). [1516] 1989. *Utopia*. New York: Cambridge University Press.

Morris, A. E. J. 1972. *History of urban form: Prehistory to Renaissance*. New York: Wiley.

Moudon, Anne Vernez. In progress. City building. Manuscript.

———. 1988. Normative/substantive and etic/emic dilemmas in design education. *Column 5 Journal of Architecture, University of Washington* (Spring): 13–15.

———. 1987. The research component of typomorphological studies. Paper presented at the AIA/ACSA Research Conference, Boston, November.

———. 1986. *Built for change: Neighborhood architecture in San Francisco*. Cambridge, Ma.: MIT Press.

Mumford, Lewis. 1961. *The city in history: Its origins, its transformations, and its prospects*. New York: Harcourt, Brace & World.

Muratori, Saverio. 1959. *Studi per una operante storia urbana di Venezia*. Rome: Instituto Poligrafico dello Stato.

Muratori, Saverio, Renato Bollati, Sergio Bollati, and Guido Marinucci. 1963. *Studi per una operante storia urbana di Roma*. Rome: Consiglio nazionale delle ricerche.

Muthesius, Stefan. 1982. *The English terraced house*. New Haven, Ct.: Yale University Press.

Myers, Barton, and George Baird. 1978. Vacant lottery. *Design Quarterly 108* (special issue).

Nasar, Jack L., ed. 1988. *Environmental aesthetics: Theory, research, and applications*. Cambridge: Cambridge University Press.

Newman, Oscar. 1980. *Community of interest*. Garden City, N.Y.: Anchor Press/Doubleday.

———. 1972. *Defensible space: Crime prevention through urban design*. New York: Macmillan.

Norberg-Schulz, Christian. 1985. *The concept of dwelling*. New York: Rizzoli.

———. 1980. *Genius loci: Toward a phenomenology of architecture*. London: Academic Editions.

Odum, Eugene P. 1971. *Fundamentals of ecology*. Philadelphia: W.B. Saunders.

Oxman, Robert M. 1987. *Urban design theories and methods: A study of contemporary researches*. Occasional paper. Sidney, Australia: Department of Architecture, University of Sidney.

Panerai, Philippe, Jean-Charles Depaule, Marcelle Demorgon, and Michel Veyrenche. 1980. *Eléments d'analyse urbaine*. Brussels: Editions Archives d'Architecture Moderne.

Passonneau, Joseph R., and Richard S. Wurman. 1966. *Urban atlas: 20 American cities, a communication study notating selected urban data at a scale of 1:48,000*. Cambridge, Ma.: MIT Press.

Perin, Constance. 1977. *Everything in its place: Social order and land use in America*. Princeton, N.J.: Princeton University Press.

———. 1970. *With man in mind: An interdisciplinary prospectus for environmental design*. Cambridge, Ma.: MIT Press.

Poëte, Marcel. [1929] 1967. *Introduction à l'urbanisme*. Paris: Editions Anthropos.

Rapoport, Amos. 1990. *History and precedents in environmental design*. New York: Plenum.

——. 1982. *The meaning of the built environment: A nonverbal communication approach*. Beverly Hills, Ca.: Sage.

——. 1977. *Human aspects of urban form: Towards a man-environment approach to form and design*. Oxford: Pergamon.

Rasmussen, S.E. 1967. *London: The unique city*. Cambridge, Ma.: MIT Press.

Rational architecture: The reconstruction of the European city. 1978. Bruxelles: Editions des Archives de l'Architecture Moderne.

Relph, Edward. 1987. *The modern urban landscape*. Baltimore, Md.: Johns Hopkins University Press.

——. 1984. Seeing, thinking and describing landscape. In *Environmental perception and behavior: An inventory and prospect*, T. Saarinen *et al.* eds. Research Paper No. 29. Chicago: Department of Geography, University of Chicago.

——. 1976. *Place and placelessness*. London: Pion.

Reps, John W. 1965. *The making of urban America: A history of city planning in the United States*. Princeton, N.J.: Princeton University Press.

Rittel, Horst, and Melvin M. Webber. 1972. *Dilemmas in a general theory of planning*. Working Paper No. 194. Berkeley: Institute of Urban and Regional Development, University of California.

Rodwin, Lloyd, and Robert M. Hollister, eds. 1984. *Cities of the mind*. New York: Plenum.

Rossi, Aldo. 1982. *The architecture of the city*. Cambridge, Ma.: MIT Press. [First Italian edition, 1966].

——. 1964. Aspetti della tipologia residenziale a Berlino. *Casabella* 288 (June): 10–20.

Rowe, Peter. 1991. *Making a middle landscape*. Cambridge, Ma.: MIT Press.

Rudofsky, Bernard. 1969. *Streets for people: A primer for Americans*. Garden City, N.Y.: Anchor Press/Doubleday.

Schlereth, Thomas J., ed. 1985a. *Material culture: A research guide*. Lawrence: University of Kansas Press.

——. 1985b. *US 40: A roadscape of the American experience*. Indianapolis: Indiana Historical Society.

——. ed. 1982. *Material culture studies in America*. Nashville, Tn.: American Association for State and Local History.

Schneider, Kenneth R. 1979. *On the nature of cities: Toward enduring and creative human environments*. San Francisco: Jossey-Bass.

Seamon, David. 1987. Phenomenology and environment/behavior research. In *Advances in environment, behavior, and design*, E.H. Zube and G.T. Moore, eds. New York: Plenum.

Seamon, David, and Robert Mugerauer, eds. 1989. *Dwelling, place, and environment*. New York: Columbia University Press.

Sennett, Richard, ed. 1969. *Nineteenth-century cities: Essays in the new urban history*. New Haven, Ct.: Yale University Press.

Sharp, Thomas. 1946. *The anatomy of the village*. Harmondsworth, Middlesex: Penguin.

Sime, Jonathan D. 1986. Creating places or designing spaces? *Journal of Environmental Psychology* 6, 1: 49–63.

Sitte, Camillo. [1889] 1980. *L'art de bâtir les villes: L'urbanisme selon ses fondements artistiques*. Paris: Editions de l'Equerre.

Slater, T.R., ed. 1990. *The built form of Western cities*. London: Leicester University Press.

Sommer, Robert. 1969. *Personal space: The behavioral basis of design*. Englewood Cliffs, N.J.: Prentice-Hall.

Spirn, Anne Whiston. 1984. *The granite garden: Urban nature and human design*. New York: Basic Books.

Spreiregen, Paul. 1965. *Urban design: The architecture of towns and cities*. New York: McGraw-Hill.

Steadman, Philip. 1983. *Architectural morphology: An introduction to the geometry of building plans*. London: Pion.

Stilgoe, John R. 1982. *Common landscape of America, 1580 to 1845*. New Haven, Ct.: Yale University Press.

Stokols, D., and Irwin Altman, eds. 1987. *Handbook of environmental psychology*. New York: Wiley.

Sutcliffe, Anthony, ed. 1984. *Metropolis 1890–1940*. Chicago: University of Chicago Press.

Thiel, Philip. 1986. *Notations for an experimental envirotechture*. Seattle: College of Architecture and Urban Planning, University of Washington.

Thompson, D'Arcy W. [1917] 1961. *On growth and form*. Cambridge: Cambridge University Press.

Todd, Nancy Jack, and John Todd. 1984. *Bioshelters, ocean arks, city farming*. San Francisco: Sierra Club Books.

Tuan, Yi-Fu. 1977. *Space and place: The perspective of experience*. Minneapolis: University of Minnesota Press.

———. 1974. *Topophilia: A study of environmental perceptions, attitudes and values.* Englewood Cliffs, N.J.: Prentice-Hall.

Unwin, Raymond. 1909. *Town planning in practice: An introduction to the art of designing cities and suburbs.* New York: B. Blom.

Upton, Dell, and John Michael Vlach, eds. 1986. *Common places: Readings in American vernacular architecture.* Athens: University of Georgia Press.

Urban Morphology Research Group. 1987 to present. *Urban morphology newsletter.* Department of Geography, University of Birmingham.

Van der Ryn, Sim, and Peter Calthorpe. 1986. *Sustainable communities: A new design synthesis for cities, suburbs, and towns.* San Francisco: Sierra Club Books.

Vance, James E., Jr. 1990. *The continuing city: Urban morphology in Western civilization.* Baltimore, Md.: Johns Hopkins University Press.

———. 1977. *This scene of man: The role and structure of the city in the geography of Western civilization.* New York: Harper's.

Venturi, Robert, Denise Scott Brown, and Steven Izenour. 1977. *Learning from Las Vegas: The forgotten symbolism of architectural form.* Cambridge, Ma.: MIT Press.

Vidler, Anthony. 1976. The third typology. *Oppositions* 7: 28–32.

Villeco, Margo, and M. Brill. 1981. *Environmental design research: Concepts, methods and values.* Washington, D.C.: National Endowment for the Arts.

Violich, Francis. 1983. *An experiment in revealing the sense of place: Subjective reading of six Dalmatian towns.* Berkeley: Center for Environmental Design Research, College of Environmental Design, University of California, Berkeley.

Walter, Eugene Victor. 1988. *Placeways: A theory of the human environment.* Chapel Hill: University of North Carolina Press.

Warner, Sam Bass. 1968. *The private city: Philadelphia in three periods of its growth.* Philadelphia: University of Pennsylvania Press.

———. 1962. *Streetcar suburbs: The process of growth in Boston, 1870–1900.* Cambridge, Ma.: Harvard University Press.

Warner, W. Lloyd. 1963. *Yankee city.* New Haven, Ct.: Yale University Press.

Webber, Melvin M. 1964. *Explorations into urban structure.* Philadelphia: University of Pennsylvania Press.

Weiss, Mark Allan. 1987. *The rise of the community builders: The American building industry and urban land planning.* New York: Columbia University Press.

Whitehand, J. W. R., ed. 1981. *The urban landscape: Historic development and management, papers by M.R.G. Conzen.* Institute of British Geographers, Special Publication No. 13. New York: Academic Press.

Whyte, William H. 1988. *City: Rediscovering the center.* New York: Doubleday.

———. 1980. *The social life of small urban spaces.* Washington, D.C.: Conservation Foundation.

Wohlwill, Joachim F. 1985. *Habitats for children: The impacts of density.* Hillsdale, N.J.: Lawrence Erlbaum Associates.

———. 1981. *The physical environment and behavior: An annotated bibliography.* New York: Plenum.

Wolfe, M.R. 1965. A visual supplement to urban social studies. *Journal of the American Institute of Planners* 31, 1: 51–1.

Wolfe, M.R., and R.D. Shinn. 1970. *Urban design within the comprehensive planning process.* Seattle: University of Washington.

Wright, Gwendolyn. 1981. *Building the dream: A social history of housing in America.* New York: Pantheon.

Wurman, Richard Saul. 1974. *Cities: A Comparison of form and scale: Models of 50 significant towns.* Philadelphia: Joshua Press.

———. 1972. *Man-made Philadelphia: A guide to its physical and cultural environment.* Cambridge, Ma.: MIT Press.

———. 1971. *Making the city observable.* Minneapolis: Walker Art Center.

Yaro, Robert D., Randall G. Arendt, Harry L. Dodson, and Elizabeth A. Brabec. 1988. *Dealing with change in the Connecticut River Valley: A design manual for conservation and development.* Amherst: Center for Rural Massachusetts, University of Massachusetts.

Zube, Ervin H., and Gary T. Moore. 1987. *Advances in environment, behavior, and design.* New York: Plenum.

"Urban Design as a Discipline and as a Profession"

from *Urban Design: The American Experience* (1994)

Jon Lang

Editors' Introduction

While literature on urban design aesthetics, history, and theory is relatively plentiful, material on the profession and practice of urban design is relatively rare, but growing as the field evolves. As is evident throughout the selections in this reader, the field began materializing in the academies and practice in the late 1960s. Its meteoric rise during the 1990s is a result of a number of factors, including: recent demographic shifts back to the city, renewed interest in place-making, postmodern emphasis on the city as a site of consumption and profit, attempts to solve urban growth problems, and increasing stakeholder voice brought by participation. As a field of study, urban design is experiencing a massive growth in the academies, with the rise of new programs and courses that capitalize on the field's recent popularity. A number of new academic journals have materialized recently as well, including the *Journal of Urban Design*, *Morphology*, *Urban Design Quarterly*, and *Urban Design International*. These have provided venues for writers to define the field, its subject matter, and the activities of its practitioners. Reasons for the late arrival of professional material on the field have been attributed to competitive interests between the varied environmental design professions (architecture, landscape architecture, planning, public health, civil and transportation engineering), as well as the difficulty in describing its holistic and ever-increasing knowledge base. While other design professions are controlled by professional governing bodies, internal standards, and statements of ethical behavior, the urban design field has yet to coalesce into a viable profession, lacks these organized structures, and typically borrows from other professions.

This reading from Jon Lang's *Urban Design: The American Experience* discusses the evolution of the field, its knowledge, and the practical roles played by urban designers in the public and private sectors. He highlights several emerging roles in the field involved with place-making, ecological responsibility, community activism, participatory design, and creative leadership. Lang implicitly attributes great importance to values in design decision-making and highlights the often contradictory roles in which urban designers find themselves. He differentiates the roles of urban designers who work for paying clients – responding to the proclivities of profit, politics, and job security – as well as those who work with non-paying clients – responding to the needs and desires of substantive users and the public interest more generally. He also suggests less obvious roles that respond to peers, awards programs, professional publications, and the self-driven ego – all of which embody a very different set of values in practice. The future of the field, he suggests, requires a more integrative profession where designers act more as midwives and collaborators, utilizing a broader base of knowledge than is used in the narrow traditional roles of practice.

Prior to becoming a Professor at the University of New South Wales in 1990, where he teaches urban design and architectural theory, Jon Lang taught at the University of Pennsylvania, 1970–1990, chairing the program in urban design during the 1980s. He has written extensively on urban design practice and theory, in addition to a number of texts on architectural history, urbanism, and design in India. His other works include: *Creating Architectural Theory: The Role of the Behavioral Sciences in Environmental Design* (New York: Van Nostrand Reinhold, 1987), a co-edited text, *Designing for Human Behavior: Architecture and the Behavioral Sciences* (Stroudsburg, PA: Dowden, Hutchinson & Ross, 1974), and *Urban Design: A Typology of Procedures and Products* (London: Elsevier and Architectural Press, 2005).

As a general text on urban design, Lang's *Urban Design: The American Experience* provides a comprehensive review of the field, its history, processes, theoretical framework and perspectives on practice. Other texts that describe the field of urban design and its practice include: Paul D. Spreiregen, *Urban Design: The Architecture of Towns and Cities* (New York: McGraw-Hill for the American Institute of Architects, 1965); Spiro Kostoff (ed.), *The Architect: Chapters in the History of the Profession* (New York: Oxford University Press, 1977); Ann Ferebee (ed.), *Education for Urban Design* (Purchase, NY: Institute for Urban Design, 1982); Jonathon Barnett, *An Introduction to Urban Design* (New York: Harper & Row and Icon, 1982); Tamas Lukovich (ed.), *Urban Design and Local Planning: An Interdisciplinary Approach* (Kensington, NSW, Australia: University of New South Wales, 1990); Ali Madanipour, *Design of Urban Space: An Inquiry into a Socio-Spatial Process* (Chichester, UK: John Wiley, 1996); and Matthew Carmona, Tim Heath, Taner Oc and Steve Tiesdell, *Public Places – Urban Spaces: The Dimensions of Urban Design* (London: Elsevier and Architectural Press, 2003).

■ ■ ■ ■ ■ ■

"*Urban design* is a relatively new term for an activity of long standing" is the way this book began. As an activity it will continue, although its name may change in response to new challenges. The focus of attention of what we now call urban design has been with the age-old activities of consciously shaping and reshaping (or forming and reforming) human settlements directly through physical design or indirectly through the establishment of rules that others must follow.

In the late 1960s, *urban design*, as the name for a field, replaced *civic design*, a label associated with the urban concerns of the School of Architecture at Liverpool University. Civic design implies perhaps a primary design focus on major municipal buildings, city halls, opera houses, and museums and their relationship to open spaces in the manner of the City Beautiful tradition. *City design* has also been used, especially by Kevin Lynch, but it implies a primary focus on the overall nature of the city. Philip Thiel's *envirotecture* may be a better label to describe the broad concerns of the design fields as a group (Thiel [1997]).

This book has described the intellectual concern about the nature and goals of urban design in the United States since the beginning of the century but primarily since World War II. Led by an interdisciplinary group of people, but predominantly architects, the growth of urban design as a field of intellectual inquiry and professional action over the last thirty years has been remarkable. Urban designing is an increasingly complex task as the range of human activities grows in size, communications processes become more diverse, new ways of putting geometries together are found, and the rate of physical changes being made in cities speeds up. The growth of knowledge makes the task of design considerably more difficult because it both opens up new options and highlights the ambiguities and contradictions in the tasks we face. Moreover, except in a few instances, the traditional ordering principles such as the cosmological rules or even the compositional principles of Rationalist architecture used historically to justify design ends and means have little intellectual weight in arguments about what constitutes good design. New urban design principles have to be generated. In the United States today, urban designers' work is and has to be seen as forming part of democratic decision-making process shaping future cities, suburbs, and other human habitats.

Since the early 1980s the scope of concern of the practicing architect as urban designer has narrowed in the United States (as elsewhere). The focus has been on large-scale site planning for development projects in the private sector, with the primary goal being to enhance the developer's profitability. Many much-admired and lively schemes have resulted in a profit for the private sector but have also been clearly in the public interest. At the same time, there are many dull, even antisocial schemes, and much of the social and public interest concerns of past urban design efforts have unfortunately been lost. This loss parallels the general decrease in concern of society as a whole with social equity issues in the broader policy-making arena. On the other hand, the concern of urban designers with design of design guidelines and linkage programs maintains the public policy line of thought that initiated the concern with environmental quality issues and that in turn initiated urban design thirty years ago. In 1977, John M. McGinty, president of the American Institute of Architects, echoed this concern:

> I honestly don't think that individual buildings matter any more. Pennzoil Place is beautiful, but who cares? You get into your car and drive past ugly parking lots and junky furniture stores on Main Street to get there. You drive into a subterranean parking lot and walk through grimy tunnels. Or you walk above ground on sidewalks that are not wide enough. By the time you get to the Pennzoil building you don't care whether it is beautiful or not. How you get there and what you look at on the way are as important as having beautiful buildings in a city. Houston is a sorry place to be in and walk in. It's like a museum of buildings. It's the total urban fabric of a city that's important.
>
> cited in Thiel, [1997]

Columbus, Indiana, with its stock of buildings by many of the world's leading architects is a pleasant enough city, but it too has learned that the finest buildings in the world, piece by piece, leave many opportunity costs in terms of the overall quality of a city. Its civic leaders have become more concerned with urban design quality.

Certainly the central areas of most American cities tend to be in better condition now than in 1977, when McGinty made his comments. Much of this is due to the economic climate of the 1980s, the filling in of gaps resulting from the premature demolition of buildings during the 1950s, 1960s and 1970s, and the improvement in the quality of many individual buildings, but much has been due to concerted city planning and urban design efforts. In thinking about the future in the United States, a special question arises: Are individual property rights so important that the benefits of overall cooperation and a concern for social equity in the development of human settlements can only be measured in enhanced property values for tax collectors and return on capital invested to developers?

The answers to such questions are not empirical ones. They are ideological, although they can be informed by empirical knowledge. They are the kinds of questions that will always arise. In order to be able to address them sensibly, urban design theory, and practice as its behavioral correlate, need to be organized in a manner that can be best brought to bear on the problems of society. The time has come to ask whether urban design needs to be seen as a discipline and a profession in its own right if it is to promote quality of life issues in the built environment – or if urban design is to be the public interest concern of architecture. If it needs to be, can it be?

DISCIPLINES AND PROFESSIONS

A discipline is a branch of knowledge. A profession is either an occupation or a group of people in that occupation. It also implies that the group of people have a specialized set of skills and a unique body of knowledge (Larson 1979). Architecture is both a discipline and a profession (see Anderson 1991). Much knowledge about architecture has little relevance to the design of buildings today but remains quite rightly part of the discipline. Also, much of an architect's professional activity (many architects would claim most of it) draws very little on the disciplinary knowledge of architecture, but instead draws on that of other fields – the management sciences, for instance. Even the act of building design itself draws heavily, in practice if not in academia, on the various branches of engineering and their basis in the discipline of physics.

The same kinds of observations can be made about civil engineering, landscape architecture, and city planning – the other major environmental design fields. They are acknowledged to be both disciplines and professions. The emergence of urban design as a professional activity raises questions about its legitimacy as a field in its own right and about whether it exists as a discipline. Certainly the other design professions regard it as part of their own domains. Whether or not it benefits society to have another specialization as a discipline and a profession in its own right is another question. Urban design certainly cannot be an exclusive profession.

URBAN DESIGN AS AN INTEGRATIVE DISCIPLINE

Paralleling the growth in empirical research on the nature of cities and urban places has been a growth in substantive urban and urban design theory based on an enhanced understanding of the person–environment relationship. There has also been a concomitant growth in our understanding of the designing processes – procedural theory. Much of this growing body of knowledge and ideas is shared with other fields – urban studies and environmental psychology on the positive theory side, and urban planning, landscape architecture, and to a lesser extent architecture on the normative side.

The theoretical content of the field of urban design, while overlapping those of other fields, is, as a synthesis, unique. Whether it should be regarded as a discipline in its own right is an open question. Much depends on the future scope of concern of other disciplines, particularly the environmental design disciplines of architecture, civil engineering, city planning, and landscape architecture. Traditionally urban design has been most closely allied with architecture and city planning by filling the intellectual and professional gap between them. As architecture reduces the domain of its concerns and methods of working, and as city planning continues to be concerned mainly with transportation and land-use planning, to the extent that it deals with other than social and economic issues, so urban design has become increasingly an entity on its own.

Architecture as a discipline, as a branch of knowledge, has focused on the nature of buildings and on the forming principles used by members of the profession in designing them. In recent years, the profession seems if anything to have narrowed its range of concerns in response to the increased complexity in building that arises from new uses and new means of construction. Programming has been taken over by independent specialists. In many places the developing field of building is taking over the construction, construction supervision, and management role traditionally held by the architect. The focus of concern of the discipline of architecture has remained firmly on the *design* aspect, architecture's unique contribution, and thus mainly on intellectual aesthetic issues. Architecture has not developed a broad environmental concern, although individual architects are becoming increasingly concerned with environmental issues in site design and the selection of materials. It has been reluctant to embrace new empirical knowledge about the physiological or psychological consequences of its work except where forced to do so by critics, clients, laws, regulations, and design guidelines.

A broad array of citizens and professionals has been concerned with the nature of the biogenic environment, particularly those living in areas of the United States where there is a shortage of land. As population pressures have increased in places that hitherto had wide open spaces, so has the concern with environmental health. Regarding urban designing as part of the process of environment change, of environmental adaptation, is a recent attitude that grows out of the basic changes in the landscape architecture profession and the growth in its scope of concern from garden design to include regional planning on an ecological and biological basis. Concomitant with this development has been the concern with the social ecology of the patterns of both agricultural landscapes and human settlements, and with the search for new forms of settlement patterns and legal tools to protect the biogenic environment. There is a sense of urgency about these issues as populations increase and more land is turned from rural to urban uses, and also, paradoxically, as abandoned agricultural land returns to woodland or grassland in a number of areas of the United States.

The bodies of knowledge of the design fields overlap considerably, and their positive substantive theory draws heavily on both the natural and social sciences. The natural sciences provide insights into the nature of materials, structures, and technology, and the social sciences into the understanding of activity systems, the distribution of activities in space within cultural frameworks, and aesthetic theories. Design methodology, the study of design processes, is common to all decision-making

anything, it is too political, the presentation require-ments are too stringent, the process needs stream-lining, there are too many agencies involved. While acknowledging these issues in the following questions, I do not consider them overwhelming arguments against design review. It is not that they are trivial, but rather that reasonably obvious solutions exist for them.

Design review is time-consuming and expensive. Architects considered delay to be the number two flaw of design review. (The lack of design experi-ence on the part of the reviewers was cited as the primary flaw.) It definitely costs more in pro-fessional fees. Of those surveyed, 66 percent esti-mated the billable hours spent on design review to be between 5 and 25 percent of their time, a percentage that compares to the time spent on the entire preliminary project design. For the client, design review undoubtedly adds to the time and cost of projects. It adds also to the cost of government, which must administer and maintain design review apparatus in the form of additional professional staff, commissions, printed materials, lawsuits, hearings, and appeals. The additional cost and time factors make the process of design review even more sub-ject to the vagaries of politics: when times are good, government can easily demand design review; when times are bad, clients can no longer afford design review and government is forced to back down or risk losing important construction projects.

Design review is easy to manipulate through persuasion, pretty pictures, and politics. Since the judgment of design is essentially discretionary and inherently difficult, it is easy to use mumbo jumbo design talk to defend decisions that are patently political (pro or con of the proposal) without letting the public become much the wiser. The political tendency is to use aesthetic control for growth control or growth encouragement, or to extract non-design-related amenities in exchange for design approval. Whatever aesthetic purpose design review may have enjoyed becomes com-pletely subordinate to the political agenda in many cases.

Design review is being performed by overworked and inexperienced staff. In the law, the wisest, most

experienced minds are called to judge. In design review, the primary reviewer is far more likely to be a junior planner without design background or an unregistered young designer or a politically appointed committee with the common thread of community prestige and power, not design ex-pertise. The staff planners around the country that I have met are tremendously sincere individuals – they study the issues, they work hard to make the right decisions, and they receive very little guidance or reward. They are often overwhelmed by the complexity of design review, which may be the leading cause in their cry for more and better design guidelines – number one reform of design review suggested by planners who review projects.

Design review is not an efficient mechanism for improving the quality of the built environment. Aside from being time-consuming and unpredictable, design review is usually limited to certain areas, uses, or sizes of projects. It is also limited, obviously, to projects undergoing change or being newly built. It is no more effective than zoning in controlling bulk, height, and setbacks (very important elements of urban design), but it is more complicated than zoning and more subject to interpretation and politics.

THE ENDEMIC PROBLEMS

I have separately organized the following sets of issues because they are much more difficult to describe fully and much more difficult to solve than the regulatory issues just mentioned. As it turns out, solving one of them tends to cause problems in another; for example, making design less arbit-ary and more objective tends to reduce the flexi-bility to make discretionary decisions that are a necessary element of aesthetic judgment. I have organized them around the robust topics of power, freedom, justice, and aesthetics.

POWER

The fundamental question in the issue of power is *who* – who will judge, whose tastes will matter, whose interest it is to control the aesthetic quality of building. Many people will support design review because they believe that it gives more community control over the environment, and in many places this is true. But does the design of urban buildings belong with the community (or rather, with their appointed planning representatives) or with those who are design experts involved in solving the whole building problem?

Design review is the only field where lay people are allowed to rule over professionals directly in their area of expertise. It seems odd that we as a society believe that the improvement of the physical environment can be made by reducing the influence of architects and increasing the influence of planners and lay appointees. As architects, we owe it to ourselves to investigate how this serious turn of events could occur. Are we being punished for the International Style? Are we seen as lackeys of the greedy developer/builder? Have we lost the respect of the public because we no longer even try to defend design excellence in the face of our clients' wishes? Are we elitist, in making projects that only we can understand and interpret, without attempting to educate the public or even reach them?

It is certain that architects – even those who approve of design review – are not willing to concede the judgment of design to lay persons. The number one complaint of architects who answered our survey about design review was that the reviewers were not trained professionals with experience in designing buildings. Nearly every architect who cited an exemplary process told us that what made it exemplary was the presence of knowledgeable professionals as reviewers. Even the city agency planners complained about non-professional members of review boards. Yet about 45 percent of all bodies that review project design do not have even one architect on them. Architects whose experience includes being reviewed by other designers are more likely to accept design review, although they may still find it flawed. Several respondents lamented the lay reviewer by making comparisons to the medical world, where lay people are not permitted to interfere with professional judgments.

Design review is grounded in personal – not public – interest. Perhaps if there were a public realm, a sense of public responsibility about the environment that led to design review, it would be a more legitimate process. For now, it is recognizably not so, being more a matter of protecting private property values from "offending" intrusions rather than a genuine public-spirited activity (Scheer, 1992). When neighbors attend design review sessions, their comments, even the fact of their attendance and concern, have more to do with the desire to stop someone from diminishing the view from their deck or to halt the construction of nearby apartment buildings or shopping centers in their backyards. While these are legitimate concerns, they are essentially self-centered, not public-centered. Neighbors seem to realize the inappropriateness of these self-centered concerns, because their rhetoric (as is the developers' rhetoric) is often disguised as protection of the public. Design review is not even effective at controlling the self-centered problems, since the common result of review will be to put a pretty face on a problem. Zoning is a much more powerful and direct tool to address size, layout, and location, but public officials are reluctant to use it. Reducing the size of buildings or denying a permit does not add to the tax base or economic growth, and promoters of large projects tend to wield political influence.

Community aesthetic input seems most legitimate when a public space is involved. Cincinnati's Fountain Square, for instance, is the subject of much public debate about its design, most of it by people who have a special interest, but at least some of which is genuine concern for the symbolic and public role that it has.

FREEDOM

The flip side of power is freedom. Unlike some of our international friends, the spirit of community in this country is heavily tempered by the belief in the rights of the individual. A somewhat related concept is the view that diversity – taken to mean varying perspectives, disagreements, and cultural differences – is a strength for society as a whole because it provides a wealth of criticism and a wealth of ideas: it keeps us on our toes. The constitution protects the individual from the power of the collective government and allows diversity to flourish.

Is design review a violation of the First Amendment right to free speech? The answer rests on two questions: (1) Are architecture and other aspects of the built environment protected as "speech" under the Constitution? (2) Can the government show a legitimate interest that would override the protection afforded to free speech in this case?

Although there has not been a single case adjudicated on the specific issue of architecture and the First Amendment, nearly all legal theorists who have approached the subject of aesthetic legislation (notably Williams, 1977; Poole, 1987; and Costonis, 1982) agree that architecture should be given the protection afforded to most forms of symbolic expression. In what appears to be an interesting contradiction, recent cases have expanded First Amendment protection to cover "commercial speech" such as signs and advertising, while at the same time the courts have overwhelmingly supported the increase in the regulation of design.

Although the language of the First Amendment clearly states that "Congress shall make no law . . . abridging the freedom of speech," there are many examples of laws in the United States that make it clear that freedom of speech is limited. In order to demonstrate that regulations and practices of design review are legitimate limits on First Amendment freedoms, theoretically a jurisdiction would need to define a very powerful public interest that would override the protection of free speech. It seems to be a dubious assertion to claim that the public interest is substantially served by controlling the color of awnings or requiring that the style of new construction is compatible with existing buildings. Even if the test requiring a substantial government interest could be met, this interest would have to be justified on grounds (such as public safety) that are not related to the suppression of an aesthetic message. In other words, it seems clear that laws that have as *their primary purpose* the curtailing of aesthetic styles or the forcing of homogeneity (known in architecture as "contextuality") would encounter First Amendment problems.

Why is it important to concern ourselves with extending First Amendment protection to architectural expression? One of the purposes of the First Amendment is to protect the individual from the tyranny of the majority. Design review/design guidelines can be interpreted as a way of reinforcing a majority-based, cultural bias (i.e., historic, white, European), especially in a threateningly pluralistic architectural and cultural milieu. Architecture is like a beacon, announcing the status, values, and interests of its culture, its creators, and its inhabitants. It could even be argued that the communicative message of architecture is so strong that community leaders, in formulating design controls, are simply trying to control the message. By excluding certain culturally diverse architectural languages or unpopular architectural styles, we literally suppress a minority viewpoint and prevent those with a different, even critical, perspective from speaking. Thus, if you believe that cosmetic imitation of quaint New England village architecture is false and damaging to the authenticity of place, you will have to express that belief without utilizing its clearest language – architecture. And the places where meaningful architecture of this nature can be explored are rapidly vanishing.

Design review rewards ordinary performance and discourages extraordinary performance. This has come to be known as the "Dolby" effect: a review that cuts out the highs and the lows. Although it is frequently cited as a criticism, it is probably less an issue in actual practice, where the excellent, exceptional, and original design proposed is often treated pretty well by design reviewers, especially if it has a famous name attached to it, and especially if the reviewers have design training. A much more severe and insidious problem, however, is related to the perception of the Dolby effect, because designers begin to anticipate the range of acceptability of particular reviewers and therefore rarely waste their clients' time proposing something original or exceptional. Of 170 architects who answered our survey, 80 percent felt that their proposals were somewhat or strongly influenced by what they knew to be acceptable to a design reviewer. Some architects told us that they liked design review because it brought them more clients who were impressed with their ability to design projects that were approved quickly. When contemplating the cumulative effects of this tendency, one can only become fearful of the mediocre quality of the future built environment and the dwindling potential for truly exceptional works of architecture in this era.

JUSTICE

Some forms of design review are more "fair" than others; that is, the rules are clearer and more objective, and the procedures are more predictable and consistent. It may seem that we should move this issue to the "solvable" side of the column, chalking it up to the newness of design review and the lack of tested processes and model codes. We must keep in mind, however, that the purpose of design review is not to deliver justice to the players, but to deliver the best environment to the community. Because of the slippery nature of design, a less discretionary system may not be flexible enough to work. Therefore, the explicit and fair process might not be the one that delivers the best environment. What follows is a discussion of the issues associated with justice and protection of the individual in design review, but the foregoing problem must be recalled while we explore these.

Design review is arbitrary and vague. Many areas of the law fall under discretionary ruling; in fact, making orderly discretionary decisions is one of the purposes of the judicial system. A police officer exercises discretion in deciding whether to arrest someone or to let him or her go. When discretion gets out of hand, as it sometimes does with the police, more rules and guidelines are laid down to limit the discretion. Just as there is no way to create a rule for every possible circumstance confronting a police officer, there is no way to formalize every rule about design. Therefore, even the most "objective" design review rests on discretionary judgment. This is not the essential legal objection, however; it is the degree to which these discretionary judgments are made consistent and nonarbitrary. Guidelines help, but many cities don't have them. Even where guidelines exist they may essentially be so vague as to be meaningless, insisting, for example, on "appropriate" scale or "compatible" design. Architects consistently complain of being sabotaged by the unclear language and unclear intentions of design review, which are clarified only in response to a specific proposal.

Design review judgments are not limited. Even though a city or town has guidelines, it is rare that the process of design review is limited to reviewing those items covered by guidelines; rather, the guidelines seem to represent a starting point, after which reviewers are relatively free – to critique whatever they like or dislike about a project. There are limits, but these seem to be drawn from a political consensus about how much power the reviewers may exert. In exemplary cases, design reviewers must not only adhere to guidelines explicitly and exclusively, but must also publish "findings" that denote their critique in terms of the guidelines. Unfortunately, the more common pattern is a free-for-all, where the designer can be attacked for any aesthetic or conceptual decision and where no official document records the review criticisms.

Design review lacks due process. Because there are usually no limitations on what is reviewed, the designer is completely at the mercy of the power of the design reviewer. Also, not all projects are subject to the same process, since the process varies from district to district and use to use, and the rules and players are constantly changing. (Only 15 percent of cities have review systems unchanged from ten years ago.) In 12 percent of cities with design review, there is no appeal of a review body's decision. Most important, in most places design review is inconsistently applied. There are no provisions for referencing earlier cases or building up case law that would limit the interpretation of guidelines or judgments and help designers and interested citizens defend their positions.

Design review is difficult to protest on aesthetic grounds. Consider the situation of an architect whose building design is severely altered, but not rejected, by the design review body. He or she has two choices: carry out the alterations and get on with the project (a choice the client is likely to support), or mount a time-consuming and expensive battle, possibly losing the client and commission in the process, as well as alienating a design board that he or she must seek approvals from on a regular basis. Thus the very nature of the design review process (use of "negotiated" coercion, discretionary decisions, uneven power balance, client/architect relationship) works against an individual's ability or desire to try to fight for aesthetic decisions.

Unless the developer finds it to his or her monetary advantage, cases about design seldom go to court. So, while "takings" suits, which claim monetary loss, are common, First Amendment suits, which claim the right of free expression, are nonexistent. Coupled with the tendency of clients to select architects on the basis of their ability to make it through the review process quickly, this may mean that an architect with thoughtfulness, creativity, and design integrity is at a distinct disadvantage.

AESTHETICS

A design reviewer must sooner or later face up to the difficulty of deciding what is right and what is wrong – in short, making judgments. Some have argued that design review could simply drop the idea of beauty, since it is too slippery to be legal, and focus instead on "shared values" (Costonis, 1989). It is clear that many aesthetic decisions are complicated by moral issues (values). We may share the belief, for example, that mowed lawns are attractive. On the other hand, mowed lawns are not good for the environment because they waste water and provide no shelter for wildlife. Fields of native flowers may not only be better in a moral sense, they may also be more beautiful. Or maybe not. It doesn't help that these decisions are relative: one man's wildflowers are another's weed-infested lawn. Clapboard is fine here, but not there. Sign variety is desired in Times Square but not on Court House Square.

Design review is reluctant to acknowledge that there are no rules to create beauty. Architecture today admits of no reference standards, no abstract principles, no Vitruvius or Alberti or even Le Corbusier to dictate propriety. Principles of good design, for today's architects, are not universal, they are specific to the problem, place-centered, expressive of time and culture. For design review to be consistent, on the other hand, principles must be harder, broader, and applicable across the board. The arbitrariness of design review is a result of the vagueness of the guidelines, and the inconsistency of the reviewers. The solution would seem to be more definite guidelines, more precise rules, judgment tempered by precedent. The tendency to increase the use of objective criteria bears this out. Yet, design excellence is not

easily defined by hard and fast principles, beauty is not subject to objective criteria, and judgments are necessarily dependent on the aesthetic response to singular, particular case, not a universal abstraction. A conflict between the increasing objectivity of design review guidelines and the very nature of postmodern architectural thought is inevitable.

Planners do not seem to be morally conflicted at the prospect of making objective criteria, on the other hand. Perhaps it is because that, in the haste to draw up the sign control standards or the contextual controls, the important questions are not being asked. What makes cities well designed or beautiful? Is making a consistent place the same as making a beautiful place? What makes a building beautiful? How can design review take heed of the different aesthetic responses that people have? Shall design review view the building as an object, to be judged without reference to its meaning or use or place in the larger site? Shall design review judge only those superficial aspects of the object such as its style or roofline? Shall design review only concern itself with contextual issues like massing and relationship to streets and leave meaning or style alone? How about the message, the "reading" of buildings – if it contributes to our response to the building, can design review judge that as well? If so, how can we give the architect freedom in his or her message? What can possibly serve as criteria for judgment? No wonder it is such a tangle.

Design review principles tend to be abstract and universal, not specific, site-related, or meaningful at the community scale. Along with the use of contextual patterns as design criteria, my survey of cities and towns with design review revealed nearly universal agreement on the elements that cities review: more than 90 percent of towns review fences and buffers, parking lot location and landscaping, signs, screening of loading and trash areas and building height. The most popular principles of good design (with at least 80 percent of towns agreeing) are directed at simple "neatening up": screening service areas and parking lots, reducing the variety of signs, and recreation and infill of contextual patterns. Ironically, the least popular or irrelevant, according to the planners who responded, were design principles that were more specifically related to building or urban design, for example, encouraging public spaces or fountains. Other than those popular principles directed at the desire to protect a site's natural environment (a finding that slightly conflicts with the same planners' admission that they do not actually review a project's response to microclimate, sunlight and shadows, the generation of pollution, or energy efficiency), most design principles being used extensively are extremely general and transferable from one place to another.

Design review encourages mimicry and the dilution of the authenticity of place. By simplifying the rules and guidelines, by encouraging banal imitations, by denying originality, creativity, or expression of difference in any way, the design review system eventually creates a dead place, a place without surprises or exigencies of site or landmarks. Fortunately, the city's uncontrollable actors (age, events, change) take care of such superficiality by immediately beginning the process of writing over it. And fortunately, too, design review is usually not that effective and is almost never followed up after a few years. But what of places that are effectively controlled for long periods of time? Some cities that have had stringent design review for long periods of time, like Cincinnati's Mariemont (a village designed in 1921 by John Nolen), *are* completely distinct from their chaotic neighbors, with a serenity that comes only from common architectural expression and homogeneity. It could be argued that the excellent quality of Nolen's original plan for Mariemont, the coherent and consistent design of the original buildings, and the respect that this excellence inspired affected later developments a great deal more than design controls. Nevertheless, Mariemont has resisted any changes through the offices of its design review. It is as if it is frozen in time. The price of its homogeneity is fossilization, an inability to change. In a tiny town like Mariemont, the price is undoubtedly worth it. But in a large, functioning, active city, such rigidity could be functionally, morally, and socially dangerous.

Outside of special historic enclaves like Charleston, South Carolina, Mariemont, or Boston's Beacon Hill, places where extreme control is exerted have a kinship to theme park perfection or urban fantasy and embody an idea that life lived here is not real life fraught with pain and crisis and emotion, but an artificial one, cleaned up, predictable, and safe. Thus the overcontrolled Battery Park City is the Disneyland equivalent of the real New York City – it is New York rendered as a stage set, spooky and unreal because it lacks the scars of urbanity: street people, vendors, handmade signs, noise, and bustle (Russell, 1992). Sadly, this approach also dilutes the meaning of the real space it imitates or preserves under glass. The camouflage of new "old" buildings resulting from misguided design review makes the authentic old buildings disappear and lose their importance and distinction.

Design review is the poor cousin of urban design. Ideally, design review's purpose would be to serve an urban design vision specifically developed for the place, the processes, and the public will. Of particular focus and importance for urban design implementation would be the public investment: streets, sidewalks, plazas, public buildings, maintenance, and parks. The use of design review for this purpose is relatively rare. Of the cities with design review, less than 30 percent subject public buildings to design review and only 18 percent review public infrastructure for design.

Design review generally focuses on single projects rather than working from an urban design program. Sometimes, design review is performed in a vacuum, operating as a studio jury, with judgments and critiques rendered on the design merits of a single project, without a concern for its place in the urban ensemble or its impacts on the nature of the surrounding space. (Of those with design review, 26 percent did not use contextualism in any way as a measure of design quality.) More often, design review is concerned with surroundings, specifically *context*, which has become confused in meaning. At the current time, planners who use context as a measure agree strongly that contextual fit means that (1) new buildings and rehabs should respect the existing pattern of buildings and open space and (2) designs that diverge widely from surroundings should not be allowed. This, too, though, is not an urban design vision or plan, but simply the recognition of an old, existing pattern that in itself constitutes too simplistic a view of urban design. Planners without physical training may find this a comforting and completely adequate approach to urban design but it negates the importance of design to create urban space, connect places, and create hierarchy and meaning. If urban design were simply a matter of the repetition of old patterns, as it seems the practice of design review encourages, there would be no opportunity to design new responses to changes in the world, like the advent of computer communication and shopping malls.

Design review is a superficial process. Of course, the effectiveness of design review is limited by the type of things commonly reviewed: reviewers focus on the surface materials and stylistic quality of buildings, and the concealment of cars and of signs. Yet the condition of the urban and sub-urban environment has more to do with the use of ubiquitous and automobile-scaled typologies – K-Marts, strip shopping centers, gas stations, fast food chains, endless pavement – than whether K-Mart has blue metal or yellow awnings or even tasteful signs. Landscaping, buffers, fences, and other popular design review requests are just ways of hiding the problem, not fixing it. The catalog of what is wrong with our environment is a catalog of what is wrong with our culture: the dominance of greed and consumption, the lack of public responsibility (on the part of both residents and builders), the deterioration of the inner city from poverty and crime, the energy waste of sprawl and automobile domination, and the abuse of the natural setting. To the extent that government is allowed to think that it is "taking care" of the "ugly" problem through the institution of design review, it is a diversion of political energy from environmental, social, and economic problems and, not insignificantly, it is a diversion from the necessity for genuine urban design. The design review solution is in fact reminiscent of the urban renewal solution: urban renewal postulated that the solution to the unsightly and deteriorating inner city was to tear it down and build new office buildings and high-priced housing.

THE INVITATION TO DEBATE

This is a fascinating topic because there seems to be no end to the ideas it engages: power, freedom, beauty, morality, justice, discretion, authenticity. After five years of being a design reviewer and five years subsequently of studying it, I have come to be concerned with the enormous effect that widespread design review will have on our cities and towns, on the profession of architecture, and on the public life and freedom of our people. These effects are just beginning to be clear. What is not clear is whether design review, a very powerful government tool, can be directed in a way that answers some of the problems addressed above. Its potential for abuse and misdirection is very strong, and even dangerous. Yet the need for thoughtful urban design in American places grows every day, and the rights of the community to expect local government to contribute to good design is unquestionable. Our task in this book is to bring the best minds to bear on the issue of design review, to look at how it is done in various places, and to offer criticism that will bring about better ways of bringing good design to the urban setting.

REFERENCES

Costonis, John. 1982. "Law and Aesthetic Regulation: A Critique and Reformation of the Dilemma." *Michigan Law Review* 80: 355.

Costonis, John. 1989. *Icons and Aliens: Law, Aesthetics, and Environmental Change.* Champaign: University of Illinois.

Gordon, Doug. 1992. "Guiding Light or Backseat Driver." *AIA Memo*, December, p. 28.

Poole, Samuel, III. 1987. "Architectural Appearance Review Regulations and the First Amendment: The Good, the Bad and the Consensus Ugly." *Urban Lawyer* 19 (Winter): 287–344.

Russell, Francis. 1992. "Battery Park City: An American Dream of Urbanism." *Proceedings of the International Symposium on Design Review*, p. 315.

Scheer, David. 1992. "Design Performance." *Proceedings of the International Symposium on Design Review*, p. 133.

Williams, Stephen. 1977. "Subjectivity, Expression, and Privacy: Problems of Aesthetic Regulation." *Minnesota Law Review* 62 (November): 1–58.

"Design Guidelines in American Cities: Conclusions"

from *Design Guidelines in American Cities: A Review of Design Policies and Guidance in Five West Coast Cities* (1999)

John Punter

Editors' Introduction

What constitutes best practice urban design policy? What are ways of doing urban design and making it effective within different political and regulatory contexts? Substantively and process-wise, what has worked over periods of time and what hasn't? What can urban designers in one city or country learn from those practicing elsewhere? On the one hand, each city is different from every other one, particularly in regards to governance, bureaucratic structure, economic bases, physical, climatic, social, and cultural contexts, and the roles of citizens in decision-making, so it seems difficult to gain meaningful lessons from how one city does things that apply to another. On the other hand, detailed knowledge about how urban design initiatives are practically accomplished in different cities, particularly when their physical, social, economic, and governmental contexts are well explained, can inspire practitioners to transcend entrenched ways of doing things and explore new possibilities.

In this regard, John Punter's extensive comparative research on design policy is very important, not least of which because he is very thorough at explaining how things work in different cities in relation to their contexts. Here, in the conclusion to his book *Design Guidelines in American Cities: A Review of Design Policies and Guidance in Five West Coast Cities*, Punter summarizes and synthesizes the findings that come from analyzing the urban design public policy of five cities on the west coast of the United States – Seattle, Portland, San Francisco, Irvine, and San Diego – cities that he posits are generally understood to have put in place best practice urban design guidance. From the analysis, which includes review of legal issues related to planning regulatory controls, Punter distills a framework for how design guidance should be structured that he argues is applicable to cities elsewhere, including in different countries. He finds that design policies should be based on careful study of the locality and full consultation with the community. Design review should be part of a comprehensive coordinated effort at design regulation and integrated into the planning process. The design review process should be efficient, fair, and effective, and that design guidelines should be precise but not overly prescriptive, underpinned by design principles, and backed by implementation advice.

John Punter's other writings on design policy include *The Design Dimension of Practice: Theory, Content, and Best Practice for Design Policies*, co-authored with Matthew Carmona (London: E & FN Spon, 1997), *From Design Policy to Design Quality: The Treatment of Design in Community Strategies, Local Development Frameworks and Action Plans*, co-authored with Matthew Carmona and David Chapman (London: Thomas Telford, 2002), and *The Vancouver Achievement: Urban Planning and Design* (Vancouver, BC: UBC Press, 2003). The first book describes and analyzes British design policy. The third is a detailed look at Vancouver's unique discretionary zoning system and approach to design regulation, including the use of urban design framework plans that shape building types and set the form and character of public spaces, that have resulted in what some consider one of the best examples of recent city-building. Vancouver has built new high-density residential neighborhoods on former industrial lands in and around its downtown core while at the same time building a vibrant, pedestrian-oriented public realm and whole communities. The unique building type that has emerged from Vancouver's design guidance is what has come to be known as a "point-tower-over-podium-base," where the base contains townhouses with individual entries facing onto streets and pedestrian walkways. Elizabeth Macdonald's article "Street-Facing Dwelling Units and Livability: The Impacts of Emerging Building Types in Vancouver's New High-Density Residential Neighborhoods," *Journal of Urban Design* (vol 10, no. 1, pp. 13–38, 2005) provides a look at on-the-ground impacts of these new buildings.

One of the best ways for students and practitioners to learn how to be effective design regulators and to write well-articulated guidelines is to look directly at examples of design guidelines and design review systems. This is becoming easier and easier to do because many cities are starting to post their urban design and planning policies on websites in easily downloadable form. The cities of Portland, Oregon (http://www.portlandonline.com/planning/) and Vancouver, British Columbia (http://www.vancouver.ca/commsvcs/planning/) both have highly sophisticated design guidance systems and their planning departments have well-structured websites that make their design guidance documents easily accessible.

AUTHOR'S INTRODUCTION

The task of this monograph was to illustrate and explain 'best practice' American design review to an international audience, providing as many illustrations, plan excerpts, and samples of guidance as possible in the belief that this would stimulate thought and encourage innovation in design guidelines and review processes in countries beyond the USA. West coast cities of the USA were selected to illustrate this 'best practice', four of them major metropolitan areas and two of them suburban municipalities. Collectively these were considered to display a long-standing commitment to design quality; a full range of design visions, strategies, goals, objectives and policies at different scales from city-wide to the neighbourhood; a sophisticated review process; a high degree of public consultation and public ownership of policies; and a wide range of implementation devices and investment programmes.

The evolution of the respective planning and design policies has been discussed and their strengths and weaknesses assessed. Their outcomes have been assessed to a much lesser extent, and it has rarely been possible to assess the effectiveness of policies in any depth. In the opening chapter design review was carefully located in the planning and permit-granting process in the largest cities, and comparisons were made with Canadian, British and French planning systems to illustrate where commonalities exist. It was seen that

there are more similarities between American planning and continental European systems than there are between American and British systems; the latter have no zoning controls and no clear development entitlements. Instead, the latter's design controls on all planning applications operated with a very large measure of professional and political discretion (notwithstanding Central Government's strong policy constraints). For the British reader the key observation is that while many of the American design documents look familiar in terms of their goals, objectives, principles, policies and guidelines, it must never be forgotten that these are backed by detailed control on bulk and use that in themselves exert a major impact on built form. For the continental European, the zoning controls will be familiar, as will the issues about how to make such controls more flexible, more responsive to development interests, and more qualitatively sophisticated. However, the question as to how new guidelines, documents or review processes might be built into a more plan-led system will loom large for the British reader.

Some time was spent explaining the key criticisms of design review in the USA in a bid to provide a conceptual framework with which to review each city's endeavours. It was seen that there were problems with the process, its efficiency and effectiveness; with the competence of planners, politicians and review boards; with the abuse of power, going beyond legal powers to impose additional constraints and obligations on developers;

with issues of freedom in terms of rights to self-expression and cultural identity, and of justice in terms of treating all applicants fairly and reasonably; and finally of aesthetics in terms of the subjectivity of design judgment (Scheer and Preiser 1994, pp. 1–10). Legal observers focused their criticisms on the process of control and the rules/principles/policies that are applied. Blaesser, in particular, focused on the legality of design review within state planning law; the need to derive policy from analysis of area character, the need to ensure that guidelines remain non-mandatory and non-prescriptive, but detailed and precise rather than vague or visionary; the need to underpin guidelines with both design principles and implementation advice, and to explain the weight to be attached to each (Blaesser 1994, pp. 49–50). Lai took a broader and more general view, on the one hand looking at general weaknesses of American planning and zoning, and on the other arguing that design review must be part of a 'comprehensive coordinated effort' to promote design quality in which other public and private agencies participate (Lai 1988).

A FRAMEWORK FOR SYNTHESIS

These criticisms provided the framework for drawing together the findings of the study in terms of the comprehensiveness of the pursuit of design quality; how policies were derived (particularly the analytical studies and consultation that underpin them); their level of precision; their basis in design theory; the extent to which they prescribe solutions; and finally the efficiency and effectiveness of the review process. Particular attention was devoted to the policy hierarchy, to the relationship between urban design at the citywide scale and the level of the individual plot, and to the relationship between goals, objectives, principles, policies and guidelines.

In these conclusions the findings of the study will be assembled around six issues which can help to organise the key arguments that have arisen in the analysis. The key issues may be encapsulated as politics; public participation; the review process; policy hierarchy, policy generation and the levels of prescription; implementation; and comprehensive co-ordination. These are not mutually exclusive issues – politics, participation and policy generation

are particularly closely related – and some aspects could be explored under several of these categories, but these issues can help structure the general conclusions of the study.

THE POLITICS OF URBAN DESIGN

The west coast cities clearly demonstrate that urban design in it broadest sense is a politically contested arena. This is most evident in the citizen revolts against the 1984 Downtown Seattle Plan and the 1985 Downtown San Francisco Plan where there was strong opposition to further large-scale, high-rise commercial development on the basis of its impact on the character of downtown, and its broader effects on congestion, transportation and housing. In San Diego the concern was suburban sprawl and the spread of new communities up the coast and into the arid interior; a balanced growth strategy soon disintegrated when existing suburban communities began to experience intensification, congestion and higher demands on existing services.

Portland provides the most positive example of how urban design goals and strategies have won powerful political support that has ensured their continuous implementation over 25 years. Carl Abbott has detailed the connection between *what* the city has accomplished in design terms and *how* it has accomplished this through its politics. He emphasises Portland's traditions of conservatism, conservation, and consensus politics (non-partisan government) developed through extensive public-private partnerships and thoroughgoing community planning as the platform for a long-term urban growth management and downtown development strategy (Abbott 1997). Others see support for these strategies being built through economic self-interest in higher property values and more stable neighbourhoods, in more profitable development opportunities and more affordable housing, and through widely shared interests in the commercial and cultural vitality of the compact city, and the protection of valuable agricultural land and natural landscapes (Richmond 1997).

The same kind of political consensus building was evident in Seattle with the 1994 Comprehensive Plan and its urban villages strategy, which tried to accommodate growth and maintain housing affordability at the same time as improving neighbourhood amenities, increasing accessibility and reducing congestion, and increasing neighbourhood design control while nonetheless allowing intensification. It remains to be seen if Seattle will succeed where San Diego failed, but certainly the Comprehensive Plan provides the basis for accommodating the growth that San Diego's mature suburbs have resisted.

Another example of the importance of politics is provided by San Diego's failure to build a political consensus for design-led planning downtown, in the face of powerful development interests and a powerful City Center Development Corporation driven by the tax increments provided through redevelopment. A further problem was a political leadership that, at least until recently, regarded design control as arbitrary, expensive and off-putting to developers.

The political complexion of the city, the extent to which it can pick and choose between developers and developments (particularly the level of economic competition between developers), the mutuality of interest between developers and residents, between business and environmental groups are all critical to the effective implementation of long-term design strategies and enhancement programmes. It is remarkably easy for policy to get out of step with public aspirations, particularly in periods of rapid growth (as in Seattle and San Francisco). One of the best ways of preventing this is to develop a very high level of public participation and neighborhood/community input into plan making and design development/regulation, but even this is not an absolute guarantee of harmony.

PUBLIC PARTICIPATION IN FORMULATING DESIGN POLICY

Issues of the politics of urban design are inseparable from issues of public participation in planning and design regulation. However, the level and extent of this participation constitutes, at one extreme, mere publicity for a plan or guidelines, and at the other extreme genuine empowerment of the community (Arnstein 1969). A wider range of different levels of participation is evident in west coast cities, from extremely limited public involvement in subdivision development and planning in Irvine (land owner planning through market research) to an extremely high level of involvement and devolved resources in Portland.

Portland is the exemplar in terms of its neighbourhood participation programme which funds neighbourhood associations, trains activists, gives associations early notification of impending development, and allows them to participate in development briefing. The 1988 Center City Plan was led by a voluntary citizen steering committee, and both this and the 1993 Albina plan had huge budgets to ensure high levels of public participation. It is noticeable that there were no citizen revolts over the 1988 Downtown Plan in Portland as there were in Seattle and San Francisco, although there was some modification of its content (a refocus on land-use issues) when the Planning Bureau took over from the citizens' committee to complete it.

Other cities have been much less ambitious with their participation initiatives, but there are some interesting experiments in the devolution of both guidance production and design control to the local level in both San Francisco and Seattle. In San Francisco, at least in part, this has been a product of financial exigencies constraining planning department initiatives, so that communities wishing to undertake closer control have had to write their own guidance or finance its production themselves. In Seattle, it is part of a pact that will allow the city to intensify development in a series of urban villages, while allowing the neighbourhoods to control the detailed design of new development, and to participate in pre-development discussions with prospective developers with a firm prospect of having their views incorporated into the final design.

There is a clear trend not just towards public consultation in design matters, but towards the public defining the principles of control and contributing to the administration of the control process itself. Such ventures provide a mechanism for managing disputes between the community and development interests, and for giving the community far greater 'ownership' of the control mechanism. The Seattle experiment in empowering neighbourhood groups will be especially interesting as the intensification process generated by zoning changes in the 1994 plan gathers pace. It will provide an interesting comparison with San Diego's experiences.

The management or neighbourhood change through elected neighbourhood associations and codes, covenants and restrictions [CC&Rs] is another form of empowerment, often dismissed as a purely negative and exclusionary process by those who see design control as a purely public activity. Baab argues that these can be devices for subordinating individual property rights to community values, and is very positive about both their effectiveness in maintaining environmental quality and their ability to translate community goals into action (Baab 1994). Taking the example of Woodbridge in Irvine, it is difficult to see anything but enforced conformity and claustrophobia, but a more positive role is evidenced in Westwood Park, San Francisco, where a less rigid set of CC&Rs has been supplemented by tailored guidelines to ensure the retention of neighbourhood qualities.

With the obvious exception of Irvine, west coast cities have made an especial effort to consult the public on design matters, and a number are now going a stage further to devolve both the production of guidance and first stage design controls to the local community. However, these efforts are at odds with the general trend identified by Southworth (1989, p. 345) to allow less participation, and to concentrate upon elite business and professional interests to capture key decision-makers and to save money. They are also at odds with Habe's research findings which found that only one-quarter of the communities surveyed promoted 'active' participation in design matters (Habe 1989, pp. 204–6). This is one of the reasons why these west coast cities continue to be exemplars of enlightened design control.

THE PROCESS OF DESIGN REVIEW

Opponents of design review, particularly developers and architects, focus upon the nature of the processes of design review and the extent to which they are subject to professional and political discretion. Lawyers are especially concerned about the abuse of discretionary power (see Blaesser 1994) and the tendency of applicants to succumb to its requirements rather than to challenge it. They have made a variety of suggestions about how the process might be made efficient, fair and effective (Lai 1988). A number of these relate to the way guidelines are developed, their basis in the nature of the locality and public values, their relationship to established

principles of design, and their level of prescription, all of which will be discussed in subsequent sections. Some of their suggestions relate specifically to the administrative procedures and the need for written opinions, principles derived from precedents, tests of the reasonableness of the decision and the right of appeal.

Abbott noted how in Portland the city routinised and depoliticised design review early on by establishing clear guidelines, trained officers, treating each case on its merits, and appointing design and landmark commissions to give decisions (Abbott 1997). The review process itself has written reports, hearings (on more complex developments) and appeal procedures all with strict timetables that will yield a decision in 11 or 17 weeks depending on the size and complexity of the proposal, even after appeal. A key aspect of Portland's system is that state legislation demands that decisions be based on demonstrable findings, hence the emphasis on clear and precise guidelines and checklists against which an application can be assessed and can be seen to be systematically evaluated. This same approach is evident in Seattle's neighbourhood design review where it is extended to emphasize pre-application negotiations and community agreement on what the decisive issues are, so that these can be used in a broader evaluation of the eventual planning application. Due process was seen to be lacking in San Francisco where discretionary review could be initiated even where a developer had met all the requirements of the zoning ordinance, and where politicians used the review mechanism to respond to citizen pressures against particular developments.

The evidence from Seattle, Bellevue, Portland and San Francisco is that design review has been fully integrated into the planning process, and it has been systematised, made transparent, democratised and professionalised, the latter by virtue of planning officers' advice and expert design or landmarks' commissions' judgments. In San Diego design review has yet to be fully established, but communities have set out their design requirements in their zoning ordinances and community plans. Meanwhile Irvine demonstrates the power of landowner control that can be exercised on developers when there is a high demand for development, and how these controls can be perpetuated through the imposition of CC&Rs on the title deeds.

THE POLICY HIERARCHY AND THE WRITING OF GUIDELINES

At the heart of the examination of the design guidelines used in west coast cities has been the question of how to write policies or guidelines that are clear, meaningful and easy for lay people to understand, and easily applicable to the control of development. Throughout the case studies the relationship between community goals, design objectives and design guidelines has been re-peatedly discussed, compared and contrasted one to another in the search for well-articulated, con-cise and comprehensive policy frameworks. To re-emphasise the importance of the task, a recent court case in Washington State overturned permit refusals which stated that development proposals were 'incompatible' and 'non-harmonious', because such judgments were not based on properly re-searched and explained policies and guidelines, The court in the case of Anderson v. Issaquah ruled that the use of such adjectives in zoning ordinances was not acceptable, that property owners must know what they are expected to do in advance, and that decision-making cannot just be turned over to a board or committee without clear guidelines being established to support their decisions (Hinshaw 1994, p. 288)

[. . .]

The value of design appraisal

One of the key characteristics of the best American urban design planning is that they are based upon thoroughgoing analysis of the charac-ter of the locality. The best example anywhere is provided by San Francisco where the 1968–70 design studies undertaken by the City Planning Department established the character of the city, the key qualities that needed to be protected, and the principles and policies that could help achieve this. Other cities undertook similar appraisals, but not on the same scale – the Lynch and Appleyard study of San Diego being one of the most interesting and comprehensible (Lynch and Appleyard 1974).

The Portland studies were among the most par-ticipative – involving various architectural and conservation groups, using Kevin Lynch method-ologies in the former and identifying all potential landmark buildings in the latter. Portland has been proceeding area by area with detailed analyses of design character through all downtown, inner city and historic districts, developing detailed guidelines from each appraisal (City of Portland 1993). The analyses make particular use of axonometrics and maps at the large scale to distil and communicate ideas, moving on to detailed analysis of architec-tural and street character at the micro-scale.

Southworth's research reveals that such thor-oughgoing appraisals are now much less common than they used to be in the period 1960–73, and further reveals that 20 percent of the newer design plans surveyed have no such analytical base (Southworth 1989). Furthermore, he is critical of many professional field surveys which he regards as vague and unstructured. His prescription for a good survey is that it sets 'clear goals and categories of analysis, and establishes a system for covering the survey area so that all areas receive equal attention [and] provides for multiple opinions in subjective analyses to reduce personal biases' (Southworth 1989, p. 376). His research reveals a rich array of analytical techniques that provide a good basis for area appraisals anywhere, but emphasises that significant public consultation remains a vital component in order to establish public, as opposed to professional, values.

Residents' views were an important component of the San Francisco Urban Design Study in 1970. In Portland since 1974 the Office of Downtown Neighborhood Associations has ensured major public inputs into all planning, zoning and guide-line reviews, including most notably the 1988 Center City Plan which was largely prepared under the auspices of a citizen steering committee with its own budget. Current initiatives in Seattle attempt to kill two birds with one stone by encour-aging communities to conduct their own area appraisals, and thereafter to develop their own objectives and guidelines, thereby ensuring that it is *their* values which are expressed in the controls rather than those of the professionals or spe-cialised amenity or development interests. It has already been seen that quite ordinary communities are capable of writing their own design guidelines, and that these can be quite original, unprescriptive and refreshingly neutral about matters of elevational treatment and architectural style (e.g. Bernal Heights in San Francisco). Certainly Seattle's manual, *Preparing Your Own Design Guidelines*, will be of value to many small communities worldwide as they contemplate the task of expressing what it is about the physical character of their settlement or neighborhood they wish to maintain.

Objectives and principles

The appropriate development of goals, objectives, principles, policies and guidelines is problematic in the sense that many of these terms are used interchangeably, and loosely, by different designers and indeed different critics and planners. In the west coast cities the three best examples seem to rely on a few key objectives each split into a number of design principles. The 1972 San Francisco Master Plan relies on four design objectives, each with nine to twelve policies (design principles); the 1985 Downtown Plan is similar but with nine objectives each split into two to five policies (also essentially principles). In the latter, further guidance is offered in a couple of paragraphs on each, with much more detailed guidelines and standards for open space and for pedestrian improvement standards.

Portland's 1988 Downtown Plan has three objectives and 26 principles (called guidelines). They are very broad and general – reinforce the pedestrian system, protect the pedestrian, bridge pedestrian obstacles, provide pedestrian stopping places, make open spaces successful – and are barely elaborated in the four sentences which accompany each guideline. They are brought together in a checklist so that planners can determine whether or not the guideline is applicable to the proposed development or not, and whether it complies or not. A very similar approach has been adopted in Seattle with 27 principles (also called guidelines) for site planning, bulk, architectural treatment, pedestrian relationship and landscaping. These are described as guidelines and the checklist is used to establish which are the priority considerations from the neighbourhood's viewpoint. Both checklists provide valuable ways of briefing developers, or articulating community wishes, or

of evaluating proposals, because they can identify which broad but widely supported principles, collectively, will be critical to design quality.

Urbanistic principles

One of the features of the best design principles is the emphasis they place upon the proposed building's relationship to the public realm and the pedestrian experience. In the most progressive authorities these urbanistic criteria receive more attention than architectural or townscape factors. Payton, relating an experience in Virginia, draws a clear line between these two kinds of factors and makes an important recommendation:

> Urbanistic criteria relate to the relationship of buildings to other buildings (vis-à-vis height relative to street width and other buildings), to set back lines, to parks etc. In essence all of those characteristics that determine the walls of the urban room. Architectural criteria are those that relate to the buildings themselves, or objects within the urban milieu. In an ideal world buildings would be successful urbanistically and architecturally. However, if only one were possible, the greatest effort should be applied to the former, consistently throughout the entire locale.
>
> Payton 1992, p. 238

Portland's design guidelines clearly illustrate some resolution of these architectural (townscape) and urbanistic (public realm) criteria, seeking to ensure first that a project is consistent with the city's character, and the broad urban design framework; secondly that it makes a contribution to the pedestrian environment; and finally that the detailed design is sensitive to the character of the locality and creates appropriate amenities. All these examples illustrate the important broadening of the concept of design beyond visual-architectural considerations. They emphasise the importance, stressed by Buchanan, Habe, and other design critics, of employing definitions of context that embrace patterns of use, activity and movement in an area. However, as recent American research demonstrates, such perspectives tend to be more exceptions than the general rule (Southworth 1989).

Guidelines and their elaboration: appropriate levels of prescription

How then should guidelines elaborate these basic principles? Guidelines can be divided into two kinds – those that are prescriptive in terms of prescribing the form of the development scheme; and those that are performance related, which seek to ensure that the development 'performs' in a certain way by responding to a particular issue. A good example of the two approaches is provided by the San Francisco Zoning Code, which is very prescriptive, and the San Francisco Residential Guidelines (1989) which try hard to be performance related. The Zoning Code specifies the provision of bay windows, their angles and overhangs, the maximum spacing of pedestrian entrances, the maximum proportion of garage doors vis-a-vis the facade, and detailed setback and landscaping provisions. The Residential Design Guidelines (1989) articulate a series of detailed design principles and how these might be applied, with key questions that the controller/applicant can ask themselves, and analytical devices to establish appropriate responses. There is no doubt that the application of these principles places significant constraints on the client and designer; it is obviously the intention of the guide to ensure that new development responds to the quality of townscape that it is placed in, and the less uniform the context the less binding the principles. These guidelines do not prescribe solutions, rather they encourage full consideration of the design issues at stake and demonstrate clearly to applicants what their designs are expected to achieve. Compatibility, not conformity, is the watchword.

The same is true of Seattle's multi-family housing guidelines (1993). Here, compatibility was defined more broadly to embrace aspects of landscape as well as architecture and urban form. The guidelines directly embrace the relationship to the public realm and seek to retain existing qualities of visibility, surveillance and private/public space, promoting both pedestrian safety and a rich pedestrian experience. In Seattle a concept like human scale is not just interpreted as an aspect of elevational treatment or building size. It embraces the social and functional aspects of the relationship between buildings and space, and in this sense it operates more like a performance standard against which the proposal can be evaluated. Most of the guidelines are similarly sophisticated. The Albina plan in Portland offers developers the alternatives of meeting a set of prescriptive supplemental compatibility standards or of submitting the project to a design review process in which the community will play an important role.

The level of prescription in design policies remains a fundamental issue that greatly exercises architects, designers and often clients. Most commentators favour policies and guidelines which do not prescribe solutions or particular built forms, but which set out principles or performance criteria leaving the designer to be free to use his or her creativity to resolve the design problem. They recognise that without full use of the designer's skills there will be no quality, and even those who wish to write detailed codes to control the form of development still try not to propose architectural solutions.

In the United States it is still quite common for both local government and landowners/developers to become obsessed with matters of style and design detail, rather like the community of Golden Hill in San Diego. Habe's research reveals that architectural considerations are the key focus of control in 98 per cent of all authorities surveyed (Habe 1989, p. 202), while nearly a third of all authorities specify certain architectural styles; others are preoccupied with architectural details often without reference to specific contexts. The west coast cities illustrate a much broader and less prescriptive approach in which architectural issues are integrated with urbanistic and landscape isssues to achieve a holistic approach to urban environmental quality. Careful study of the context (broadly interpreted) is required of every applicant, but design principles rather than design solutions still allow the developer and designer the opportunity to respond creatively to the carefully defined constraints.

Legal experts concerned with the potential and actual abuse of discretionary powers have particularly emphasised the need for a clear division between mandatory controls (which are limited to judicially accepted parameters like height, bulk, density, building line, setback) and design guidelines (Blaesser 1994). Portland and Seattle provide two good examples where this has been achieved successfully.

Visions and strategies – conveying the desired future form

One of the most striking features of the American experience is the way that design thinking has recently come back to permeate planning at the metropolitan, district and neighbourhood levels. Mark Hinshaw (1994, pp. 287–8) has argued that through most of the 1970s and 1980s urban design fell into severe disfavour as planners moved into policy planning. Other relevant factors are likely to have been the deep economic recessions of the 1970s and 1980s, public disenchantment with the results of urban renewal and redevelopment, while increasing competition between cities in the 1990s has tended to revive design concerns. Hinshaw notes that urban design is now making a strong comeback in the United States as community image, community design and environmental quality become more widely discussed, and as its potential to express desired qualities in built form and environmental regulation is realised. He considers urban design to be particularly relevant as growth management is developed in the United States and the reshaping of suburban development becomes an urgent necessity.

In the west coast cities, most notably perhaps in Portland and most recently in Seattle, the importance of thinking strategically in urban design terms is palpable in the attempts to convey city wide future urban form. This embraces the areas for major intensification and concentration of commercial development; the patterns of infrastructure investment especially transportation; the accessible areas for residential intensification; the townscapes to be conserved and the agricultural and natural areas to be protected. Seattle's new urban villages strategy set out in the 1994 Comprehensive Plan is a fine example of how a generalised urban design concept can express a city-wide vision of the future that is comprehensible to the public and the development industry. Developed through two years of debate with broadly-based discussion groups and community forums, the vision expresses with great simplicity what Seattleites want for their city in terms of reduced congestion and improved transit, protection of the environment and neighbourhood quality, living compactly but ensuring housing affordability, improving suburban services and ensuring economic vitality and employment growth.

One might criticise Seattle for not clearly presenting this vision with the kinds of maps, diagrams, sketches and axonometrics that make such ideas accessible to a wide constituency, but no such criticism can be made of Portland's design strategy for its downtown. Established in 1972 and refined in 1988, this design strategy is expressed largely through maps and axonometric line drawings, starting with a concept plan emphasising the areas of intensive development and infrastructure investment. Then it expresses each planning policy in a spatial way on a map – economic development, riverfront, historic preservation and, finally, urban design. The latter then attempts to integrate these different aspects into an overall design framework. The key concepts employed, and the notation for the strategy have been the subject of much thought and refinement, but they are still not entirely satisfactory in the way that they integrate aspects of built form, public space, and activities, both current and projected. They are, however, more complete and more sophisticated than other known examples.

These detailed spatial strategies are backed up by a set of action proposals, and supplemented by a set of programmes with timings and relevant implementation agencies identified. The whole provides a clear framework for private development decisions and acts as a corporate document to guide public investment and initiative. Like the Seattle plan, it sets out a vision for the future of the central city developed in conjunction with business and resident groups. What is striking about the strategy is that the 1988 version, while extended, elaborated and more detailed than its 1972 forebear, is still essentially promulgating the same vision and approach, and it is this continuity which is a testament to the robustness of the original concept and a key to the sustained positive impacts upon downtown itself.

Of course, design strategies necessarily precede the writing of design policies and guidelines, and should emerge from the whole process of visioning and goal setting citywide, at the district or neighbourhood levels. They are a key element in ensuring that urban design thought plays a much more prominent role in the coordination, integration and modification of systematic planning policies. They provide a spatial framework for developing enhancement programmes and other forms of direct public action to ensure that the

valuable. For example, *Urban Design in Australia* (Government of Australia 1994), the report by the Prime Minister's Urban Design Task Force, considered the whole issue of the hierarchy of policy from visions through strategies, through briefs, performance codes, prescriptive codes and guidelines, and the value of 'future character statements' (pp. 46–52). Similarly the intention was to provide constructive guidance to numerous local authorities across the country struggling to come to grips with the problems of effectively managing urban change and ensuring environmental quality. Meanwhile in cities in Asia, there are concerted attempts to inject design guidelines into zoning systems which have hitherto largely failed to deliver an acceptable level of environmental quality and activity (Cheng Wu forthcoming).

The relevance to British practice

To conclude, however, the author feels duty bound to consider how these 'best practice' ideas might be translated into the British system. Since this issue has been very much the sub-text of the research undertaken, and indeed was in many senses its *raison d'être*, it would be remiss to ignore the opportunity to comment (see Punter 1996). As has been stressed at numerous points in the foregoing, the British system, based as it is upon a discretionary system of decision-making, at both the technical advice and political decision-making stages, is the exception rather than rule in the world system, so the applicability of British experience to other countries is severely limited. Nonetheless, looking at the issue the other way round, it is clear that the American system of design review has much to teach the British system of design control.

Many of these lessons are embedded in the critiques of and responses to design control, that have already been reviewed. They embrace the need for clear principles, the need for guidelines to be based on study of the locality, the problems of high levels of subjectivity, lack of appropriate review skills and discrimination against the new, different and minority taste. The British system has attempted to respond to these issues over the last twenty years. Many local planning authorities, especially some of the London Boroughs, larger provincial cities as well as a number of historic cities, have developed a sophisticated range of policies and guidance, other advisory agencies and design initiatives to respond to these difficulties. Since 1994 Central Government has taken a much more positive attitude to design, and the personal interests of the then Secretary of State, the Rt. Hon. John Gummer, led to the sponsorship of a number of valuable research projects and experiments which have raised the level of debate and knowledge.

This is not the point at which to explain in any detail recent developments in British practice, though it is important to know that urban design has undergone a major revival in the 1990s, just as it has in the USA, due to a complex of changes in environmental consciousness (sustainability), European competitiveness, a desire to counter standardization / globalization in urban forms and retain local distinctiveness in town and country.

Having miraculously survived the deregulatory tide of the 1980s, design control and urban design were revived by the campaigns of the Prince of Wales and the previous Secretary of State, and eagerly exploited by local planning authorities, design consultants and the Urban Design Group, and now enjoy a position of considerable prominence in the planning agenda. Whether they can retain that prominence, particularly under a new Labour Government, is a moot point, and it depends in no little measure upon the ability of the profession to articulate the kinds of visions, strategies, clear objectives and principles that we have discussed in this book, and to win public support for them. These visions embrace quite different scales of planning from the national questions of where new housing should be located (4.4 m new houses in England by 2015), to the sub-regional questions of suburban expansion and new settlements, down to the local level in terms of intensification, inner-city regeneration and revitalisation.

Unlike most other developed planning systems, Central Government maintains tight control on local initiative in British planning, especially in the area of design where, interestingly, it has maintained a critical eye on 'overprescriptive policies', rooting them out of plans and dismissing them when they have led to an appeal by a developer. Until very recently it has discouraged local authorities from preventing all but the very worst designs ('outrages'), and told them to concentrate upon basic issues of height, bulk, massing, scale, layout, access and landscape. In 1997, however, three new ideas were added to Government advice, at least partly in response to studies of US experience (Delafons 1990; Punter 1996):

- the importance of urbanistic criteria;
- the importance of design appraisal;
- the public's role in guidance preparation.

In his 'Quality in Town and Country' initiative the previous Secretary of State provoked a national debate about how design quality might be promoted through the planning system. The main outcome was an urban design campaign which has led to a wide range of design brief/framework/strategy experiments in towns and cities across the country, and which has particularly encouraged others to think more ambitiously about city-wide design strategies and frameworks as additional documents to the development plan (Cowan 1997).

Whether these initiatives are going to continue depends essentially, as we have seen, on whether local government is prepared, and has the resources, to undertake a sustained effort to put design at the heart of its planning efforts, and to take it beyond into other spheres of corporate activity. That in turn depends on establishing a constituency for good design, political will and a process by which local authorities can continually involve significant numbers of people in the planning process, and win the confidence of local business and the development industry. Local government continues to struggle with minimal resources and often absurd boundaries, factors which work against the kind of initiatives that are necessary to take forward large-scale sustainable urban design initiatives. (Perhaps this is a task for the new integrated Government Regional Offices.)

The lessons from America about resources indicate that major design policy initiatives, thorough analysis, plan and guideline preparation, are very expensive activities. The breadth and depth of Portland's consultation was explained in large part by the huge proportion of the planning budget allocated to plan preparation, neighbourhood group support and the like. Seattle's recent planning efforts have been underwritten by a grant of $500,000 from the Federal Government, while San Francisco's urban design plan was paid for by the Federal Government out of the urban renewal budget. Limited planning budgets mean limited initiatives. Neighbourhood groups can be encouraged to undertake appraisals and develop and operate guidelines, but design initiatives demand money, whether they be staff resources, consultants' expertise, or budgets for environmental enhancement. It is a sign of the times that Portland can no longer afford to produce its *Developer's Handbook* or to update it. As a recent symposium on *The New Agenda for Urban Design* has recognised, resources and funding, and Central Government support for the lead role of the local authority, lie at the heart of more innovative and more strategic design interventions in British cities (Cowan 1997, p. 23).

To return to the relevance of American models to British practice design guidelines, or design objectives and principles as we have come to perceive them, is to enter the complex debate about the nature of British development plans and development control (Booth 1996). In 1990 the Government introduced a new system of district-wide development plans which would be given greater weight than hitherto in the determination of planning applications. References to a plan-led system are somewhat misleading because the plan remains only one, albeit important, consideration in development control decisions in the British discretionary system. Comprehensive plan coverage has yet to emerge – after seven years only half the English local authorities have completed the long and cumbersome process of analysis, drafting, government review, consultation, redrafting, objection, public inquiry, (Government) inspector's report, redrafting and adoption. There are plenty of opportunities for public involvement in this process, but limited resources for focus groups, 'planning for real' and other activities that might generate a true sense of plan ownership.

The design policies in British development plans have two major difficulties from an American perspective. First of all they have to respond to all developments, not just the major schemes – 80 percent of planning applications are house extensions, minor residential developments, and small alterations and extensions to commercial premises. Second the policies are the only controls available to a planning authority, and there are no zoning maps, use or dimensional controls to reinforce the design dimension. Density and plot ratio controls have largely disappeared from plans, driven out by deregulatory tendencies to give more freedom to house builders and developers, and by dissatisfaction with their design outcomes which often frustrate sensitive design. (By 1998 density policies were being readvocated by Central Government to support more sustainable forms of development.)

Both these 'difficulties' place serious constraints on design policies and invest them with a much weightier and more comprehensive role. This must embrace aspects of density, use, open space, and car parking, etc., as well as issues of site and context, architectural character and relationships with the public realm. Perhaps inevitably it makes the policies more cautious, negative, contested and legalistic, and much more difficult to absorb as a whole.

Neither of these factors should prevent policies from being clearly related to vision statements, design objectives and plan strategy, and from being very well structured and organised with checklists of criteria and considerations against which applications can be evaluated. Nor should it prevent these policies being supplemented by valuable design guidance on common problems such as the design of residential layout, shop fronts or car parking. Some plans particularly those of the London Boroughs which have had twenty years to perfect them, are models of a well-organised and disciplined approach to policy making, but many others, prepared for the first time, are quite the opposite. Almost all the policies suffer from the problems of being inaccessible and unassimilable by the lay person; being too long, too complex, and too poorly presented, so that it becomes better not to contemplate whether anyone (including the control officer) actually reads and uses them (Punter and Carmona 1997).

One of the major questions American practice raises for British planning is whether it should persist with its attempts to offer detailed design control on the minutiae of development. Lewis Keeble once remarked that British planning 'often swallowed the camels and strained at the gnats', by which he meant that it was preoccupied with minor development issues at the expense of major, and often strategic, decisions. Could planners devise 'supplemental compatibility rules' (Portland) to allow minor development to be largely self-regulating? Could they go a stage further and abandon householder control to neighborhour agreements? Would the more selective and strategic approach to design

intervention and control practised in America be more successful and more effective? Or is design control at this level a vital part of the public expectations of planning as a neighbourhood protection service, and would the myriad of less regulated developments seriously erode the quality of the British built environment?

Thinking the unthinkable is one of the benefits of cross-comparative work, but the new system of development plans with comprehensive, district-wide coverage should not be abandoned without a proper trial and evaluation of its effectiveness. Perhaps new design documents are already emerging alongside statutory plans to achieve some of those things achieved by American guidelines-quarters studies, design frameworks, city-wide strategy documents, character assessments (for conservation areas) and landscape character assessments on the urban fringes, urban nature conservation strategies, and legibility studies. In the meantime design appraisal, thoroughgoing public consultation, clear goals, objectives and design principles, supported by a wide range of design guidance that is accessible and comprehensible to the average applicant, all consolidated into a clear policy hierarchy, will offer important advances to many local planning authorities. However, the experience of the most design-progressive American and European cities (Portland, Seattle, Barcelona, Berlin, etc.), also demonstrates that the public sector has to lead by example in investing in design quality whenever it builds facilities or infrastructure, and whenever it modifies streets and public spaces.

REFERENCES

Abbott, C., "The Portland Region: Where City and Suburbs Talk to Each Other – and Often Agree", *Housing Policy Debate*, 8(1), 1997, pp. 11–52.

Arnstein, S., "A Ladder of Citizen Participation", *American Institute of Planners Journal*, July, 1969, pp. 216–24.

Baab, D.G., "Private Design Review in Edge City" in B.C. Scheer and W. Preiser, eds, *op. cit.*, 1994, pp. 187–96.

Banerjee, T. and Southworth, M., eds, *City Sense and City Design: Writing and Projects of Kevin Lynch*, Cambridge Mass, MIT Press, 1990.

Blaesser, B.W., "The Abuse of Discretionary Power" in B.C. Scheer and W. Preiser, eds, *op. cit.*, 1994, pp. 42–55.

Booth, P., *Controlling Development, Certainty and Discretion in Europe, the USA and Hong Kong*, London, UCL Press, 1996.

City of Portland, *Adopted Albina Community Plan*, Portland, OR, Bureau of Planning, 1993.

Cowan, R., "The New Urban Design Agenda", *Urban Design Quarterly*, 63, 1997, pp. 18–37.

Delafons, J., *Aesthetic Control: A Report on Methods used in the USA to Control the Design of Buildings*, Berkeley, University of California, Institute of Urban and Regional Development, Monograph 41, 1990.

Government of Australia, *Urban Design in Australia*, Report by the Prime Minister's Urban Design Task Force, Canberra, Government of Australia, 1994.

Habe, R., "Public Design Control in American Communities", *Town Planning Review*, 60(3), 1989, pp. 195–219.

Hinshaw, M., "The New Legal Dimensions of Urban Design" in A.V. Moudon, ed., *op. cit.*, 1994, pp. 287–9.

Kropf, K., "An Alternative Approach to Zoning in France: Typology, Historical Character and Development Control", *European Planning Studies*, 4(6), 1996, pp. 717–38.

Lai, R.T-Y., *Law in Urban Design and Planning: The Invisible Web*, New York, Van Nostrand Reinhold, 1988.

Lai, R.T-Y., "Can the Process of Architecture Review Withstand Legal Scrutiny?" in B.C. Scheer and W. Preiser, eds, *op. cit.*, 1994, pp. 31–40.

Lightner, B.C. and Preiser, W., eds, *Proceedings of the International Symposium on Design Review*, Cincinnati, University of Cincinnati, 1992.

Lynch, K. and Appleyard, D. "Temporary Paradise? A Look at the Special Landscape of the San Diego Region" in T. Banerjee and M. Southworth eds, *op. cit.*, 1974, pp. 720–63.

Moudon, A.V., ed., *Urban Design: Reshaping our City*, Seattle, University of Washington (Conference Proceedings), 1994.

Nelissen, N. and de Vocht, C.L.F.M., "Design Control in the Netherlands", *Built Environment*, 20(2), 1994, pp. 142–57.

Nyström, L., "Design Control in Planning: The Swedish Case", *Built Environment*, 20(2), 1994, pp. 113–26.

Payton, N.I., "Corrupting the Masses with Good Taste" in B.C. Lightner and W. Preiser, eds, *op. cit.*, 1992, pp. 235–42.

Punter, J.V., "Developments in the Urban Design Review: The Lessons of West Coast Cities of the United States for British Practice", *Journal of Urban Design*, 1(1), 1996, pp. 23–45.

Punter, J.V. and Carmona, M., *The Design Dimension of Local Plans: Theory, Content and Best Practice*, London, Chapman and Hall, 1997.

Richmond, H.R., "Comment . . . on the Portland Region . . .", *Housing Policy Debate*, 8(1), 1997, pp. 53–64.

Samuels, I., "The Plan d'Occupation des Sol for Asnièressur Oise: a Morphological Design Guide" in R. Hayward and S. McGlynn, eds, *Making Better Places: Urban Design Now*, Oxford, Butterworth, 1993, pp. 113–21.

Scheer, B.C. and Preiser, W., eds, *Design Review: Challenging Urban Aesthetic Control*, New York, Chapman and Hall, 1994.

Southworth, M., "Theory and Practice of Contemporary Urban Design – a Review of Urban Design Plans in the United States", *Town Planning Review*, 60(4), 1989, pp. 369–402.

Whyte, W.H., *City: Rediscovering the Center*, New York, Doubleday, 1988.

EPILOGUE

The urban design field has grown substantially since the mid-1990s. Many university-based urban design programs and concentrations have been created or are expanding. A growing number of professional firms reference themselves as urban designers. Accompanying this has been an increased amount of research and writing about what the design of cities and of their parts may try to achieve. As well, increased information about what urban design practitioners around the world are actually designing – projects, plans, design guidance, and visually-based design decision-making processes – is available in print and on-line. With all this activity, the field seems poised to define itself more clearly than has been the case until now; articulating the unique contribution of urban design that distinguishes it from the allied fields of architecture, landscape architecture, and urban planning. Dimensions that seem to be emerging are a commitment to drawing on a broadly interdisciplinary theoretical base, a focus on shaping physical urban form – most particularly the public realm – to meet everyday human needs and aspirations, a concern with integrating across all scales of design (region, city, neighborhood, and site), and an understanding that the process of city building is a collaborative effort that takes place over time. In practice, urban design framework plans are emerging as the field's distinctive design product. Framework plans are a mechanism for designing cities and their parts rather than detailed building design. A distinctive emerging process is the growth of innovative participatory design charrettes. The urban design field will, of course, always be intertwined with the fields of architecture, landscape architecture, urban planning, and transportation planning, and many urban designers will continue to train in these fields, gain professional credentials within them, and practice these professions as well as urban design. Because of its holistic focus, the nature of the urban design field is that it will always be interdisciplinary.

From the wealth of readings assembled in this reader, it is clear that there is much literature that relates to the urban design field. This book has focused on the most salient of what is considered the best literature. But each of the readings comes from larger contexts. The selections in this reader are just samples. Serious students will want to engage with the wider literature – to read the whole of Jane Jacobs and Kevin Lynch, and if for nothing else than to realize how wrongly focused design fields can become the whole of Le Corbusier.

Just as important as being grounded in the literature, is learning from observed reality. Directly looking at and experiencing cities is perhaps the best way to learn about the relationship between urban form and human activity in all of its beautiful complexity. Experience of cities and their elements allows an urban designer to build up a repertoire of precedents and form possibilities that are not abstract but directly known. Immersed in the experience of the city, the urban connoisseur more deeply understands the success and limitations of what has come before and what is possible. As well, valuable knowledge about form dimensions and relationships comes from experiencing spatial forms in relation to one's own bodily capacities and one's own perceptions. In other words, it is important to get out there and experience cities – the more the better.

ILLUSTRATION CREDITS

Every effort has been made to contact copyright holders for their permission to reprint illustrations in this book. The publishers would be grateful to hear from any copyright holder who is not here acknowledged and will undertake to rectify any errors or omission in future editions of this book. Following is copyright information for the illustrations that appear in this book.

PART ONE: HISTORICAL PRECEDENTS FOR THE URBAN DESIGN FIELD

BACON. Figure 1: Florence. Illustration by Alois K. Strobl, from *Design of Cities* by Edmund Bacon. Copyright © 1967, 1974 by Edmund N. Bacon. Used by permission of Penguin, a division of Penguin Group (USA) Inc.
Figure 2: The Capitoline Hill before reconstruction by Michelangelo; and Figure 3: The Campidoglio by Michelangelo. Illustrations by Joseph Aronson, from *Design of Cities* by Edmund Bacon. Copyright © 1967, 1974 by Edmund N. Bacon. Used by permission of Penguin, a division of Penguin Group (USA) Inc.
MUMFORD. All images from *Garden Cities of To-morrow*, by Ebenezer Howard (Cambridge, MA: MIT Press, 1965). Copyright © 1965 by MIT Press. Reprinted by permission of MIT Press.
PERRY. Figures 1–4 from "The Neighborhood Unit," from *Neighborhood and Community Planning: Regional Survey Volume VII – Regional Plan of New York and Its Environs*, by Clarence A. Perry (New York: Russell Sage Foundation, 1929). Reprinted by permission of the Russell Sage Foundation.

PART FOUR: DIMENSIONS OF PLACE-MAKING

LYNCH. All images from *Image of the City*, by Kevin Lynch (Cambridge, MA: MIT Press, 1960). Copyright © 1960 by MIT Press. Reprinted by permission of MIT Press.
CULLEN. Serial Vision from *The Concise Townscape* (London: Van Nostrand Reinhold, 1961). Copyright © 1961 by Elsevier. Reprinted by permission of Elsevier.
KELBAUGH. Illustration from *Repairing the American Metropolis: Common Place Revisited*, by Douglas Kelbaugh (Seattle, WA: University of Washington Press, 2002). Copyright © 2002 by the University of Washington Press. Reprinted by permission of the University of Washington Press.
ELLIN. The Axes of Postmodern Urbanism from *Postmodern Urbanism*, by Nan Ellin (New York: Princeton Architectural Press, 1999). Copyright © 1999 by Princeton Architectural Press Inc. Reprinted by permission of Princeton Architectural Press.

PART FIVE: TYPOLOGY AND MORPHOLOGY IN URBAN DESIGN

KRIER. All illustrations from *Houses, Palaces, Cities*, by Léon Krier (London: Architectural Design AD Editions Ltd., Volume 54, 1984). Copyright © 1984 by Architectural Design AD Editions Ltd. Reprinted by permission of Architectural Design AD Editions Ltd and the author Léon Krier.

MOUDON. Figure 1: Modularity in the Built Landscape. "Getting to Know the Built Landscape: Typomorphology," by Anne Vernez Moudon, in Karen A. Franck and Lynda H. Schneekloth (eds), *Ordering Space: Types in Architecture and Design* (New York: Van Nostrand Reinhold, 1994). Copyright © 1994 by Anne Vernez Moudon. Reprinted by permission of Anne Vernez Moudon. Originally published in A.V. Moudon, *Built for Change: Neighborhood Architecture in San Francisco* (Cambridge, MA: MIT Press, 1986, p. 124).

Figure 2: Elements of U.S. suburban residential forms: houses, lots, and streets; and Figure 3: Elements of U.S. suburban residential forms: plan units. "Getting to Know the Built Landscape: Typomorphology," by Anne Vernez Moudon, in Karen A. Franck and Lynda H. Schneekloth (eds), *Ordering Space: Types in Architecture and Design* (New York: Van Nostrand Reinhold, 1994). Copyright © 1994 by Anne Vernez Moudon. Reprinted by permission of Anne Vernez Moudon. Originally published in A.V. Moudon, "The Evolution of Twentieth-Century Residential Forms: An American Case Study," in J.W.R. Whitehand and P.J. Larkham (eds), *Urban Landscapes: An International Perspective* (London: Routledge, 1992, pp. 173–6).

PART SIX: CONTEMPORARY CHALLENGES AND RESPONSES

FREY. Figure 1: Alternative forms of sustainable cities according to Haughton and Hunter (1994). "Compact, Decentralised or What? The Sustainable City Debate," in *Designing the City: Towards a More Sustainable Urban Form* (London: E & FN Spon, 1999). Copyright © 1999 by Hildebrand Frey. Reprinted by permission of Hildebrand Frey and the Taylor & Francis Group.

PART SEVEN: ELEMENTS OF THE PUBLIC REALM

WHYTE. Typical Sighting Map. *The Social Life of Small Urban Spaces*, by William H. Whyte (New York: Project for Public Spaces, 2001; original 1980). Copyright © 1980 by William H. Whyte. Reprinted by permission of the estate of William H. Whyte and The Project for Public Spaces.

GEHL. Graphic representation of the relationship between the quality of outdoor spaces and the rate of occurrence of outdoor activities. From "Three Types of Outdoor Activities" and "Life Between Buildings," from *Life Between Buildings: Using Public Space*, 5th edn (Copenhagen: Danish Architectural Press, 2001). Copyright © 2001 by Jan Gehl. Reprinted by permission of Jan Gehl.

HESTER. All illustrations from *Neighborhood Space*, by Randolph Hester (Stroudsburg, PA: Dowden, Hutchinson & Ross, 1976). Copyright © 1976 by Randolph Hester. Reprinted by permission of Randolph Hester.

SUCHER. All illustrations from *City Comforts: How to Build an Urban Village*, by David Sucher (Seattle, WA: City Comforts Inc., 2003. www.citycomforts.com). Copyright © 2003 by David Sucher. Reprinted by permission of David Sucher.

METRO. All illustrations from *Green Streets: Innovative Solutions for Stormwater and Stream Crossings*, by Metro (Portland, OR: Metro, 2002). Copyright © 2002 by Metro. Reprinted by permission of Metro, Portland, Oregon.

PART EIGHT: PRACTICE AND PROCESS

COPYRIGHT INFORMATION

PART FIVE: TYPOLOGY AND MORPHOLOGY IN URBAN DESIGN

PART SIX: CONTEMPORARY CHALLENGES AND RESPONSES

PART SEVEN: ELEMENTS OF THE PUBLIC REALM

PART EIGHT: PRACTICE AND PROCESS

Index

Figures in **bold** type refer to passages included in the text.

THE URBAN DESIGN READER

The Urban Design Reader brings together some of the most influential writing on the historical development and contemporary practice of urban design. Emerging as a distinct field of environmental design practice in the late 1950s, urban design bridges the fields of architecture, planning, landscape architecture, civil engineering, urban development, and social science – with a focus on physical form and the social use of space. Among university programs, the design professions, interest groups and city governments around the world, the practice of urban design is recognized as a means of addressing twenty-first-century urban challenges. As planning and development processes have become more participatory in recent years, the number of people interested in improving the design of their cities and neighborhoods has also grown. The timeliness of *The Urban Design Reader* parallels recent public interest in making better cities and urban places.

This anthology includes forty-one selections that illuminate the history, theory and practice of urban design. In addition to classic writings from the field's luminaries, such as Jane Jacobs, William H. Whyte, and Kevin Lynch, *The Urban Design Reader* provides recent material on the urban design aspects of contemporary urbanism, place-makig, density, sustainability, neighborhood planning, traffic calming, green infrastructure, and the public realm. The readings are organized into eight topical parts beginning with historical precedents, continuing with a variety of theoretical and pragmatic concerns, and concluding with material on current professional practice. The parts begin with introductory essays that contextualize and situate major themes within the field. In addition, introductions for each selection highlight important lessons emanating from the literature, as well as biographical information on the authors and suggested supplemental reading.

The Urban Design Reader provides a set of essential readings that will be valuable to design and planning students, public decision-makers and design professionals. For those interested in the design of our neighborhoods, towns and cities, these selections offer an indispensable intellectual and practical foundation for the urban design field.

Michael Larice is Associate Professor of Urban Design and City Planning at the University of Pennsylvania. He is a licensed architect, planner and urban designer.

Elizabeth Macdonald is Assistant Professor of Urban Design and City Planning at the University of California, Berkeley. She is a principal in the firm Jacobs Macdonald: Cityworks in San Francisco, California.

THE ROUTLEDGE URBAN READER SERIES

Series editors

Richard T. LeGates,

Professor of Urban Studies, San Francisco State University

Frederic Stout,

Lecturer in Urban Studies, Stanford University

The Routledge urban reader series responds to the need for comprehensive coverage of the classic and essential texts that from the basis of intellectual work in the various academic disciplines and professional fields concerned with cities.

The readers focus on the key topics encountered by undergraduates, graduates and scholars in urban studies and allied fields, the contributions of major theoreticians and practitioners and other individuals, groups and organizations that study the city or practice in a field that directly affects the city.

As well as drawing together the best of classic and contemporary writings on the city, each reader features extensive general, section and selection introductions prepared by the volume editors to place the selections in context, illustrate relations among topics, provide information on the author and point readers towards additional related bibliographic material.